Alberto FERRERO - Filippo GAZZOLA - Maurizio ZANOTTI

Elements of Advanced Mathematical Analysis for Physics and Engineering

$$0.\overline{9} \geq 1 \qquad \int_{\mathbb{R}^n} e^{-i\xi\cdot x} e^{-|x|^2} dx = \pi^{n/2} e^{-|\xi|^2/4}$$

$$\log z + \log z \neq 2 \log z \qquad -\int_{\mathbb{R}^n} u \Delta \Phi = u(0)$$

$$e^{i\pi} = -1 \qquad u_t - \eta \Delta u + (u \cdot \nabla) u + \nabla p = f$$

Società Editrice
ESCULAPIO

Alberto Ferrero
Universita del Piemonte Orientaleà
15121 Alessandria - Italy
alberto.ferrero@mfn.unipmn.it

Filippo Gazzola
Department of Mathematics
Politecnico di Milano
20133 Milano - Italy
filippo.gazzola@polimi.it

Maurizio Zanotti
Structural and Machine designer
Maurizio zanotti@alice.it -

ISBN 978-88-7488-645-6

Società Editrice Esculapio s.r.l.
Via Terracini, 30 – 40131 Bologna
www.editrice-esculapio.com – info@editrice-esculapio.it

Reprint 2015

Editorial Staff: Giancarla Panigali, Carlotta Lenzi
Translated by Ilaria Lucardesi

Società Editrice
ESCULAPIO

Contents

Preface

Deep comprehension of applied sciences requires a solid knowledge of Mathematical Analysis. For most of high level scientific research, the good understanding of Functional Analysis and weak solutions to differential equations is essential. This book aims to deal with the main topics that are necessary to achieve such a knowledge. Still, this is the goal of many other texts in advanced analysis; and then, what would be a good reason to read or to consult this book? In order to answer this question, let us introduce the three Authors.

Alberto Ferrero got his degree in Mathematics in 2000 and he is presently researcher in Mathematical Analysis at the Università del Piemonte Orientale. Filippo Gazzola got his degree in Mathematics in 1987 and he is now full professor in Mathematical Analysis at the Politecnico di Milano. Maurizio Zanotti got his degree in Mechanical Engineering in 2004 and presently he is structural and machine designer and lecturer professor in Mathematical Analysis at the Politecnico di Milano.

The three Authors, for the variety of their skills, decided to join their expertise to write this book. One of the reasons that should encourage its reading is that the presentation turns out to be a reasonable compromise among the essential mathematical rigor, the importance of the applications and the clearness, which is necessary to make the reference work pleasant to the readers, even to the inexperienced ones.

The range of treated topics is quite wide and covers the main basic notions of the scientific research which is based upon mathematical models. We start from vector spaces and Lebesgue integral to reach the frontier of theoretical research such as the study of critical exponents for semilinear elliptic equations and recent problems in fluid dynamics. This long route passes through the theory of Banach and Hilbert spaces, Sobolev spaces, differential equations, Fourier and Laplace transforms, before which we recall some appropriate tools of Complex Analysis. We give all the proofs that have some didactic or applicative interest, while we omit the ones which are too technical or require too high level knowledge.

This book has the ambitious purpose to be useful to a broad variety of readers. The first possible beneficiaries are of course the second or third year students of a scientific course of degree: in what follows they will find the topics that are necessary to approach more advanced studies in Mathematics and in other fields, especially Physics and Engineering. This text could be also useful to graduate students who want to start a Ph.D. course: indeed it contains the matter of a multidisciplinary Ph.D. course given by Filippo Gazzola for several years at Politecnico di Milano. Finally, this book could be addressed also to the ones who have already left education far-back but occasionally need to use mathematical tools: we refer both to university professors and their research, and to professionals and designers who want to model a certain phenomenon, but also to the nostalgics of the good old days when they were students. It is precisely for this last type of reader that we have also reported some elementary topics, such as the properties of numerical sets and of the integrals; moreover, every chapter is provided with examples and specific exercises aimed at the involvement of the reader. Let us start immediately inviting the reader to find an "anomaly" among the six formulas appearing in the cover.

Exercises for the topics covered in the present book may bee found in [31] from the same Editor. This book is the translation from Italian of the book "Elementi di Analisi Superiore per la Fisica e l'Ingegneria". The translation is due to Ilaria Lucardesi.

Chapter 1

Banach and Hilbert spaces

In this chapter we present the basic notions of Functional Analysis that will be useful in the following chapters. Abstract functional spaces and their properties are an essential tool in the study of integral transforms and partial differential equations.

The first part of the chapter is devoted to the definitions and the properties of Banach spaces. Such spaces are nothing else than vector spaces in which is defined a positive function, called norm. The norm allows to measure the distance between two vectors of the space. In this way, thanks to the acquired structure of metric space, on a Banach space it will be possible to introduce the notions of convergence of a sequence, of limit and of continuity of a functional. Banach spaces are endowed with a further fundamental property, the completeness, whose definition can be found in this chapter.

In the remaining part of the chapter we analyze more in detail Hilbert spaces and the linear mappings between them. Hilbert spaces are Banach spaces on which is possible to define a scalar product. Particular emphasis is also given to the spectral properties of the linear mappings from a Hilbert space into itself.

1.1 Vector spaces and norms

Let us begin with a list of basic notions of Linear Algebra.

Definition 1.1.1. *Let $\mathbb{K}$ be the real field $\mathbb{R}$ or the complex field $\mathbb{C}$. The elements of the field $\mathbb{K}$ will be called scalars. A* **vector space** *(or linear space) over the field $\mathbb{K}$ is a non empty set X such that:*

(i) for every pair of elements $x, y \in X$ there exists a third element of X, denoted by $x + y$ and called sum of x plus y,

(ii) for every $x \in X$ and $\lambda \in \mathbb{K}$ there exists an element of X, denoted by λx and called product of λ and x.

Such operations must satisfy the following properties:

$\diamond$ *For every $x, y, z \in X$ there holds $(x + y) + z = x + (y + z)$.*

◇ *There exists an element* $0 \in X$, *called zero vector or zero, such that for every* $x \in X$ *there holds* $0 + x = x + 0 = x$.

◇ *For every* $x \in X$ *there exists an element* $-x$, *called the opposite of* x, *such that* $x + (-x) = (-x) + x = 0$.

◇ *For every* $x, y \in X$ *there holds* $x + y = y + x$.

◇ *For every* $x, y \in X$ *and for every* $\lambda \in \mathbb{K}$ *there holds* $\lambda(x + y) = \lambda x + \lambda y$.

◇ *For every* $x \in X$ *and for every* $\lambda, \mu \in \mathbb{K}$ *there holds* $(\lambda + \mu)x = \lambda x + \mu x$.

◇ *For every* $x \in X$ *and for every* $\lambda, \mu \in \mathbb{K}$ *there holds* $(\lambda\mu)x = \lambda(\mu x)$.

◇ *For every* $x \in X$ *there holds* $1x = x$.

The simplest idea of vector in geometry is the one of the equivalence classes of oriented segments in the Euclidean space $\mathbb{R}^n$. Two oriented segments belong to the same equivalence class if they have the same direction, orientation and length. Definition 1.1.1, which is suitable for a more general notion of vector, summarizes the properties of the vectors of the space $\mathbb{R}^n$.

In what follows, we will omit to cite the field $\mathbb{K}$ over which the considered vector space is defined, but we will specify it for clearness just when necessary.

We recall also the following definitions:

Definition 1.1.2. *Let* X *be a vector space. A non empty subset* Y *of* X *is said* **vector subspace** *of* X *if:*

(i) for every $x, y \in Y$ *there holds* $x + y \in Y$,

(ii) for every $x \in Y$ *and for every* $\lambda \in \mathbb{K}$ *there holds* $\lambda x \in Y$.

In other words, Y is a subset of X which turns out to be a vector space as well, with the operations of addition of vectors and product between vectors and scalars induced by X itself.

Definition 1.1.3. *Let* X *be a vector space. Given* n *vectors* $x_1, ..., x_n \in X$, *it is called* **linear combination** *of* $x_1, ..., x_n$ *with coefficients* $\lambda_1, ..., \lambda_n \in \mathbb{K}$ *the following sum*

$$\lambda_1 x_1 + ... + \lambda_n x_n.$$

Definition 1.1.4. *Let* X *be a vector space. Then the vectors* $x_1, ..., x_n \in X$ *are called* **linearly independent** *if whenever* $\lambda_1, ..., \lambda_n \in \mathbb{K}$ *are scalars such that*

$$\lambda_1 x_1 + ... + \lambda_n x_n = 0$$

then

$$\lambda_1 = ... = \lambda_n = 0.$$

We say that the vectors $x_1, ..., x_n \in X$ *are linearly dependent if they are not linearly independent.*

Definition 1.1.5. *Let X be a vector space. Given n vectors $x_1, ..., x_n \in X$ we say that they* **generate** *X if for every $x \in X$ there exist $\lambda_1, ..., \lambda_n \in \mathbb{K}$ such that*

$$x = \lambda_1 x_1 + ... + \lambda_n x_n.$$

We shall denote this situation by writing $X = \text{span}\{x_1, ..., x_n\}$. If the vectors $x_1, ..., x_n$ are also linearly independent then the set $\{x_1, ..., x_n\}$ is called a **basis** *of X.*

We can now define the dimension of a vector space.

Definition 1.1.6. *Let X be a vector space and assume that X admits a basis of n vectors; then X is said to have* **dimension** *n and it is set $\text{dim}X = n$.*

If X does not admit any basis composed by a finite number of vectors, X is said to have infinite dimension, $\text{dim}X = \infty$.

It can be shown that the dimension of a vector space does not depend on the choice of the basis. For the proof and other notions about the theory of vector spaces see [63].

Let us see some simple examples of vector spaces:

Example 1.1.7. For $n \geq 1$ integer number, let us consider the space $\mathbb{K}^n$ of ordered n-tuples of real or complex numbers. Let the sum of two vectors be defined as follows: for every $(x_1, ..., x_n), (y_1, ..., y_n) \in \mathbb{K}^n$ let

$$(x_1, ..., x_n) + (y_1, ..., y_n) := (x_1 + y_1, ..., x_n + y_n).$$

For every $\lambda \in \mathbb{K}$ and for every $(x_1, ..., x_n) \in \mathbb{K}^n$ we define the product of a scalar and a vector:

$$\lambda(x_1, ..., x_n) := (\lambda x_1, ..., \lambda x_n).$$

We leave to the reader to check that all the properties listed in Definition 1.1.1 are satisfied; therefore, $\mathbb{K}^n$ is a vector space over $\mathbb{K}$ with the operations defined above. Moreover, it can be verified that $\text{dim}\mathbb{K}^n = n$. Indeed, it is enough to choose the basis

$$\{(1, 0, ..., 0), (0, 1, 0, ..., 0), ..., (0, 0, ..., 1)\}. \tag{1.1}$$

Note that the set $\mathbb{C}^n$, as well as a vector space over the field $\mathbb{C}$, can also be considered as a vector space over the field $\mathbb{R}$. But in the latter case the dimension of $\mathbb{C}^n$ is $2n$. Indeed (1.1) is not any more a basis, while the set of the $2n$ vectors

$$\{(1, 0, ..., 0), ..., (0, 0, ..., 1), (i, 0, ..., 0), ..., (0, 0, ..., i)\}$$

is a basis. □

Example 1.1.8. Given a non empty set $A \subset \mathbb{R}^n$ consider the set X of functions $f : A \to \mathbb{K}$. For every $f, g \in X$ and for every $\lambda \in \mathbb{K}$ set

$$(f+g)(x) := f(x) + g(x) \ , \quad (\lambda f)(x) := \lambda f(x) \qquad \forall x \in A.$$

Also in this case it can be easily verified that the set X, endowed with these two operations, becomes a vector space over $\mathbb{K}$.

Furthermore, we can define the subset $C^0(A)$ of X of the continuous functions in A, equipped with the same operations introduced in X. Since the sum of continuous functions is a continuous function and the product of a continuous function with a scalar is again a continuous function, the space $C^0(A)$ is a vector subspace of X. □

In the next definition we introduce the notion of norm in a vector space. Thanks to it, it will be possible to define a distance between vectors, making the vector space a metric space.

Definition 1.1.9. *Let X be a vector space. A* **norm** *in X is a function*

$$\| \cdot \| : X \to [0, \infty)$$

such that for every scalar λ and for every $x, y \in X$, the following properties hold true:

(i) $\|x\| = 0 \Leftrightarrow x = 0$ *(non degeneracy)*
(ii) $\|\lambda x\| = |\lambda| \cdot \|x\|$ *(homogeneity)*
(iii) $\|x + y\| \leq \|x\| + \|y\|$ *(triangle inequality).*

A vector space on which is defined a norm is called **normed space.**

Let us present some examples of normed vector spaces.

Example 1.1.10. In the space $\mathbb{R}^n$ consider the Euclidean norm defined as follows: given $x = (x_1, ..., x_n) \in \mathbb{R}^n$, let

$$\|x\| := \left(\sum_{k=1}^{n} x_k^2 \right)^{1/2} .$$

Similarly, we can define the norm in $\mathbb{C}^n$. □

Example 1.1.11. Let $a, b \in \mathbb{R}$ and let $C^0([a,b])$ be the vector space of continuous functions in $[a,b]$. We define on it the function

$$\|u\|_{C^0} := \max_{x \in [a,b]} |u(x)|$$

for every $u \in C^0([a,b])$. Such function turns out to be a norm in $C^0([a,b])$, called **uniform norm**. The first two properties of Definition 1.1.9 are straightforward. The third property is satisfied since the following chain of inequalities hold true

$$\|u+v\|_{C^0} = \max_{[a,b]} |u+v| \leq \max_{[a,b]} (|u| + |v|) \leq \|u\|_{C^0} + \|v\|_{C^0} \qquad \forall u, v \in C^0([a,b]).$$

We leave to the reader the proof that also the following **integral norm**

$$\|u\| := \int_a^b |u(x)|dx \qquad \forall u \in C^0([a,b])$$

is a norm in $C^0([a,b])$. □

A normed space $(X, \|\cdot\|)$ becomes a metric space if it is equipped with the metric

$$d(x,y) := \|x-y\| \qquad \forall x,y \in X.$$

Therefore, on the normed spaces it is possible to introduce all the topological notions, such as for instance the concepts of open and closed sets, the continuity of functions, the convergence of sequences, etc. For the benefit of the reader, below we recall the definition of convergence of a sequence.

Definition 1.1.12. *Let X be a normed vector space. A sequence $\{x_n\} \subset X$ is said to be* **convergent** *to $x \in X$ if $d(x_n, x) = \|x_n - x\| \to 0$ as $n \to \infty$.*

1.2 Banach spaces

The notion of completeness of a space is related to the behavior of particular sequences that we define as follows:

Definition 1.2.1. *Let X be a normed vector space. A sequence $\{x_n\} \subset X$ is called a* **Cauchy sequence** *if:*

$$\forall\, \varepsilon > 0 \quad \exists\, N \in \mathbb{N} \quad s.t. \quad n,m > N \Rightarrow \|x_m - x_n\| < \varepsilon.$$

Basically, Cauchy sequences "do not diverge nor oscillate". In particular, converging sequences are Cauchy sequences. Indeed, if $\{x_n\}$ converges to x then for every $\varepsilon > 0$ there exists $N \in \mathbb{N}$ such that

$$\|x_n - x\| < \frac{\varepsilon}{2} \qquad \forall n > N.$$

Therefore, exploiting the triangle inequality, we get

$$\|x_m - x_n\| \le \|x_m - x\| + \|x_n - x\| < \varepsilon \qquad \forall n,m > N.$$

Nevertheless, in general it is not true that Cauchy sequences are convergent. Thus it makes sense the following

Definition 1.2.2. *A normed space in which all the Cauchy sequences converge is said to be* **complete**. *A complete normed vector space is called* **Banach space**.

The notion of completeness can be extended also to the subsets of a normed space:

Definition 1.2.3. *Let X be a normed space. A subset $A \subseteq X$ is said to be* **complete** *if all the Cauchy sequences contained in A converge to some element of A.*

In Banach spaces compact sets play an important role:

Definition 1.2.4. *Let X be a normed space. A subset $A \subset X$ is said to be* **compact** *if every sequence $\{x_n\} \subset A$ admits a subsequence converging to some element $x \in A$.*

Exercise 1.2.5. Prove that every compact set in a Banach space is closed and bounded. If the space has finite dimension, show that also the viceversa holds true (and that this is false in infinite dimension). □

The first example of complete space is related to elementary numerical sets:

Theorem 1.2.6. *The set of rational numbers $\mathbb{Q}$ is not complete. The set of real numbers $\mathbb{R}$ is complete.*

We do not provide here the proof of this result since it can be found in every basic textbook of Mathematical Analysis. We would rather discuss the meaning of Definition 1.2.2: a complete space must include the limits of all its Cauchy sequences. Noticing that this kind of check is unfeasible, we propose an alternative method that we formalize in the

Remark 1.2.7. Since the Cauchy sequences are the only "candidates" to converge and since the completeness is, actually, a condition of *closure*, in order to prove the completeness of a space it is necessary to show that the limit of a converging sequence belongs to the space itself. For example, in order to show that the space of rational numbers $\mathbb{Q}$ is not complete, it is enough to consider the sequence of rational numbers $a_n = (1 + \frac{1}{n})^n$, which converges to $e \notin \mathbb{Q}$. □

Our second example of complete space is important as well and concerns continuous functions:

Theorem 1.2.8. *Let $K \subset \mathbb{R}^n$ be a non empty compact set and let $C^0(K)$ be the space of continuous functions on K with values in $\mathbb{R}$. Equipped with the uniform norm (see Example 1.1.11)*

$$\|u\|_{C^0(K)} = \max_K |u|, \tag{1.2}$$

the space $C^0(K)$ is a Banach space.

Proof. The fact that (1.2) is a norm can be proved as in Example 1.1.11.

Let us proceed as suggested in Remark 1.2.7. Let us consider a sequence $\{u_m\} \subset C^0(K)$ and let us assume that it converges to a certain function u, with respect to the norm (1.2) (uniform convergence); we have to prove that $u \in C^0(K)$. Let us fix a generic $x_0 \in K$ and a generic $\varepsilon > 0$; by the assumption of uniform convergence, we know that

$$\exists M = M(\varepsilon) \in \mathbb{N} \text{ s.t.} \qquad m \geq M \Rightarrow |u_m(x) - u(x)| < \varepsilon \quad \forall x \in K.$$

Thus, by the triangle inequality in $\mathbb{R}$, we infer

$$\begin{aligned} |u(x) - u(x_0)| &\leq |u(x) - u_M(x)| + |u_M(x) - u_M(x_0)| + |u_M(x_0) - u(x_0)| \\ &< 2\varepsilon + |u_M(x) - u_M(x_0)| \end{aligned}$$

and, passing to the limit as $x \to x_0$ (we recall that $u_M \in C^0(K)$), we obtain

$$\lim_{x \to x_0} |u(x) - u(x_0)| \leq 2\varepsilon;$$

by the arbitrariness of ε, this implies that

$$\lim_{x \to x_0} u(x) = u(x_0).$$

Hence, $u \in C^0(K)$. □

We now show that completeness is a feature of the *pair* (space,norm):

Example 1.2.9. Let $I = [0,1]$ and let us consider in $C^0(I)$ the integral norm introduced in Example 1.1.11:

$$\|u\| = \int_0^1 |u| \qquad \forall u \in C^0(I).$$

For $m \geq 3$, let us consider the sequence $\{u_m\} \subset C^0(I)$ defined by

$$u_m(x) = \begin{cases} 1 & \text{if } x \in [0, \frac{1}{2}] \\ 1 + \frac{m}{2} - mx & \text{if } x \in [\frac{1}{2}, \frac{1}{2} + \frac{1}{m}] \\ 0 & \text{if } x \in [\frac{1}{2} + \frac{1}{m}, 1] \,. \end{cases}$$

In addition, let us consider the function

$$u(x) = \begin{cases} 1 & \text{if } x \in [0, \frac{1}{2}) \\ 0 & \text{if } x \in (\frac{1}{2}, 1] \,. \end{cases}$$

An easy computation leads to

$$\lim_{m \to \infty} \|u_m - u\| = \lim_{m \to \infty} \int_0^1 |u_m - u| = \lim_{m \to \infty} \frac{1}{2m} = 0.$$

In view of Definition 1.1.12, this means that the sequence $\{u_m\}$ converges with respect to the integral norm to u, which however is not continuous. Thus, the space $C^0(I)$ is not complete with respect to the integral norm. □

We conclude this section with the notion of equivalent norms:

Definition 1.2.10. *Two norms $\|\cdot\|_1$ and $\|\cdot\|_2$ are said to be* **equivalent** *in a vector space X if:*

$$\exists\, \alpha, \beta > 0 \quad s.t. \quad \alpha\|u\|_1 \leq \|u\|_2 \leq \beta\|u\|_1 \qquad \forall u \in X.$$

Therefore, two equivalent norms "control" each other, meaning that $\|\cdot\|_1 \to 0$ if and only if $\|\cdot\|_2 \to 0$; in particular, $(X, \|\cdot\|_1)$ defines a Banach space if and only if $(X, \|\cdot\|_2)$ defines a Banach space as well. As already proved in Theorem 1.2.8 and in Example 1.2.9, in infinite dimensional spaces not all the norms are equivalent and the property of completeness depends on the norm considered. On the other hand, in finite dimensional spaces, all the norms are equivalent.

Exercise 1.2.11. Show that the following norms are equivalent in $\mathbb{R}^2$:

$$\|(x,y)\|_1 = |x| + |y|, \qquad \|(x,y)\|_2 = \sqrt{x^2+y^2}, \qquad \|(x,y)\|_\infty = \max\{|x|,|y|\}. \quad \square$$

1.3 Linear operators and dual spaces

In this section we study the linear operators between Banach spaces. Given a linear operator L, we will denote by Lx (instead of $L(x)$) the image of the vector x. An operator L between two Banach spaces X_1 and X_2 (over the field $\mathbb{K}$) is linear if

$$L(x+y) = Lx + Ly \quad \forall x, y \in X_1\,, \qquad L(ax) = aLx \quad \forall x \in X_1\,,\ \forall a \in \mathbb{K}.$$

Moreover we shall say that L is continuous if for every $x_0 \in X_1$ the following implication holds true:

$$\|x - x_0\|_{X_1} \to 0 \;\Rightarrow\; \|Lx - Lx_0\|_{X_2} \to 0.$$

The **kernel** $\ker(L)$ of a linear operator $L : X_1 \to X_2$ is the pre-image of the zero vector of X_2:

$$\ker(L) = \{x \in X_1\,;\ Lx = 0\}$$

and it turns out to be a subspace of X_1.

The **image** (or range) $R(L)$ of a linear operator $L : X_1 \to X_2$ is the set of images:

$$R(L) = \{y \in X_2\,;\ \exists\, x \in X_1,\ Lx = y\}$$

and it turns out to be a subspace of X_2.

Definition 1.3.1. *A linear operator $L : X_1 \to X_2$ is said to be* **bounded** *if there exists a constant $C > 0$ such that:*

$$\|Lx\|_{X_2} \leq C\,\|x\|_{X_1} \qquad \forall x \in X_1.$$

This relation reveals that the ball of radius R in X_1 is transformed by L in a subset of the ball of radius CR in X_2. Thus, the constant C can be interpreted as a control of the expansion rate of L. If $C < 1$, then L would be a contraction of distances.

The following result holds true:

Theorem 1.3.2. *A linear operator $L : X_1 \to X_2$ is bounded if and only if it is continuous.*

Proof. The easy check is left to the reader. □

Exercise 1.3.3. Find an example of linear discontinuous operator (unbounded). We suggest to take the space ℓ^2 introduced in Definition 1.4.8 below, to consider its canonical (orthonormal) basis $\{e_1, e_2, ...\}$ and to choose the linear operator $L : \ell^2 \to \ell^2$ satisfying $Le_n = ne_n$. □

The vector space of linear continuous (or bounded) operators from a Banach space X_1 to another Banach space X_2 is denoted by the symbol $\mathcal{L}(X_1, X_2)$. Such space is usually equipped with the norm

$$\|L\|_{\mathcal{L}(X_1,X_2)} := \sup_{\|x\|_{X_1}=1} \|Lx\|_{X_2}. \tag{1.3}$$

Exercise 1.3.4. Show that the vector space $\mathcal{L}(X_1, X_2)$ with the norm (1.3) is a Banach space. □

If X_1 is a real (or complex) space and $X_2 = \mathbb{R}$ (o $X_2 = \mathbb{C}$), instead of using the word operator we use the word **functional**. In this case, instead of Lx, we will write $\langle L, x\rangle$ and we will read it duality (or bracket) between L and x.

Definition 1.3.5. *The set of linear continuous functionals on a Banach space X is called* **dual space** *of X and is denoted by X'.*

When considering the dual space X', the norm (1.3) reads

$$\|L\|_{X'} = \sup_{\|x\|_X=1} |\langle L, x\rangle|. \tag{1.4}$$

It can be shown that X', equipped with this norm, is a Banach space, see Exercise 1.3.4. As an immediate consequence of the definition of the norm (1.4) we obtain the following Schwarz inequality:

$$|\langle L, x\rangle| \le \|L\|_{X'}\|x\|_X \qquad \forall x \in X,\ \forall L \in X'. \tag{1.5}$$

1.4 Hilbert spaces

Let us begin this section with the following important

Definition 1.4.1. *Let X be a vector space. It is called* **scalar product** *in X a function*

$$(\cdot,\cdot) : X \times X \to \mathbb{K}$$

such that for every $x, y, z \in X$ and for every $\lambda, \mu \in \mathbb{K}$ there hold:
(i) $(x,x) \geq 0$; $(x,x) = 0 \Leftrightarrow x = 0$ (non degeneracy)
(ii) $(x,y) = \overline{(y,x)}$ (symmetry)
(iii) $(\mu x + \lambda y\ ,\ z) = \mu(x,z) + \lambda(y,z)$ (linearity with respect to the first argument).

In the property (ii), we denote by $\overline{(y,x)}$ the complex conjugate of (y,x). Clearly, if $\mathbb{K} = \mathbb{R}$, such conjugate coincides with (y,x), being a real number; in that case, by property (iii) we deduce that the scalar product is linear not only with respect to the first argument, but also with respect to the second one: thus we shall say that $(\cdot,\cdot)$ is a *non negative symmetric bilinear form* from X^2 to $\mathbb{R}$. If in turn $\mathbb{K} = \mathbb{C}$, the (complex) scalar product is linear with respect to the first component, but in view of (ii)-(iii) we infer that

$$(z, \lambda x + \mu y) = \overline{\lambda}(z,x) + \overline{\mu}(z,y)$$

for every $x, y, z \in X$ and for every $\lambda, \mu \in \mathbb{C}$.

The scalar product can be estimated thanks to the

Theorem 1.4.2. (Schwarz inequality). *Let X be a vector space and let $(\cdot,\cdot)$ be a scalar product in X. Then for every $x, y \in X$ there holds*

$$|(x,y)| \leq \sqrt{(x,x)}\sqrt{(y,y)}. \tag{1.6}$$

Proof. We can assume $(x,y) \neq 0$ (in particular $x, y \neq 0$), otherwise (1.6) is trivial.

Let $A = (x,x)$, $B = |(x,y)|$, $C = (y,y)$, $\lambda = \frac{B}{(y,x)}$. Then, $\lambda \in \mathbb{C}$ satisfies $|\lambda| = 1$. In view of the properties of the scalar product, we deduce that for every $t \in \mathbb{R}$ there holds

$$0 \leq (x - t\lambda y, x - t\lambda y) = (x,x) - t\lambda(y,x) - t\overline{\lambda}(x,y) + t^2(y,y).$$

Hence, $A - 2Bt + Ct^2 \geq 0$ for every $t \in \mathbb{R}$. But then, the discriminant of this polynomial of degree two must be non positive: $B^2 \leq AC$. By replacing A, B, C, this proves (1.6). □

Exercise 1.4.3. Show that (1.6) holds true with the equality sign if and only if x and y are linearly dependent. □

Thanks to the Schwarz inequality (1.6) we can give the following

Definition 1.4.4. *Let X be a vector space and let $(\cdot,\cdot)$ be a scalar product in X. We will call* **norm induced by the scalar product** *the function defined by*

$$\|x\| = \sqrt{(x,x)} \qquad \forall x \in X.$$

We can now verify that the function just defined is actually a norm. The first two properties of the norm, expressed in Definition 1.1.9, are immediate. For what concerns the triangle inequality, thanks to the Schwarz inequality (1.6) we have

$$\|x+y\|^2 = \|x\|^2 + (x,y) + (y,x) + \|y\|^2 \le \|x\|^2 + 2\|x\|\|y\| + \|y\|^2 = (\|x\| + \|y\|)^2$$

for every $x, y \in X$.

Exercise 1.4.5. In most of our arguments, the scalar field will be $\mathbb{K} = \mathbb{R}$. We suggest the reader to rewrite (simplifying them!) the previous definitions and proofs in such particular case. □

We now give the following

Definition 1.4.6. *It is called* **Hilbert space** *a vector space (real or complex) endowed with a scalar product, which is complete with respect to the norm induced by the scalar product.*

Thus, a Hilbert space is a vector space for which the scalar product induces a norm that makes it a Banach space.

Exercise 1.4.7. Let H be a Hilbert space. Show the so called *parallelogram law*:

$$\left\|\frac{x+y}{2}\right\|^2 + \left\|\frac{x-y}{2}\right\|^2 = \frac{1}{2}(\|x\|^2 + \|y\|^2) \qquad \forall x, y \in H.$$ □

The Euclidean space $\mathbb{R}^n$ is a Hilbert space with respect to the scalar product defined by

$$(x,y) = \sum_{k=1}^{n} x_k\, y_k \qquad \forall x = (x_1, ..., x_n),\ y = (y_1, ..., y_n) \in \mathbb{R}^n.$$

Similarly, we can define on the space $\mathbb{C}^n$ the scalar product

$$(x,y) = \sum_{k=1}^{n} x_k\, \overline{y}_k \qquad \forall x = (x_1, ..., x_n),\ y = (y_1, ..., y_n) \in \mathbb{C}^n.$$

We can extend such example to the case in which (formally) $n = \infty$:

Definition 1.4.8. *We call ℓ^2 the real vector space of the sequences $\{x_k\} \subset \mathbb{R}$ such that*

$$\sum_{k=1}^{\infty} x_k^2 < \infty,$$

equipped with the scalar product

$$(x,y)_{\ell^2} = \sum_{k=1}^{\infty} x_k\, y_k \qquad \text{for every } \ x = \{x_k\},\ y = \{y_k\} \in \ell^2 \ .$$

Similarly, we can define the complex space ℓ^2 of the sequences $\{x_k\} \subset \mathbb{C}$ such that

$$\sum_{k=1}^{\infty} |x_k|^2 < \infty,$$

equipped with the scalar product

$$(x,y)_{\ell^2} = \sum_{k=1}^{\infty} x_k\, \overline{y}_k \qquad \text{for every } \ x = \{x_k\},\ y = \{y_k\} \in \ell^2 \ .$$

This space will play a crucial role in the sequel. Therefore it is useful to understand how the vectors of ℓ^2 look like. Let us consider the two real sequences

$$x = \left\{1, \frac{1}{2}, \frac{1}{3}, \dots, \frac{1}{k}, \dots\right\}, \qquad y = \left\{1, \frac{1}{\sqrt{2}}, \frac{1}{\sqrt{3}}, \dots, \frac{1}{\sqrt{k}}, \dots\right\}.$$

Then

$$(x,x)_{\ell^2} = \sum_{k=1}^{\infty} \frac{1}{k^2} = \frac{\pi^2}{6}\,, \qquad (y,y)_{\ell^2} = \sum_{k=1}^{\infty} \frac{1}{k} = +\infty$$

and hence $x \in \ell^2$ while $y \notin \ell^2$. Basically, the sequences belonging to ℓ^2 are the ones which are "strongly" infinitesimal at infinity.

Exercise 1.4.9. Verify that ℓ^2 is a Hilbert space. For $1 < p < \infty$, show that the space ℓ^p of sequences $x = \{x_k\} \subset \mathbb{R}$ (or $x = \{x_k\} \subset \mathbb{C}$ in the case of the complex space ℓ^p) such that

$$\sum_{k=1}^{\infty} |x_k|^p < \infty$$

and equipped with the norm

$$\|x\|_{\ell^p} = \left(\sum_{k=1}^{\infty} |x_k|^p \right)^{1/p}$$

is a Banach space, but is not a Hilbert space in case $p \neq 2$. The proof that $\|\cdot\|_{\ell^p}$ is a norm is not trivial at all and is based upon the following Hölder inequality

$$\sum_{k=1}^{\infty} |x_k y_k| \leq \left(\sum_{k=1}^{\infty} |x_k|^p\right)^{1/p} \cdot \left(\sum_{k=1}^{\infty} |y_k|^{p'}\right)^{1/p'}$$

for every $x \in \ell^p$ and $y \in \ell^{p'}$ with $p' = p/(p-1)$. To have an idea of the proof see Theorem 2.4.6. □

The convergence introduced in Definition 1.1.12 is not the only possible one in infinite dimensional spaces; sometimes it is called *strong convergence* (or convergence in norm) in order to distinguish it from other kinds of convergence that we introduce now:

Definition 1.4.10. *Let H be a Hilbert space. We say that $\{x_n\}$* **converges weakly** *to x and we write $x_n \rightharpoonup x$ if, as $n \to \infty$,*

$$(x_n, y) \to (x, y) \qquad \forall y \in H.$$

It is easy to see that *strong convergence implies weak convergence*. Indeed, if $x_n \to x$ as $n \to \infty$ (namely $\|x_n - x\| \to 0$), in view of the Schwarz inequality (1.6) we deduce that

$$|(x_n - x, y)| \leq \|x_n - x\| \cdot \|y\| \to 0 \qquad \forall y \in H. \tag{1.7}$$

Thus by Definition 1.4.10, we get $x_n \rightharpoonup x$.

In general, the viceversa is not true, as we see in the following

Example 1.4.11. Let ℓ^2 be the real space introduced in Definition 1.4.8. Let us consider $\{e_n\} \subset \ell^2$ where e_n is the sequence with all the terms 0, except the n-th term which equals 1:

$$e_1 = (1, 0, 0, ..., 0, ...), \quad e_2 = (0, 1, 0, ..., 0, ...), \quad ..., \quad e_n = (0, 0, 0, ..., 1, ...), \quad ...$$

Let us then fix $x = (x_1, ..., x_n, ...) \in \ell^2$ so that

$$\sum_{n=1}^{\infty} x_n^2 < \infty$$

which, in turn, implies $\lim\limits_{n\to\infty} x_n = 0$. Hence we have $(x, e_n)_{\ell^2} = x_n \to 0$ which, by the arbitrariness of x, implies that $e_n \rightharpoonup 0$. But, since $\|e_n - 0\|_{\ell^2} = 1$, the sequence $\{e_n\}$ does not converge strongly to 0. □

Exercise 1.4.12. Show that in finite dimensional spaces, strong convergence is equivalent to weak convergence: it is enough to consider the case $\mathbb{R}^n$. □

In the next result it will be shown what is the relationship between bounded and weak convergent sequences:

Theorem 1.4.13. *Let H be a Hilbert space.*
(i) If $\{x_n\} \subset H$ and $x \in H$ are such that $x_n \rightharpoonup x$, then $\{x_n\}$ is bounded in the norm induced by the scalar product and there holds

$$\|x\| \leq \liminf_{n\to\infty} \|x_n\|.$$

(ii) If $\{x_n\} \subset H$ is bounded, then there exist a subsequence $\{x_{n_k}\}$ and an element $x \in H$ such that $x_{n_k} \rightharpoonup x$.

Proof. See [10, Proposizione III.5] and [10, Teorema III.27]. □

As a consequence of Theorem 1.4.13 we have

Theorem 1.4.14. *Let H be a Hilbert space and let $\{x_n\}, \{y_n\} \subset H$ be such that $x_n \rightharpoonup x$ and $y_n \to y$. Then*

$$(x_n, y_n) \to (x, y).$$

Proof. By exploiting the Schwarz inequality (1.6) we get

$$|(x_n, y_n) - (x, y)| \leq |(x_n, y_n - y)| + |(x_n - x, y)| \leq \|x_n\|\|y_n - y\| + |(x_n - x, y)| \to 0$$

as $n \to \infty$, since $\|x_n\|$ is bounded in view of Theorem 1.4.13, $\|y_n - y\| \to 0$ thanks to the convergence $y_n \to y$ and $(x_n - x, y) \to 0$ by the convergence $x_n \rightharpoonup x$. □

1.5 Orthogonal projections in a Hilbert space

Let H be a Hilbert space with the norm $\|\cdot\|$. A set $K \subset H$ is said to be **convex** if:

$$\forall\, x, y \in K \quad tx + (1-t)y \in K \quad \forall\, t \in [0, 1].$$

There holds

Theorem 1.5.1. (projection on a closed convex set). *Let H be a Hilbert space and let $K \subseteq H$ be a non empty closed convex set. Then, for every $x \in H$ there exists a unique $y \in K$ such that*

$$\|x - y\| = \min_{z\in K} \|x - z\|. \tag{1.8}$$

We set $y = P_K x$ and we call it projection of x on K.

Moreover, if H is a real space, $y \in K$ is characterized by the following property:

$$(x - y, z - y) \leq 0 \quad \forall z \in K. \tag{1.9}$$

Proof. We divide the proof into three parts.

Existence. Let $\{z_n\}$ be a minimizing sequence for (1.8), namely $z_n \in K$ and

$$d_n = \|x - z_n\| \to d = \inf_{z \in K} \|x - z\|.$$

By applying the parallelogram law (Exercise 1.4.7) to the vectors $x - z_n$ and $x - z_m$ we obtain

$$\left\| x - \frac{z_n + z_m}{2} \right\|^2 + \left\| \frac{z_n - z_m}{2} \right\|^2 = \frac{1}{2}(d_n^2 + d_m^2).$$

Since $\frac{z_n + z_m}{2} \in K$, we infer that $\left\| x - \frac{z_n+z_m}{2} \right\| \geq d$ and hence

$$\left\| \frac{z_n - z_m}{2} \right\|^2 \leq \frac{1}{2}(d_n^2 + d_m^2) - d^2$$

whence

$$\lim_{m,n \to \infty} \|z_n - z_m\| = 0.$$

Thus, $\{z_n\}$ is a Cauchy sequence and, thanks to the completeness of H, there exists $y \in H$ such that $z_n \to y$. Moreover $y \in K$ (being K closed and $\{z_n\} \subset K$) and

$$\|x - y\| = \lim_{n \to \infty} \|x - z_n\| = \lim_{n \to \infty} d_n = d.$$

Uniqueness. Let $y_1, y_2 \in K$ verify (1.8). Again by the parallelogram law we have that

$$\left\| \frac{y_1 - y_2}{2} \right\|^2 + d^2 \leq \left\| \frac{y_1 - y_2}{2} \right\|^2 + \left\| x - \frac{y_1 + y_2}{2} \right\|^2 = \frac{1}{2}\left(\|x - y_1\|^2 + \|x - y_2\|^2\right) = d^2$$

from which follows immediately that $\|y_1 - y_2\| = 0$ and hence $y_1 = y_2$.

Equivalence between (1.8) *and* (1.9). Let H be a real space. Let $y \in K$ verify (1.8) and let $z \in K$. Thanks to the convexity of K, we have

$$(1-t)y + tz \in K \quad \text{for every } t \in [0,1]$$

and hence

$$\begin{aligned} \|x - y\|^2 &\leq \|x - [(1-t)y + tz]\|^2 = \|(x - y) - t(z - y)\|^2 \\ &= \|x - y\|^2 - 2t(x - y, z - y) + t^2\|z - y\|^2. \end{aligned}$$

In view of the last equality we infer $2(x - y, z - y) \leq t\|z - y\|^2$ and, as $t \to 0$, we obtain (1.9). Conversely, if y verifies (1.9) then by the bilinearity of the scalar product there holds

$$\|y - x\|^2 - \|z - x\|^2 = 2(x - y, z - y) - \|y - z\|^2 \leq 0 \quad \forall z \in K$$

whence (1.8) follows. □

In $\mathbb{R}^2$, Theorem 1.5.1 has a simple geometrical interpretation:

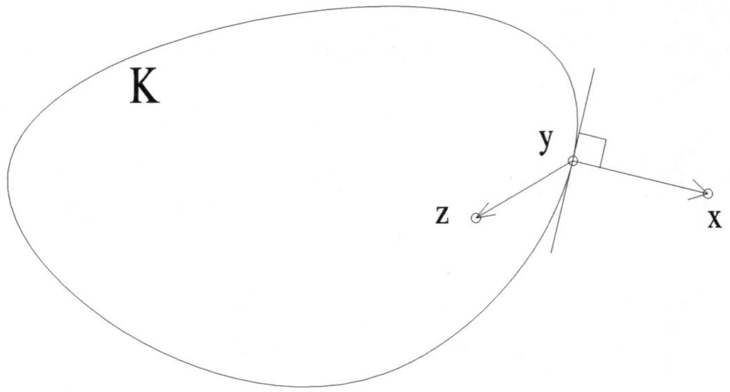

Projection on a convex set

From this picture we see that the projection y of x on K is unique and is the point at minimal distance; moreover, the vectors $(x-y)$ and $(z-y)$ form an angle greater than a right angle (negative scalar product).

We remark that Theorem 1.5.1 applies also in the particular case in which K is a (closed) subspace of H. In this case, it is useful the following

Definition 1.5.2. *Let H be a Hilbert space and let $M \subseteq H$ be a vector subspace. Then it is defined the* **orthogonal space** *to M in H and is denoted by $M^\perp$ the following vector subspace of H*

$$M^\perp := \{x \in H; (x, y) = 0 \textit{ for every } y \in M\}.$$

In the next theorem we state a result about the orthogonal projection on closed subspaces:

Theorem 1.5.3. *Let H be a Hilbert space and let $M \subseteq H$ be a closed vector subspace. The following results hold true:*

(i) for every $x \in H$ there exists a unique decomposition $x = Px + Qx$ such that $Px \in M$ and $Qx \in M^\perp$;

(ii) for every $x \in H$ the vectors Px and Qx coincide respectively with the vectors $P_M x$ and $P_{M^\perp} x$ defined in Theorem 1.5.1;

(iii) the functions $P : H \to M$ and $Q : H \to M^\perp$ are linear;

(iv) for every $x \in H$ there holds $\|x\|^2 = \|Px\|^2 + \|Qx\|^2$ (Pythagoras Theorem).

Proof. (i) We begin by showing the uniqueness of the decomposition. Let $x_1, x_2 \in M$ and $y_1, y_2 \in M^\perp$ be such that $x = x_1 + y_1$ and $x = x_2 + y_2$, then $x_1 - x_2 = y_2 - y_1$. But $x_1 - x_2 \in M$, $y_2 - y_1 \in M^\perp$ and $M \cap M^\perp = \{0\}$ from which follows that $x_1 = x_2$ and $y_1 = y_2$.

We now pass to the existence. Let us set $Px = P_M x$ with $P_M x$ as in Theorem 1.5.1 so that $Px \in M$. We prove that $Qx := x - Px \in M^\perp$. By Definition 1.5.2, we have to show that $(Qx, y) = 0$ for every $y \in M$ with $\|y\| = 1$. With such a choice of y, in view of the definition of Qx and (1.8), we have that for every $\alpha \in \mathbb{K}$

$$\|Qx\|^2 = \|x - Px\|^2 \leq \|x - (Px + \alpha y)\|^2$$
$$= \|Qx - \alpha y\|^2 = \|Qx\|^2 - (Qx, \alpha y) - (\alpha y, Qx) + |\alpha|^2\|y\|^2$$

and hence, being $\|y\| = 1$,

$$0 \leq -\overline{\alpha}(Qx, y) - \alpha\overline{(Qx, y)} + |\alpha|^2.$$

By choosing $\alpha = (Qx, y)$ we immediately obtain $(Qx, y) = 0$.

(ii) Let $y \in M^\perp$ be arbitrary. Since $Px \in M$ and $Qx - y \in M^\perp$, in view of the properties (ii)-(iii) in Definition 1.4.1 we have

$$\|x - y\|^2 = \|Px + (Qx - y)\|^2 = \|Px\|^2 + (Px, Qx - y) + (Qx - y, Px) + \|Qx - y\|^2$$
$$= \|Px\|^2 + \|Qx - y\|^2 \geq \|Px\|^2 = \|x - Qx\|^2.$$

This last equality implies that

$$\|x - Qx\| = \min_{y \in M^\perp} \|x - y\|$$

and by Theorem 1.5.1 we deduce that $Qx = P_{M^\perp}x$.

(iii) For every $x, y \in H$ and for every $a, b \in \mathbb{K}$ we have

$$P(ax + by) + Q(ax + by) = ax + by = a(Px + Qx) + b(Py + Qy)$$

and hence

$$P(ax + by) - aPx - bPy = -Q(ax + by) + aQx + bQy \in M \cap M^\perp = \{0\}$$

since $P(ax + by) - aPx - bPy \in M$ and $-Q(ax + by) + aQx + bQy \in M^\perp$.

(iv) This follows immediately by the properties (ii)-(iii) in Definition 1.4.1 and by the orthogonality of Px and Qx. □

From Theorem 1.5.3 follows the

Corollary 1.5.4. *Let H be a Hilbert space and let $M \subseteq H$ be a closed subspace. Let P_M be the projection on M defined in Theorem 1.5.1. Then there holds*

$$\|P_Mx - P_My\| \leq \|x - y\| \qquad \forall x, y \in H. \tag{1.10}$$

Proof. As in Theorem 1.5.1, let P and Q denote the projections P_M and $P_{M^\perp}$ respectively. Then, in view of Theorem 1.5.3 (iii)-(iv), we deduce that

$$\|x - y\|^2 = \|P(x - y)\|^2 + \|Q(x - y)\|^2 \geq \|Px - Py\|^2 \qquad \forall x, y \in H. \qquad \square$$

Besides the importance in itself, Theorem 1.5.3 also finds applications in the study of the dual space of a Hilbert space. Indeed we prove that every element of the dual space can be represented by means of a scalar product with a suitable element.

Theorem 1.5.5. (Riesz representation Theorem). *Let H be a Hilbert space and let H' be its dual. For every $f \in H'$ there exists a unique $u \in H$ such that*

$$\langle f, v\rangle = (v, u) \qquad \forall v \in H.$$

Moreover, there holds $\|f\|_{H'} = \|u\|$.

Proof. If $f \equiv 0$ it is enough to take $u = 0$. Let then $f \in H'\backslash\{0\}$ and let $M = \ker(f) = \{v \in H; \langle f, v\rangle = 0\}$. Let P and Q be the orthogonal projections on M and $M^\perp$ respectively. Since f is continuous, M is a closed subspace of H (prove it!) and then, given $z \notin M$, thanks to Theorem 1.5.3 the definition $w := Qz \in M^\perp$ makes sense. Moreover $w \neq 0$ since otherwise there would be $z = Pz \in M$, in contrast with the choice made on z. Let us now consider

$$u := \frac{\overline{\langle f, w\rangle}}{\|w\|^2} w \in M^\perp$$

so that

$$\langle f, u\rangle = \|u\|^2. \tag{1.11}$$

For every $v \in H$ we set

$$v' = v - \frac{\langle f, v\rangle}{\|u\|^2} u$$

so that $\langle f, v'\rangle = 0$ and hence $v' \in \ker(f) = M$. Since $u \in M^\perp$, we have

$$0 = (v', u) = (v, u) - \langle f, v\rangle .$$

Finally, we have that $\langle f, v\rangle = (v, u)$ for every $v \in H$. Moreover, by the Schwarz inequality (1.5), we derive

$$\|f\|_{H'} = \sup_{\|v\|=1} |\langle f, v\rangle| = \sup_{\|v\|=1} |(v, u)| \leq \|u\|.$$

On the other hand, in view of (1.4), (1.11) and by the linearity of f we also have

$$\|f\|_{H'} \geq \left|\left\langle f, \frac{u}{\|u\|}\right\rangle\right| = \frac{|\langle f, u\rangle|}{\|u\|} = \|u\|.$$

Thus we can conclude that $\|f\|_{H'} = \|u\|$. Finally, let us verify the uniqueness of such vector u. Indeed, let $u_1, u_2 \in H$ satisfy

$$(v, u_1) = (v, u_2) = \langle f, v\rangle$$

for every $v \in H$. Then, choosing $v = u_1 - u_2$ we immediately obtain $\|u_1 - u_2\| = 0$, that is $u_1 = u_2$. □

The proof of the Riesz Theorem shows how the vector of H with whom the generic functional $f \in H'$ is identified, is nothing else but the unique vector orthogonal to the kernel of f which satisfies the condition $\langle f, u \rangle = \|u\|^2$. Basically, the representation Theorem allows to identify a Hilbert space with its dual (for example the dual space of $\mathbb{R}^n$ is $\mathbb{R}^n$ itself and the same also for $\mathbb{C}^n$). If we introduce the operator

$$R: \; H' \to H$$

which associates to every element $f \in H'$ the unique element $u \in H$ representing it, namely such that

$$\langle f, v \rangle = (v, u) \qquad \forall v \in H,$$

then R is bijective and isometric:

$$f \leftrightarrow u \quad \text{ and } \quad \|f\|_{H'} = \|u\|.$$

The operator R takes the name of **Riesz operator** (or canonical isometry) between H' and H.

1.6 Separable spaces

We immediately start explaining what is meant by separable space, recalling the notions of countability and density:

Definition 1.6.1. *A set A is said to be* **countable** *if there exists a bijection from A to $\mathbb{N}$. Let X be a normed space and let $A \subseteq B \subseteq X$; A is said to be* **dense** *in B if:*

$$\forall\, x \in B,\ \forall\, \varepsilon > 0\ \exists\, y \in A \quad s.t. \quad \|x - y\| < \varepsilon;$$

this situation is also denoted by $\overline{A} = B$, namely the **closure** *of A coincides with B. Finally, a normed vector space X is said to be* **separable** *if it admits a countable dense subset $Y \subseteq X$.*

There holds the following:

Theorem 1.6.2. *The union $A_1 \cup A_2 \cup ... \cup A_k \cup ...$ and the Cartesian product $A_1 \times A_2 \times ... \times A_k \times ...$ of a finitely or countably infinitely many countable sets are countable.*

Thanks to Theorem 1.6.2 we can affirm that $\mathbb{Q}$ *is a countable set*: indeed, a generic element $\frac{p}{q} \in \mathbb{Q}$ can be seen as an ordered pair $(p, q) \in \mathbb{N} \times \mathbb{Z}$, thus generating a one-to-one correspondence between $\mathbb{Q}$ and the set of pairs (p, q) such that p and q are mutually prime.

A countable set can be sequentially ordered through a sequence in such a way that for each element of the set it is possible to determine the *predecessor* and the *successor. The Cantor diagonalization argument* provides a possible ordering of a sequence of the elements of $\mathbb{Q}_+$. We construct a table with infinitely many rows and columns in which the generic positive rational number $\frac{p}{q}$ (with $p, q \in \mathbb{N}$) is located in the p-th row (numerator) and q-th column (denominator):

$N\backslash D$	1	2	3	4	5	6	...
1	1/1	1/2	1/3	1/4	1/5	1/6	...
2	2/1	(2/2)	2/3	(2/4)	2/5	(2/6)	...
3	3/1	3/2	(3/3)	3/4	3/5	(3/6)	...
4	4/1	(4/2)	4/3	(4/4)	4/5	(4/6)	...
5	5/1	5/2	5/3	5/4	(5/5)	5/6	...
6	6/1	(6/2)	(6/3)	(6/4)	6/5	(6/6)	...
...	...	...	...	...	...	...	...

Then we erase the reducible fractions, that are denoted in brackets. We proceed by extracting the elements of the table: we consider the elements (not erased) of the diagonals $\nearrow$ going from the left to the right and from the bottom to the top. Going on in this way, we obtain the ordering

$$1, \quad 2, \quad \frac{1}{2}, \quad 3, \quad \frac{1}{3}, \quad 4, \quad \frac{3}{2}, \quad \frac{2}{3}, \quad \frac{1}{4}, \quad 5, \quad \frac{1}{5}, \quad 6, \quad \frac{5}{2}, \quad \frac{4}{3}, \quad \frac{3}{4}, \quad \frac{2}{5}, \quad \frac{1}{6}, \quad \dots$$

With this ordering, it is possible to determine the predecessor and the successor of every rational number: for example, the predecessor of $3/38$ is $4/37$ while its successor is $2/39$. On the other hand, it is not possible to determine the predecessor and the successor of a rational number with the usual ordering (the inequality $<$) even if... the following exercise seems to contradict what has just been said:

Exercise 1.6.3. Consider the rational numbers $r = 1$ and $s = 0.\overline{9}$. Then s is the predecessor of r since between the two it can not be inserted any other rational number. The reader try to explain the apparent contradiction. Suggestion: compute $9s = 10s - s$ and...do not be surprised! □

In regard to the field $\mathbb{R}$, there holds the following

Theorem 1.6.4. *The set $\mathbb{R}$ is not countable.*

Proof. It is enough to prove that the interval $(0, 1)$ is not countable. Let us proceed by contradiction, by assuming that $(0, 1)$ is countable, namely that there exists a one-to-one correspondence $\varphi : \mathbb{N} \to (0, 1)$ that permits to order the numbers of $(0, 1)$ in a sequence:

$$\begin{aligned} a_1 &= 0.\alpha_{11}\,\alpha_{12}\,\alpha_{13}\dots \\ a_2 &= 0.\alpha_{21}\,\alpha_{22}\,\alpha_{23}\dots \\ a_3 &= 0.\alpha_{31}\,\alpha_{32}\,\alpha_{33}\dots \\ &\vdots \end{aligned}$$

where α_{ni} is the i-th decimal digit of a_n. In order to obtain a contradiction, it is sufficient to construct a number $b \in (0,1)$ which is different from those of the sequence $\{a_n\}$. Let $b = 0.\beta_1\beta_2\beta_3...$ where the decimal digits β_n are defined in this way: $\beta_n = 7$ if $0 \leq \alpha_{nn} \leq 4$, $\beta_n = 3$ if $5 \leq \alpha_{nn} \leq 9$. Such number does not coincide with none of the a_n since $\beta_n \neq \alpha_{nn}$. □

Therefore, thanks to the Theorems 1.6.2 and 1.6.4, we can affirm that $\mathbb{R}$ is "strictly greater" than $\mathbb{Q}$. Nevertheless, even if they are "few", rational numbers can approximate real numbers:

Proposition 1.6.5. *The set of rational numbers $\mathbb{Q}$ is dense in $\mathbb{R}$.*

Proof. It is enough to observe that a real number can be approximated as good as we want with a rational number, by truncating its decimal expansion after arbitrarily many decimal digits. □

Therefore, since $\mathbb{Q}$ is countable, according to Definition 1.6.1 *the set of real numbers $\mathbb{R}$ is separable.* Similarly, in view of Theorem 1.6.2, also $\mathbb{R}^n$ and ℓ^2 turn out to be separable. The same holds true for $\mathbb{C}^n$ and the complex space ℓ^2.

We now pass to the analysis of the bases of a Hilbert space.

Definition 1.6.6. *Let H be a Hilbert space. We shall say that $\{e_n\}$* **generates** *H if the closure of the space of the finite linear combinations of $\{e_n\}$ coincides with H. In other words, for every $x \in H$ and for every $\varepsilon > 0$ there exist $N \in \mathbb{N}$ and $\lambda_1, ..., \lambda_N \in \mathbb{K}$ such that*

$$\left\| x - \sum_{n=1}^{N} \lambda_n e_n \right\| < \varepsilon.$$

If in addition $\{e_n\}$ is orthonormal, that is

$$\|e_n\| = 1 \quad \forall n, \qquad (e_m, e_n) = 0 \quad \forall m \neq n\ ,$$

we will say that $\{e_n\}$ is a **Hilbertian basis** *of H.*

If there is no ambiguity, a Hilbertian basis will be also simply called basis: however, we should not confuse this concept with that of *algebraic basis* introduced in Definition 1.1.5, namely a *finite* family $\{e_i\}$ of H such that every element of H can be written in a unique way as a (finite) linear combination of $\{e_i\}$.

Not all the Hilbert spaces admit a Hilbertian basis; nevertheless there holds the

Theorem 1.6.7. *Every separable Hilbert space admits a Hilbertian basis.*

Proof. Let $\{x_n\} \subset H$ be a countable dense subset of H; it is easy to construct a sequence $\{y_n\} \subset \{x_n\}$ of linearly independent vectors such that every element of $\{x_n\}$ is a finite linear combination of elements of $\{y_n\}$.

Consider the sequence $\{V_n\}$ of subspaces of H, where V_n is the subspace spanned by $y_1, ..., y_n$. Since the vectors of $\{y_n\}$ are linearly independent, it holds $V_n \subset V_{n+1}$ in a strict sense, moreover $\dim V_n = n$. The space $V := \cup_{n=1}^{\infty} V_n$ is a subspace of H and it is dense in H ($\overline{V} = H$) since $\{x_n\} \subset V$.

Let us set $e_1 := y_1/\|y_1\|$; then $\{e_1\}$ is an orthonormal basis of V_1. Let us assume as induction hypothesis that for a generic n the space V_n admits an orthonormal basis $\{e_1, ..., e_n\}$ and let us construct an orthonormal basis for V_{n+1}. For this purpose, let us define $w_{n+1} := y_{n+1} - \sum_{k=1}^{n}(y_{n+1}, e_k)e_k$ and $e_{n+1} = w_{n+1}/\|w_{n+1}\|$. With this choice of e_{n+1} we have that $\{e_1, ..., e_{n+1}\}$ is an orthonormal basis of V_{n+1}.

By induction, we can construct a *sequence* $\{e_n\}$ such that for every $n \geq 1$ the family $\{e_1, ..., e_n\}$ is an orthonormal basis of V_n. In view of this fact (and in view of $\overline{V} = H$) we deduce that the closure of the space generated by $\{e_n\}$ coincides with H; hence, $\{e_n\}$ is a Hilbertian basis of H. □

Recalling that a series in a normed space is defined as the limit (according to Definition 1.1.12) of the sequence of partial sums, we prove the

Theorem 1.6.8. *Let H be a separable Hilbert space. Let $\{e_n\}$ be a Hilbertian basis, whose existence is ensured by Theorem 1.6.7. Then*

$$x = \sum_{n=1}^{\infty}(x, e_n)e_n \qquad \textit{(generalized Fourier series)} \tag{1.12}$$

and furthermore

$$\|x\|^2 = \sum_{n=1}^{\infty} |(x, e_n)|^2. \tag{1.13}$$

The coefficients $c_n := (x, e_n)$ are called **Fourier coefficients** *of x with respect to the basis $\{e_n\}$.*

Proof. For every $k \in \mathbb{N}$ we define the orthogonal projection of H on the space spanned by $\{e_1, ..., e_k\}$:

$$P_k y := \sum_{n=1}^{k}(y, e_n)e_n \qquad \forall y \in H.$$

Exploiting the orthonormality of the basis $\{e_n\}$ and iterating the assertion (iv) of Theorem 1.5.3, we deduce that

$$\|y\|^2 = \|y - P_k y\|^2 + \|P_k y\|^2 = \|y - P_k y\|^2 + \sum_{n=1}^{k} |(y, e_n)|^2 \qquad \forall y \in H. \tag{1.14}$$

Let us now fix $x \in H$; then, according to Definition 1.6.6, for every $\varepsilon > 0$ there exist $N \in \mathbb{N}$ and $\lambda_1, ..., \lambda_N \in \mathbb{K}$ such that, setting

$$\overline{x} := \sum_{n=1}^{N} \lambda_n e_n ,$$

it holds $\|x - \overline{x}\| < \varepsilon$. Clearly, $P_k\overline{x} = \overline{x}$ for every $k \geq N$. But then, in view of Corollary 1.5.4 and by the triangle inequality, we also have that

$$\|P_k x - x\| \leq \|P_k x - P_k \overline{x}\| + \|\overline{x} - x\| < 2\varepsilon$$

and, by the arbitrariness of ε, we obtain (1.12). Finally, thanks to (1.14) and to the fact that $P_k x \to x$ as $k \to \infty$, we immediately obtain (1.13). □

In the next theorem we see how all the infinite dimensional separable Hilbert spaces are isomorphic and can be "identified" with ℓ^2 (see Definition 1.4.8).

Theorem 1.6.9. *Let H be an infinite dimensional separable Hilbert space. Then H is isomorphic and isometric to the space ℓ^2, namely there exists an isomorphism $L : H \to \ell^2$ such that*

$$\|Lx\|_{\ell^2} = \|x\|_H$$

for every $x \in H$.

Proof. Let $\{e_n\}$ be a Hilbertian basis of H. Thanks to Theorem 1.6.8 it is possible to define a linear mapping $L : H \to \ell^2$ which associates to each $x \in H$ the sequence $c = \{c_1, c_2, ...\} \subset \ell^2$ of its Fourier coefficients with respect to the basis $\{e_n\}$. Moreover, such function verifies

$$\|Lx\|_{\ell^2} = \|c\|_{\ell^2} = \left(\sum_{n=1}^{\infty} |(x, e_n)|^2 \right)^{1/2} = \|x\|_H$$

for every $x \in H$. The injectivity and the continuity of L immediately follow from the fact that it is an isometry. Let us verify the surjectivity: let $c = (c_1, c_2, ...) \in \ell^2$ and consider the partial sum

$$x_n := \sum_{k=1}^{n} c_k e_k .$$

In view of the convergence of $\sum_{k=1}^{\infty} |c_k|^2$ and of the orthonormality of e_k we deduce that for every $\varepsilon > 0$ there exists $N \in \mathbb{N}$ such that for every $n > m > N$

$$\|x_n - x_m\|_H^2 = \sum_{k=m+1}^{n} |c_k|^2 < \varepsilon .$$

Therefore, the sequence $\{x_n\}$ is a Cauchy sequence in H and, by the completeness of H, there exists $x \in H$ such that $x_n \to x$. It is easy to verify that the Fourier coefficients of x coincide with c_k and hence $Lx = c$. In conclusion, L is invertible and the inverse L^{-1} is continuous, being an isometry as well. □

1.7 Stampacchia and Lax-Milgram Theorems

In this section we will consider just real Hilbert spaces, therefore in the following we will omit to specify it. We want to reformulate and generalize Theorem 1.5.1 in order to make it applicable to the problems that we will study in the sequel.

Definition 1.7.1. *Let H_1 and H_2 be Hilbert spaces. A* **bilinear form** *on $H_1 \times H_2$ is a function*

$$a: H_1 \times H_2 \to \mathbb{R}$$

satisfying the following properties:

(i) for every fixed $v \in H_2$, the function $u \mapsto a(u,v)$ is linear on H_1

(ii) for every fixed $u \in H_1$, the function $v \mapsto a(u,v)$ is linear on H_2.

If $H_1 = H_2 = H$ we will simply say that $a(\cdot,\cdot)$ is a bilinear form on H, instead of $H \times H$.

Definition 1.7.2. *A bilinear form $a : H \times H \to \mathbb{R}$ is said to be*

(i) **continuous** *if there exists a constant $C > 0$ such that*

$$|a(u,v)| \leq C\|u\|\|v\| \qquad \forall u, v \in H;$$

(ii) **coercive** *if there exists a constant $\alpha > 0$ such that*

$$|a(v,v)| \geq \alpha\|v\|^2 \qquad \forall v \in H;$$

(iii) **symmetric** *if*

$$a(u,v) = a(v,u) \qquad \forall u, v \in H.$$

Let us extend Theorem 1.5.1 to a slightly more general setting:

Theorem 1.7.3. (Stampacchia Theorem). *Let H be a Hilbert space and let $a(\cdot,\cdot)$ be a continuous and coercive bilinear form on H. Let $K \subset H$ be a non empty closed convex set. Given $f \in H'$, there exists a unique $u \in K$ such that*

$$a(u, v-u) \geq \langle f, v-u \rangle \qquad \forall v \in K.$$

Moreover, if $a(\cdot,\cdot)$ is symmetric, then u is characterized by the following property:

$$\frac{1}{2}a(u,u) - \langle f,u\rangle = \min_{v\in K}\left\{\frac{1}{2}\ a(v,v) - \langle f,v\rangle\right\}.$$

Proof. It is not too different from the one of Theorem 1.5.1; see [10, Teorema V.6] for the details. □

In the case in which the convex set K coincides with the whole space we have the

Corollary 1.7.4. (Lax-Milgram Theorem). *Let H be a Hilbert space and let $a(\cdot,\cdot)$ be a continuous and coercive bilinear form on H. Then for every $f \in H'$, there exists a unique $u \in H$ such that*

$$a(u,v) = \langle f, v\rangle \qquad \forall v \in H. \tag{1.15}$$

Moreover, if $a(\cdot,\cdot)$ is symmetric, u is characterized by the property

$$\frac{1}{2}a(u,u) - \langle f,u\rangle = \min_{v\in H}\left\{\frac{1}{2}\ a(v,v) - \langle f,v\rangle\right\}. \tag{1.16}$$

Proof. The first part is a direct consequence of Theorem 1.7.3. The only change lies in (1.15): we leave this small effort to the reader as an exercise. □

The Lax-Milgram Theorem is a simple and powerful tool for the resolution of linear elliptic partial differential equations, see Section 7.1. It is worth to notice the link between the equation (1.15) and the minimization problem (1.16). This relationship has often a mechanical interpretation (principle of least action, minimization of an energy, etc.). In the Calculus of Variations, the equation (1.15) is called **Euler equation** of the minimization problem (1.16).

Remark 1.7.5. Let us assume that $H = \mathbb{R}$ and $a(v,w) = v\cdot w$ for every $v, w \in \mathbb{R}$. Thus considering $f \in \mathbb{R}$, the function defined in (1.16) becomes

$$\varphi(v) = \frac{v^2}{2} - f\cdot v$$

and a minimizer $u \in \mathbb{R}$ satisfies $\varphi'(u) = 0$ and hence $u = f$. On the other hand, if $u \in \mathbb{R}$ solves problem (1.15) then $u\cdot v = f\cdot v$ for every $v \in \mathbb{R}$ and hence, again, $u = f$. □

This remark underlines how Corollary 1.7.4 is the generalization of an elementary principle; for this very reason, it is a powerful tool to prove existence and uniqueness of solutions in generic Hilbert spaces. Our use will be precisely to apply it in solving elliptic equations.

1.8 Spectral theory for real Hilbert spaces

In this section we introduce the notions and the basic results concerning the spectral theory on real Hilbert spaces. In the following we will omit to specify that the spaces considered are real spaces.

If H is a generic Hilbert space, we simply denote by $\mathcal{L}(H)$ the space $\mathcal{L}(H,H)$, see Section 1.3.

Definition 1.8.1. *Let H be a Hilbert space and let $T \in \mathcal{L}(H)$. Then it is defined the* **resolvent set of** T, *and it denoted by $\rho(T)$, the set*

$$\{\lambda \in \mathbb{R} \; ; \; T - \lambda I : H \to H \text{ is bijective}\}$$

where $I \in \mathcal{L}(H)$ denotes the identity operator on H. The set

$$\sigma(T) := \mathbb{R} \backslash \rho(T)$$

is called the **spectrum** *of T while the set*

$$\sigma_p(T) := \{\lambda \in \mathbb{R} \; ; \; \ker(T - \lambda I) \neq \{0\}\} \subseteq \sigma(T)$$

is called **point spectrum** *of T and its elements are called* **eigenvalues of** T. *Moreover, if $\lambda \in \mathbb{R}$ is an eigenvalue, then the subspace* $\ker(T - \lambda I)$ *is called* **eigenspace** *and its elements are called* **eigenvectors** *associated to λ.*

Remark 1.8.2. Note that if $H = \mathbb{R}^n$ then the spectrum and the point spectrum coincide and the aforementioned notion of eigenvalue coincides with the usual definition.

Though, if H has infinite dimension, it may happen that the point spectrum is strictly contained into the spectrum. Here we report the following example taken from [10]: let $H = \ell^2$ and let $T \in \mathcal{L}(\ell^2)$ be the operator which associates to every vector $x = \{x_1, x_2, ...\} \in \ell^2$ the vector $Tx := \{0, x_1, x_2, ...\}$. Notice that T is injective and non surjective and hence $0 \in \sigma(T)$, but $0 \notin \sigma_p(T)$ since $\ker(T - 0I) = \ker(T) = \{0\}$. □

An important property of linear operators is expressed in the

Definition 1.8.3. *An operator $T \in \mathcal{L}(H_1, H_2)$ is said to be* **compact** *if for every bounded set $B \subset H_1$ the set $T(B) \subset H_2$ has compact closure in H_2.*

From Theorem 1.3.2 we immediately deduce that a compact linear operator is also continuous; actually, there holds a much stronger property:

Theorem 1.8.4. *Let H_1 and H_2 be Hilbert spaces. An operator $T \in \mathcal{L}(H_1, H_2)$ is compact if and only if for every sequence $\{x_n\} \subset H_1$ such that $x_n \rightharpoonup x$ in H_1 there holds $Tx_n \to Tx$ in H_2.*

Proof. First, let us assume that T is compact and let $\{x_n\} \subset H_1$ be such that $x_n \rightharpoonup x$ in H_1. In view of Theorem 1.4.13, we obtain that $\{x_n\}$ is bounded and hence, by the compactness of T, we deduce that $\{Tx_n\}$ is a set with compact closure in H_2. Therefore we can extract from the sequence $\{Tx_n\}$ a subsequence such that $Tx_{n_k} \to y$ for some $y \in H_2$. By applying Theorem 1.3.2 we infer that $Tx_{n_k} \to Tx$, from which we deduce that $y = Tx$ does not depend on the subsequence chosen. This allows to conclude that the original sequence satisfies $Tx_n \to Tx$ in H_2.

Viceversa, let us assume that for every sequence $\{x_n\} \subset H_1$ such that $x_n \rightharpoonup x$ in H_1 there holds $Tx_n \to Tx$ in H_2 and let us prove that T is compact. Let $B \subset H_1$ be a bounded set and let us consider an arbitrary sequence $\{y_n\} \subset T(B)$. For every n there exists $x_n \in B$ such that $y_n = Tx_n$ and hence $\{x_n\} \subset B$ is a bounded sequence in H_1 and, in view of Theorem 1.4.13, we can state that there exists a subsequence $\{x_{n_k}\}$ such that $x_{n_k} \rightharpoonup x$ for some $x \in H_1$. Thanks to the assumption made above, there holds $y_{n_k} = Tx_{n_k} \to Tx$. Thus we have proved that every sequence contained in $T(B)$ admits a converging subsequence. Similarly, every sequence contained into $\overline{T(B)}$ admits a subsequence converging to some element of $\overline{T(B)}$. This proves that $T(B)$ has compact closure. □

Among the compact operators, the ones that will play a crucial role in the next chapters will be the embeddings:

Definition 1.8.5. *Let X and Y be two Hilbert spaces such that $X \subset Y$. We will say that X is* **continuously embedded** *in Y if there exists a constant $C > 0$ such that*

$$\|x\|_Y \leq C\ \|x\|_X \qquad \forall x \in X. \tag{1.17}$$

The condition (1.17) *means that the embedding (i.e. the inclusion mapping) from X into Y is continuous (see Theorem 1.3.2). Moreover, we will say that the embedding $X \subset Y$ is compact if*

$$x_n \overset{X}{\rightharpoonup} x \ \Rightarrow\ x_n \overset{Y}{\longrightarrow} x;$$

referring to this situation we will use the notation $X \subset\subset Y$.

Given an operator $T \in \mathcal{L}(H)$ it is possible to define an operator $T^* \in \mathcal{L}(H)$ called the **adjoint** of T, so that

$$(Tx, y) = (x, T^*y) \qquad \forall x, y \in H. \tag{1.18}$$

Indeed for every fixed $y \in H$ it is enough to define the functional $f_y \in H'$ given by $\langle f_y, x\rangle := (Tx, y)$ for every $x \in H$. Thanks to Theorem 1.5.5 we can affirm that there exists a unique vector depending on y, that we will denote by T^*y, such that

$$(Tx, y) = \langle f_y, x\rangle = (T^*y, x).$$

Once constructed the operator T^* in this way, it is an easy exercise to verify that $T^* \in \mathcal{L}(H)$.

We now give the following

Definition 1.8.6. *Let H be a Hilbert space. An operator $T \in \mathcal{L}(H)$ is said to be* **self-adjoint** *if $T = T^*$.*

We now state some results concerning the spectral properties of compact operators.

Theorem 1.8.7. (Fredholm alternative Theorem). *Let H be a Hilbert space and let $T \in \mathcal{L}(H)$ be compact. Then for every $\lambda \neq 0$ the following conclusions hold true:*

$$\dim \ker(T - \lambda I) = \dim \ker(T^* - \lambda I) < \infty$$

and

$$R(T - \lambda I) = \ker(T^* - \lambda I)^{\perp}.$$

Proof. See [10, Teorema VI.6]. □

Theorem 1.8.8. *Let H be a Hilbert space and let $T \in \mathcal{L}(H)$ be compact. Then $\sigma(T)\backslash\{0\} = \sigma_p(T)\backslash\{0\}$. Moreover $\sigma(T)$ is bounded and has at most zero as cluster point. In particular, $\sigma(T)$ is at most countable.*

Proof. See [10, Teorema VI.8]. □

Theorem 1.8.9. *Let H be a Hilbert separable space and let $T \in \mathcal{L}(H)$ be compact and self-adjoint. If λ, μ are distinct eigenvalues of T then the corresponding eigenspaces are mutually orthogonal. Moreover, there holds*

$$H = \overline{\text{span}} \bigcup_{\lambda \in \sigma_p(T)} \ker(T - \lambda I)$$

where $\overline{\text{span}}$ denotes the closure of the space of finite linear combinations of eigenvectors of T.

Proof. See [10, Teorema VI.11]. □

Theorems 1.8.7-1.8.9 admit a "variational version" which has a fundamental importance in the study of the partial differential equations that we will treat in the next chapters.

Let us introduce some notations. Consider two Hilbert separable spaces $V \subset H$ such that the inclusion $I_1 : V \to H$ is compact (consequently, continuous). In the following we will denote by $(\cdot, \cdot)$ and $\|\cdot\|$ the scalar product and the norm in V respectively, and by $(\cdot, \cdot)_H$ and $\|\cdot\|_H$ the scalar product and the norm in H respectively. Then it is possible to embed in a suitable sense the space H into V' in order to obtain a so called **Hilbert triplet**

$$V \subset H \subset V'.$$

The inclusion $H \subset V'$ is to be meant according to the embedding $I_2 : H \to V'$ defined by

$$\langle I_2 u, v\rangle := (u, v)_H \qquad \forall u \in H \ \forall v \in V .$$

We are identifying a vector $u \in H$ with $I_2 u$, which is a linear and continuous functional on V. In the following, the functional $I_2 u \in V'$ will be simply denoted by u. We now define on V a bilinear continuous form $a : V \times V \to \mathbb{R}$. In place of coercivity we will require a less restrictive condition, called **weak coercivity**: we assume that there exist a constant $\alpha > 0$ and a constant $\lambda_0 \geq 0$ such that

$$a(v,v) + \lambda_0 \|v\|_H^2 \geq \alpha \|v\|^2 \qquad \forall v \in V\,. \tag{1.19}$$

Let us introduce the operator $L \in \mathcal{L}(V, V')$ defined by

$$\langle Lu, v \rangle = a(u,v) \qquad \forall u, v \in V\,.$$

We intend to study variational problems that are more general than the ones studied with the Lax-Milgram Theorem (Corollary 1.7.4). More precisely, we consider the problem of the existence of a vector $u \in V$ such that

$$a(u,v) = \lambda(u,v)_H + \langle f, v \rangle \qquad \forall v \in V \tag{1.20}$$

for λ varying in $\mathbb{R}$ and f in V'. Furthermore, we consider two other problems, called respectively **associated homogeneous** problem and **adjoint homogeneous** problem:

$$a(u,v) = \lambda(u,v)_H \qquad \forall v \in V \tag{1.21}$$

and

$$a_*(u,v) = \lambda(u,v)_H \qquad \forall v \in V\,, \tag{1.22}$$

where $a_*(\cdot,\cdot)$ is the bilinear form called **adjoint** of $a(\cdot,\cdot)$, defined by

$$a_*(u,v) = a(v,u) \qquad \forall u, v \in V.$$

Clearly, $u = 0$ is a solution for (1.21) and (1.22). The existence of non trivial solutions is not as clear. If the latter situation occurs for a certain $\lambda \in \mathbb{R}$, λ is said to be an **eigenvalue** and the corresponding solution is said to be an **eigenvector**. Since the set of eigenvectors E_λ is a vector subspace of V, such set is called **eigenspace** of λ and its dimension is called **geometric multiplicity** of λ. Finally, the set of eigenvalues is called **spectrum** and is denoted by $\sigma_p(a)$.

We can state the following

Theorem 1.8.10. *Let $V \subset H$ be separable Hilbert spaces and let (V, H, V') be a Hilbert triplet such that the embedding $V \subset H$ is compact and such that V is dense in H. Let $a : V \times V \to \mathbb{R}$ be a bilinear continuous form verifying condition* (1.19) *for some $\lambda_0 \geq 0$. Let $\lambda \in \mathbb{R}$ and $f \in V'$. Then the following conclusions hold true:*

(i) λ is an eigenvalue of (1.21) *if and only if λ is an eigenvalue of* (1.22) *and the eigenspaces of the two problems have finite and equal dimension;*

(ii) the set of eigenvalues $\sigma_p(a)$ has no cluster point and is contained into the half-line $(-\lambda_0, \infty)$;

(iii) problem (1.20) *admits a solution $u \in V$ if and only if $\langle f, u_* \rangle = 0$ for every solution u_* for* (1.22)*;*

(iv) if, in addition, the bilinear form $a(\cdot,\cdot)$ is symmetric and if u and v are eigenvectors of (1.21) associated to distinct eigenvalues, then there holds

$$a(u, v) = (u, v)_H = 0$$

and moreover

$$H = \overline{\text{span}}^H \bigcup_{\lambda\in\sigma_p} E_\lambda, \tag{1.23}$$

where we denote by E_λ the eigenspace corresponding to λ and by $\overline{\text{span}}^H$ the closure in H of the space of finite linear combinations of eigenvectors of $a(\cdot,\cdot)$. Finally, if we endow V with the new scalar product

$$((u, v)) := a(u, v) + \lambda_0(u, v)_H \qquad \forall u, v \in V$$

equivalent to $(\cdot,\cdot)$, then

$$V = \overline{\text{span}}^V \bigcup_{\lambda\in\sigma_p} E_\lambda \tag{1.24}$$

where by $\overline{\text{span}}^V$ we mean the closure in V of the space of finite linear combinations of eigenvectors of $a(\cdot,\cdot)$.

Proof. Here we give just an idea of the proof. Let $L \in \mathcal{L}(V, V')$ be the operator implicitly defined by

$$\langle Lu, v\rangle := a(u, v) \qquad \forall u, v \in V.$$

If we define $I = I_2 \circ I_1$, then by the Lax-Milgram Theorem we deduce that the operator $L_0 := L+\lambda_0 I : V \to V'$ is invertible and that the linear operator $K : V \to V$ defined by $K := L_0^{-1} \circ I$ is compact since L_0^{-1} is continuous and I is compact. The continuity of L_0^{-1} follows from the open mapping Theorem, see [10, Corollario II.6]. At this point, the proof is achieved by applying Theorems 1.8.7-1.8.9 to the operator K. □

Chapter 2

Lebesgue integral and L^p spaces

The theory of Lebesgue integration is an important tool in the applications, therefore, having a good knowledge of it is essential. In this chapter we give the basic definitions and the main results.

If compared to the more intuitive theory of Riemann integration, the theory of Lebesgue integration introduces a wider class of integrable functions; such extension of the concept of integrability endows the L^1 space of Lebesgue integrable functions of the structure of Banach space, when endowed with the integral norm. The same kind of structure is inherited also by the L^p space of functions u such that $|u|^p \in L^1$.

Therefore, the theoretical results described in Chapter 1 naturally apply to L^p spaces. In turn, the theory of L^p spaces finds natural applications in the theory of distributions, of integral transforms and of Sobolev spaces.

The last part of this chapter is devoted to distributions. The main reason for which the distributions were introduced is the necessity of having at disposal a set that contains all the local integrable functions and, at the same time, that allows to differentiate, in a suitable sense, its elements infinitely many times. Moreover, it is worth to observe that different physical phenomena that can be ascribed to a "pulse" are modeled by the so called "Dirac delta function", namely a sort of function which takes the infinite value at a point and equals zero elsewhere. An object of this kind, that cannot be properly considered a function, turns out to be a distribution.

2.1 Riemann integral

We first consider the case of a bounded interval Q of $\mathbb{R}^n$, namely the Cartesian product of n intervals of $\mathbb{R}$:

$$Q = [a_1, b_1] \times [a_2, b_2] \times ... \times [a_n, b_n]$$

(we remark that the intervals are closed). The measure (n-dimensional) of Q is simply the product of the measures of the intervals $[a_i, b_i]$:

$$|Q| := (b_1 - a_1)(b_2 - a_2)...(b_n - a_n). \tag{2.1}$$

Such definition is the natural extension of the notions of length of a segment ($n = 1$), area of a rectangle ($n = 2$) and volume of a right parallelepiped ($n = 3$).

Then we split the intervals $[a_k, b_k]$ (for $k = 1, ..., n$) introducing a finite number m_k (possibly depending on k) of values which we denote by c_k^i:

$$a_k =: c_k^0 < c_k^1 < ... < c_k^{m_k} < c_k^{m_k+1} := b_k.$$

In this way we obtain a partition of Q into a finite number of intervals:

$$Q = \bigcup_* Q_{i_1,...,i_n} \qquad \text{with } Q_{i_1,...,i_n} = [c_1^{i_1}, c_1^{i_1+1}] \times ... \times [c_n^{i_n}, c_n^{i_n+1}] \tag{2.2}$$

where $\cup_*$ denotes the union as the index i_k varies between 0 and m_k.

Let us now consider a bounded function $f : Q \to \mathbb{R}$ and let us set

$$M_{i_1,...,i_n} = \sup_{Q_{i_1,...,i_n}} f, \qquad \mu_{i_1,...,i_n} = \inf_{Q_{i_1,...,i_n}} f.$$

We define the upper sum and lower sum of f with respect to the subdivision (2.2):

$$S = \sum_* M_{i_1,...,i_n}(c_1^{i_1+1} - c_1^{i_1})...(c_n^{i_n+1} - c_n^{i_n}), \quad s = \sum_* \mu_{i_1,...,i_n}(c_1^{i_1+1} - c_1^{i_1})...(c_n^{i_n+1} - c_n^{i_n}).$$

Obviously, for any choice of the partition (2.2), it holds $s \leq S$. Moreover, making the partition thicker, the upper sums do not increase and the lower sums do not decrease. Thus we can conclude that *every lower sum is less or equal to every upper sum, even with respect to different partitions.*

Then we let the subdivision of Q vary and we define

$$S_-(f) := \inf S, \qquad s_+(f) := \sup s,$$

where inf and sup are computed with respect to all the possible partitions of Q. In view of what we have just said, we surely obtain that $S_-(f) \geq s_+(f)$. In case $S_-(f) = s_+(f)$ we will say that f is **Riemann integrable** in Q and we shall set

$$\int_Q f(x)\,dx = S_-(f) = s_+(f).$$

Now we pass to a general bounded subset Ω of $\mathbb{R}^n$ and we consider $f : \Omega \to \mathbb{R}$. It is possible to enclose Ω in an interval $Q \supset \Omega$ and define the function

$$\widetilde{f}(x) := \begin{cases} f(x) & \text{if } x \in \Omega \\ 0 & \text{if } x \in Q \setminus \Omega. \end{cases}$$

We will say that f is Riemann integrable in Ω if $\widetilde{f}$ is Riemann integrable in Q; in this case, we shall set

$$\int_\Omega f = \int_Q \widetilde{f} \,.$$

Here we are not interested in listing the properties of the Riemann integral, for which we refer to [56, Capitolo 5]. We limit ourselves to make some remarks which will be useful for the comparison with the Lebesgue integral.

First of all, we observe that the definition of Riemann integral is very simple and intuitive. Basically, it amounts to approximate a bounded function f with two "step" functions (piecewise constant). The first function gives an overestimation, the second one an underestimation. If we are able to provide "good" approximations (such that the overestimation and the underestimation differ by arbitrarily small quantities) we shall say that the function is Riemann integrable. This ease is definitely an advantage with respect to the Lebesgue integral, but there are some inconveniences: for its simplicity, there is the risk of not being able to integrate "strange" functions. It is easy to show (using the notion of uniform continuity) that every function which is continuous over a compact set is Riemann integrable. Though, if the function is "too discontinuous" the machinery fails. For instance, let us consider the Dirichlet function f defined in the interval $[0,1]$ by

$$f(x) := \begin{cases} 1 & \text{if } x \in [0,1] \cap \mathbb{Q} \\ 0 & \text{if } x \in [0,1] \setminus \mathbb{Q}. \end{cases} \tag{2.3}$$

It is easy to understand that for every subdivision of $[0,1]$ it holds $S = 1$ and $s = 0$; so that f is not Riemann integrable in $[0,1]$. But the main problem does not consist in the "lack" of Riemann integrable functions. In fact, the Riemann integral does not fit well into the setting of Functional Analysis introduced in Chapter 1, since the space of uniformly bounded and Riemann integrable functions is not closed with respect to the pointwise convergence: let us illustrate this fact with an example which involves again the Dirichlet function.

Example 2.1.1. We order the rational numbers in $[0,1]$ by means of a sequence $\{a_k\}$. This can be done since $\mathbb{Q}$ is countable (see Theorem 1.6.2). Then we set

$$f_k(x) = \begin{cases} 1 & \text{if } x \in \{a_1, ..., a_k\} \\ 0 & \text{if } x \notin \{a_1, ..., a_k\}\,. \end{cases} \tag{2.4}$$

The sequence f_k is bounded and has k removable discontinuity points, hence it is Riemann integrable and $\int_0^1 f_k = 0$. However, the pointwise limit of the sequence $\{f_k\}$ is the Dirichlet function (2.3), which is not Riemann integrable. □

In order to avoid these inconveniences and have an integral which is suitable for the tools of Functional Analysis, we have to change the definition of integral.

2.2 Lebesgue measure

2.2.1 Open and closed sets

We start again from the intervals $Q \subset \mathbb{R}^n$ and from their measure defined by (2.1). We call **pluri-interval** (or pluri-rectangle) the union R of a finite number of intervals $Q_1, Q_2, ..., Q_k$ with pairwise disjoint interiors:

$$R = \bigcup_{i=1}^{k} Q_i, \qquad \mathring{Q}_i \cap \mathring{Q}_j = \emptyset \quad \forall i \neq j.$$

The measure of the pluri-interval is defined as the sum of the measures of the intervals it is composed by:

$$|R| := \sum_{i=1}^{k} |Q_i|.$$

Clearly, the same pluri-interval can be composed by the union of intervals corresponding to different subdivisions; but it is clear as well that its measure does not depend on the subdivision chosen.

Exercise 2.2.1. Show that if R_1 and R_2 are pluri-intervals, we have
(i) $|R_1| \leq |R_2|$ if $R_1 \subseteq R_2$;
(ii) $|R_1 \cup R_2| \leq |R_1| + |R_2|$, with the equality sign if and only if $\mathring{R}_1 \cap \mathring{R}_2 = \emptyset$. □

After having considered the pluri-intervals, we pass to the measure of open and closed sets:

Definition 2.2.2. *Let A be an open bounded non empty subset of $\mathbb{R}^n$. The measure of A is the supremum of the measures of the pluri-intervals contained in A:*

$$|A| := \sup\{|R|;\ R\ pluri-interval,\ R \subset A\}.$$

Clearly, it holds $|A| > 0$; indeed, if $x \in A$, there exists a spherical neighborhood of x with radius $r > 0$ entirely contained into A; such neighborhood contains at least an interval: e.g., the n-dimensional cube with side $2r/\sqrt{n}$. We also remark that, if R is a pluri-interval, it holds $|\mathring{R}| = |R|$.

Definition 2.2.3. *Let C be a closed bounded non empty subset of $\mathbb{R}^n$. The measure of C is the infimum of the measures of the pluri-intervals containing C:*

$$|C| := \inf\{|R|;\ R\ pluri-interval,\ R \supseteq C\}.$$

It may happen that $|C| = 0$ also for closed non empty sets: it is enough to consider a point in $\mathbb{R}^n$ (with $n \geq 1$) or a regular curve in $\mathbb{R}^n$ (with $n \geq 2$).

2.2.2 Generic sets

After having defined the measures of open and closed sets, we are now in position to consider a generic subset of $\mathbb{R}^n$:

Definition 2.2.4. *Let E be a bounded subset of $\mathbb{R}^n$. The outer measure $|E|^*$ of E is the infimum of the measures of the open sets which contain E; the inner measure $|E|_*$ of E is the supremum of the measures of the closed sets which are contained into E:*

$$|E|^* := \inf\{|A|; A \text{ open bounded}, \ A \supseteq E\} \ , \quad |E|_* := \sup\{|C|; C \text{ closed}, \ C \subseteq E\}.$$

If E is open or closed, obviously it holds $|E|^* = |E| = |E|_*$. Moreover, for every bounded set E it holds $|E|^* \geq |E|_*$.

Definition 2.2.5. *A bounded set $E \subset \mathbb{R}^n$ is said to be* **Lebesgue measurable** *if its measures, inner and outer, coincide. The number*

$$|E| := |E|^* = |E|_*$$

is then called **n-dimensional measure** *of E.*

We have introduced these notions of measurability and measure only for bounded sets of $\mathbb{R}^n$. Nevertheless, in the applications it may be useful to extend the concept of measurability to unbounded sets, also allowing, for some of them, the measure to be infinite. To this purpose, after observing that the intersection of two (bounded) measurable sets is measurable, we give the following

Definition 2.2.6. *A set $E \subseteq \mathbb{R}^n$ (not necessarily bounded) is said to be* **Lebesgue measurable** *if for every $r > 0$ the set $E \cap B_r$ is measurable according to Definition 2.2.5, with $B_r := \{x \in \mathbb{R}^n; \ |x| < r\}$. In this case, it is defined*

$$|E| := \lim_{r \to \infty} |E \cap B_r| \in [0, \infty].$$

By choosing $r > 0$ large enough (so that $E \cap B_r = E$), we deduce that the sets that are measurable according to Definition 2.2.5 are measurable also according to Definition 2.2.6. From now on, by Lebesgue measurability we will mean the notion introduced in Definition 2.2.6.

Exercise 2.2.7. Let $E, F \subset \mathbb{R}^n$ be measurable. Prove that $E \cup F$, $E \cap F$, $E \backslash F$, $E^c := \mathbb{R}^n \backslash E$ are measurable sets according to Definition 2.2.6. □

Example 2.2.8. According to Definition 2.2.6, the set $\mathbb{R}^n$ itself turns out to be a measurable set of infinite measure. However, there exist unbounded sets with finite measure. For instance, let

$$E := \{(x_1, x_2) \in \mathbb{R}^2; \ x_1 > 1 \ , \ 0 < x_2 < 1/x_1^2\}$$

then $|E|$ is nothing but the area of the portion of the half-plane $x_1 > 1$ lying between the x_1 axis and the graph of the function $f(x_1) := 1/x_1^2$. As a consequence, it holds

$$|E| = \int_1^\infty \frac{1}{x_1^2}\, dx_1 = 1. \qquad \square$$

The main property of measurable sets is stated in the following

Theorem 2.2.9. *Let $\{E_i\}$ be a sequence of measurable subsets of $\mathbb{R}^n$; then the set $E := \bigcup_{i=1}^{\infty} E_i$ is measurable and the following inequality (countable subadditivity of the measure) holds true:*

$$|E| \leq \sum_{i=1}^{\infty} |E_i|. \tag{2.5}$$

In particular, if $E_i \cap E_j = \emptyset$ for every $i \neq j$ then (2.5) holds true with the equality sign (countable additivity of the measure). Moreover, if in addition we assume that the sequence $\{E_i\}$ is non decreasing with respect to the inclusion, namely $E_1 \subseteq E_2 \subseteq \ldots$, then

$$|E| = \lim_{i \to \infty} |E_i|.$$

Proof. See [56, Capitolo 5, Teorema 2.4]. $\square$

Exercise 2.2.10. Using Theorem 2.2.9, prove the following facts:

(i) Let $\{E_i\}$ be a sequence of measurable sets of $\mathbb{R}^n$. Then the set $G = \bigcap_{i=1}^{\infty} E_i$ is measurable.

(ii) If in addition $\{E_i\}$ is a sequence which is non increasing with respect to the inclusion, namely such that $E_1 \supseteq E_2 \supseteq \ldots$, then $|G| = \lim_{i\to\infty} |E_i|$. $\square$

In order to understand the importance of Theorem 2.2.9, observe that the intersection of countably many open sets is not in general open, and that the countable union of closed sets is not in general closed.

Remark 2.2.11. A point has measure zero; a set of countably many points, in view of (2.5) in Theorem 2.2.9, has measure zero as well; more generally, the countable union of sets of measure zero is a set of measure zero. $\square$

Basically, almost all the sets are measurable; the non measurable ones are quite "pathological" cases. A famous example is the one due to Vitali [60] and is based upon the axiom of choice and upon a suitable equivalence relation.

Example 2.2.12. Let us consider the interval $I = [0, 1)$ and let us establish the following equivalence relation r (that we will denote by $\approx$) on its points:

$$r:\ x \approx y \Leftrightarrow |x - y| \in [0, 1) \cap \mathbb{Q}.$$

We create a partition of I by splitting it in equivalence classes with respect to r. Since the set $\mathbb{Q}$ is countable, every equivalence class contains countably many elements. Since I is uncountable, in view of Theorem 1.6.2 we infer that the number of equivalence classes is uncountable.
With a suitable choice rule, we construct a set T by selecting only one element from each equivalence class. For what we have said, T is uncountable.
Suppose by contradiction that T is measurable and, for every $s \in [0,1) \cap \mathbb{Q}$, consider the sets defined as follows:

$$T_s = \{x \in [0,1) \ ; \ x = y+s, \ y \in T\} \cup \{x \in [0,1) \ ; \ x = y+s-1, \ y \in T\} = T_s^1 \cup T_s^2;$$

$$\widehat{T}_s = \{x \in \mathbb{R} \ ; \ x = y+s, \ y \in T\}.$$

Since $\widehat{T}_s$ is the translation of T by the quantity s, it has the same measure of T. We observe that $\widehat{T}_s = T_s^1 \cup T_s^3$ with

$$T_s^3 = \{x \in \mathbb{R} \ ; \ x = y+1, \ y \in T_s^2\} \subseteq [1,2) \ ,$$

moreover $|T_s^2| = |T_s^3|$, being one a shift of the other.
Since $T_s^1 \cap T_s^3 = \emptyset = T_s^1 \cap T_s^2$, we obtain

$$|T_s| = |T_s^1| + |T_s^2| = |T_s^1| + |T_s^3| = |\widehat{T}_s| = |T| \qquad \forall s \in [0,1) \cap \mathbb{Q}. \tag{2.6}$$

Since $T_s \cap T_t = \emptyset$ if $t, s \in [0,1)$ are distinct rational numbers, in view of Theorem 2.2.9 and by (2.6) we derive that

$$\left| \bigcup_{s \in [0,1) \cap \mathbb{Q}} T_s \right| = \sum_{s \in [0,1) \cap \mathbb{Q}} |T_s| = \begin{cases} 0 & \text{if } |T| = 0 \\ +\infty & \text{if } |T| > 0. \end{cases} \tag{2.7}$$

On the other hand, since $\bigcup_{s \in [0,1) \cap \mathbb{Q}} T_s = [0,1)$ (prove it!), it turns out that

$$\left| \bigcup_{s \in [0,1) \cap \mathbb{Q}} T_s \right| = 1,$$

in contradiction with (2.7). □

2.3 Lebesgue integral

2.3.1 Measurable functions, integrable functions

We begin with the following

Definition 2.3.1. *Let $\Omega \subseteq \mathbb{R}^n$ be a Lebesgue measurable set. A function $f : \Omega \to \mathbb{R}$ is said to be* **measurable** *if for every $c \in \mathbb{R}$ the set*

$$\{x \in \mathbb{R}^n;\ f(x) < c\}$$

is Lebesgue measurable.

We leave to the reader the proof of the main properties of measurable functions, which we gather in the following

Exercise 2.3.2. Let $\Omega \subseteq \mathbb{R}^n$ be a Lebesgue measurable set. Prove that:
(i) $f : \Omega \to \mathbb{R}$ is measurable if and only if for every $c \in \mathbb{R}$ the following sets are measurable as well:
$\{x \in \mathbb{R}^n;\ f(x) \leq c\}$, $\{x \in \mathbb{R}^n;\ f(x) > c\}$, $\{x \in \mathbb{R}^n;\ f(x) \geq c\}$;
(ii) if $f : \Omega \to \mathbb{R}$ is measurable then also the following functions are measurable: $f^+ := \max\{f, 0\}$, $f^- := -\min\{f, 0\}$, $|f|$, $c \cdot f$ (for every $c \in \mathbb{R}$);
(iii) if $f, g : \Omega \to \mathbb{R}$ are measurable then also the functions $f + g$, $f \cdot g$, $\min\{f, g\}$, $\max\{f, g\}$ are measurable;
(iv) if $\{f_k\}$ is a sequence of measurable functions defined on Ω which converges pointwise to a function f, then f is measurable. □

Definition 2.3.3. *Let $\Omega \subseteq \mathbb{R}^n$ be measurable and let $f, g : \Omega \to \mathbb{R}$ be measurable. We say that f equals g* **almost everywhere** *in Ω, and we write $f = g$ a.e. in Ω, if*

$$|\{x \in \Omega;\ f(x) \neq g(x)\}| = 0. \tag{2.8}$$

Moreover, we say that a certain property is true for a.e. $x \in \Omega$ if the set of $x \in \Omega$ for which such property does not hold is a set with measure zero.

We remark that, actually, the set considered in (2.8) is measurable thanks to the properties stated in Exercise 2.3.2.

Let $\Omega \subset \mathbb{R}^n$ be a set of finite measure and let $f : \Omega \to \mathbb{R}$ be a bounded measurable function, whose image is contained into the interval (m, M):

$$m < f(x) < M \qquad \forall x \in \Omega.$$

We introduce the lower and upper sums (according to Lebesgue) of f as follows. Consider a partition $\mathcal{D} = \{\alpha_0, \alpha_1, ..., \alpha_N\}$ of the interval $[m, M]$ with $m = \alpha_0 < \alpha_1 < ... < \alpha_N = M$; define the (disjoint) sets

$$\Omega^i = \{x \in \Omega : \alpha_{i-1} \leq f(x) < \alpha_i\} = f^{-1}([\alpha_{i-1}, \alpha_i)) \quad i = 1, ..., N.$$

Such sets are all measurable, see Exercises 2.2.7 and 2.3.2. Thus we set

$$S(\mathcal{D}, f) = \sum_{i=1}^{N} \alpha_i |\Omega^i|\,, \qquad s(\mathcal{D}, f) = \sum_{i=1}^{N} \alpha_{i-1} |\Omega^i|.$$

Clearly, for every partition $\mathcal{D}$, it holds

$$m|\Omega| \leq s(\mathcal{D}, f) \leq S(\mathcal{D}, f) \leq M|\Omega|. \tag{2.9}$$

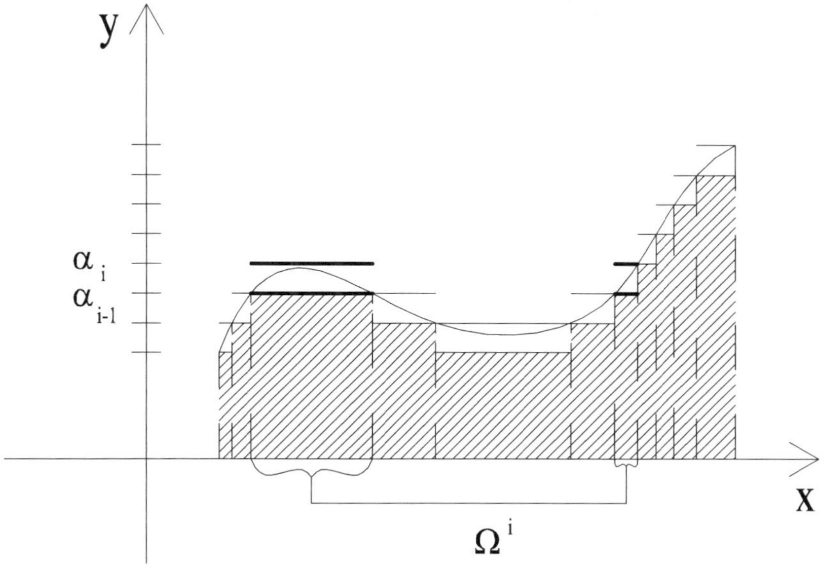

Lower and upper sums according to Lebesgue

Formula (2.9) compares the lower and upper sums of f corresponding to the same partition. It is easy to see that, making the partition $\mathcal{D}$ thicker, the upper sums do not increase and the lower sums do not decrease. Thus we may conclude that *every lower sum is less or equal to every upper sum, even with respect to different partitions.*

In other words, for $\mathcal{D}$ varying among all the possible partitions of $[m, M]$, we have

$$\sup_{\mathcal{D}} s(\mathcal{D}, f) \leq \inf_{\mathcal{D}} S(\mathcal{D}, f). \tag{2.10}$$

Now we observe that for every partition $\mathcal{D}$ it holds

$$S(\mathcal{D}, f) - s(\mathcal{D}, f) = \sum_{i=1}^{N} (\alpha_i - \alpha_{i-1})|\Omega^i| \leq \delta|\Omega|\,, \tag{2.11}$$

where we have denoted by δ the size of the subdivision, namely $\delta := \max_i(\alpha_i - \alpha_{i-1})$. Since δ can be chosen arbitrarily small, by combining (2.10) and (2.11) we infer that, for every bounded and measurable function f,

$$\sup_{\mathcal{D}} s(\mathcal{D}, f) = \inf_{\mathcal{D}} S(\mathcal{D}, f). \tag{2.12}$$

Finally, we can give the

Definition 2.3.4. *Let $\Omega \subset \mathbb{R}^n$ have finite measure and let $f : \Omega \to \mathbb{R}$ be bounded and measurable. The number defined in* (2.12) *is called* **Lebesgue integral** *of f on Ω and is denoted by*

$$\int_\Omega f = \int_\Omega f(x)\,dx = \sup_{\mathcal{D}} s(\mathcal{D}, f) = \inf_{\mathcal{D}} S(\mathcal{D}, f).$$

Starting from Definition 2.3.4, which is valid for bounded measurable functions on sets of finite measure, it is possible to introduce the notions of integrability and Lebesgue integral for more general functions and sets. First, we consider a measurable set $\Omega \subseteq \mathbb{R}^n$ with finite measure and a measurable function $f : \Omega \to \mathbb{R}$ which is non negative (not necessarily bounded). For every $\lambda > 0$, we define the truncated function

$$f_\lambda(x) := \begin{cases} f(x) & \text{if } f(x) \le \lambda \\ \lambda & \text{if } f(x) > \lambda. \end{cases} \tag{2.13}$$

For every $\lambda > 0$, $\int_\Omega f_\lambda$ exists and is an increasing function with respect to λ; thus the limit as $\lambda \to \infty$ exists. If such limit is infinity, we shall say that f is not integrable over Ω and we will write

$$\int_\Omega f = +\infty\ . \tag{2.14}$$

If instead such limit is finite, we shall say that f is integrable over Ω and we will set

$$\int_\Omega f = \lim_{\lambda\to\infty} \int_\Omega f_\lambda\ .$$

Let us now consider a measurable set $\Omega \subseteq \mathbb{R}^n$ (not necessarily of finite measure) and a measurable function $f : \Omega \to \mathbb{R}$ which is non negative (not necessarily bounded). We proceed as in Definition 2.2.6, that is, for every $r > 0$, we consider the set $\Omega_r = \Omega \cap B_r$. If there exists $r > 0$ such that f is not integrable on Ω_r, we will say that f is not integrable on Ω; also in this case we may write (2.14). If instead f is integrable on Ω_r for every $r > 0$, we notice that the function

$$r \mapsto \int_{\Omega_r} f$$

is increasing and hence it admits a limit as $r \to \infty$. If such limit is infinity we shall say again that f is not integrable over Ω. If on the contrary such limit is finite, we will say that f is integrable over Ω and we will set

$$\int_\Omega f = \lim_{r\to\infty} \int_{\Omega_r} f\ .$$

Finally, we give the definition of integrability and of Lebesgue integral for functions having arbitrary sign and defined on generic measurable sets.

Definition 2.3.5. *Let $\Omega \subseteq \mathbb{R}^n$ be a Lebesgue measurable set. A measurable function $f : \Omega \to \mathbb{R}$ is said to be Lebesgue integrable if the (non negative) functions $f^+ := \max\{f, 0\}$ and $f^- := -\min\{f, 0\}$ are Lebesgue integrable. In this case it is set*

$$\int_\Omega f := \int_\Omega f^+ - \int_\Omega f^-.$$

The class of Lebesgue integrable functions over Ω will be denoted by $L^1(\Omega)$.

2.3.2 Main properties of the Lebesgue integral

We first list some elementary properties of Lebesgue integrable functions. We leave the easy proof to the reader as an exercise. In all the properties listed below, $\Omega \subseteq \mathbb{R}^n$ is a Lebesgue measurable set.

1. *Homogeneity.* If $f \in L^1(\Omega)$ and $c \in \mathbb{R}$, then $cf \in L^1(\Omega)$ and
$$\int_\Omega cf = c\int_\Omega f.$$

2. *Additivity with respect to the integrand function.* If $f, g \in L^1(\Omega)$, then also $f + g \in L^1(\Omega)$ and
$$\int_\Omega (f+g) = \int_\Omega f + \int_\Omega g.$$

3. *Monotonicity.*
   (a) If $f, g \in L^1(\Omega)$ and $f \leq g$, then $\int_\Omega f \leq \int_\Omega g$.
   (b) $|\int_\Omega f| \leq \int_\Omega |f|$ for every $f \in L^1(\Omega)$.
   (c) If $m \leq f \leq M$ and $|\Omega| < \infty$, then
   $$m \leq \frac{1}{|\Omega|}\int_\Omega f \leq M.$$
   (d) If $|f| \leq g$ and $g \in L^1(\Omega)$, then also $f \in L^1(\Omega)$.

4. *Countable additivity with respect to the domain of integration.* If $f \in L^1(\Omega)$ and $\Omega = \bigcup_{k=1}^{\infty} \Omega_k$, $\Omega_k \cap \Omega_j = \emptyset$ for $k \neq j$, then
$$\int_\Omega f = \sum_{k=1}^{\infty} \int_{\Omega_k} f.$$

5. *Sets of measure zero.* Let $f \in L^1(\Omega)$

(a) If $|\Omega| = 0$, then $\int_\Omega f = 0$.
(b) If $g = f$ a.e. in Ω, then also $g \in L^1(\Omega)$ and $\int_\Omega f = \int_\Omega g$.
(c) If $f \geq 0$ and $\int_\Omega f = 0$, then $f = 0$ a.e. in Ω.

We continue the list of properties by stating a result about the reduction of multiple integrals and the interchange of the order of integration.

Theorem 2.3.6. (Fubini Theorem). *Let $\Omega \subseteq \mathbb{R}^{n+m}$ be a measurable set and let $f \in L^1(\Omega)$; if $\Omega \neq \mathbb{R}^{n+m}$ we consider f defined over all $\mathbb{R}^{n+m}$, by extending it to 0 in $\mathbb{R}^{n+m} \setminus \Omega$. Then, denoting by $(x, y) \in \mathbb{R}^{n+m}$ a generic point (with $x \in \mathbb{R}^n$, $y \in \mathbb{R}^m$), we have:*

(i) for a.e. $\xi \in \mathbb{R}^n$, the restriction of f to $\Omega_\xi := \Omega \cap \{x = \xi\}$ is integrable with respect to y;

(ii) for a.e. $\eta \in \mathbb{R}^m$, the restriction of f to $\Omega^\eta := \Omega \cap \{y = \eta\}$ is integrable with respect to x;

(iii) there hold the reduction formulas

$$\int_\Omega f(x,y)dxdy = \int_{\mathbb{R}^n} \left(\int_{\Omega_\xi} f(\xi, y)dy \right) d\xi = \int_{\mathbb{R}^m} \left(\int_{\Omega^\eta} f(x,\eta)dx \right) d\eta \; .$$

Finally we state what is probably the most important property, a theorem of passage to the limit under the integral sign.

Theorem 2.3.7. (Lebesgue or dominated convergence Theorem). *Let $\{f_k\}$ be a sequence of measurable functions from Ω (assumed to be measurable) to $\mathbb{R}$. If f_k tends to f a.e. in Ω and there exists $g \in L^1(\Omega)$ such that $|f_k| \leq g$ for every k, then $f \in L^1(\Omega)$ and*

$$\lim_{k \to \infty} \int_\Omega f_k = \int_\Omega f.$$

Proof. See [56, Teorema 3.4, Capitolo 5]. □

From the point of view of Lebesgue integration, the sequence defined in (2.4) is the zero sequence, being $f_k = 0$ a.e. in $[0, 1]$, and it converges to the Dirichlet function (2.3), which vanishes a.e. as well. On the other hand, Example 2.1.1 shows how the analogous of Theorem 2.3.7 *does not hold* if we replace the Lebesgue integral with the Riemann integral.

2.4 L^p Spaces

According to Definition 2.3.3, in this section we shall say that a certain property holds **almost everywhere** (a.e.) in a measurable set $\Omega \subseteq \mathbb{R}^n$ if it holds true for every $x \in \Omega$ except at most in a set of measure zero. The sets with measure zero will be called **negligible**.

We give the following definitions:

Definition 2.4.1. *Let $p \in \mathbb{R}$ be such that $1 \leq p < \infty$ and let $\Omega \subseteq \mathbb{R}^n$ be measurable. We set*

$$L^p(\Omega) = \{f : \Omega \to \mathbb{R};\ \ f \ \textit{measurable and}\ |f|^p \in L^1(\Omega)\}$$

or equivalently

$$L^p(\Omega) = \left\{f : \Omega \to \mathbb{R};\ \ f \ \textit{measurable and} \int_\Omega |f(x)|^p\, dx < \infty\right\}.$$

Moreover, we define over $L^p(\Omega)$ the function

$$\|f\|_{L^p} := \left(\int_\Omega |f(x)|^p\, dx\right)^{\frac{1}{p}} \tag{2.15}$$

for every $f \in L^p(\Omega)$. Only if necessary, we will specify the set of integration and write $\|\cdot\|_{L^p(\Omega)}$ instead of $\|\cdot\|_{L^p}$.

Definition 2.4.2. *Let $\Omega \subseteq \mathbb{R}^n$ be a measurable set. For every non negative measurable function $f : \Omega \to \mathbb{R}$ we set*

$$\operatorname*{ess\,sup}_{x\in\Omega} f(x) := \inf\{\alpha \geq 0;\ |f^{-1}(\alpha, \infty)| = 0\} \in [0, \infty]$$

and we call such value **essential supremum** *of f on Ω. Then we set*

$$L^\infty(\Omega) = \left\{f : \Omega \to \mathbb{R};\ \ f \ \textit{measurable and} \operatorname*{ess\,sup}_{x\in\Omega}|f(x)| < +\infty\right\}.$$

On the set $L^\infty(\Omega)$ we define the function

$$\|f\|_{L^\infty} := \operatorname*{ess\,sup}_{x\in\Omega}|f(x)|$$

for every $f \in L^\infty(\Omega)$.

The functions belonging to the space L^∞ are bounded outside a set of measure zero, namely if $f \in L^\infty(\Omega)$ then there exists a constant $C > 0$ such that $|f(x)| \leq C$ for almost every $x \in \Omega$. More precisely, we can affirm that

$$|f(x)| \leq \|f\|_{L^\infty} \qquad \text{for a.e. } x \in \Omega.$$

Exercise 2.4.3. (i) Prove that for every non negative measurable function $f : \Omega \to \mathbb{R}$ there holds

$$\operatorname*{ess\,sup}_{x\in\Omega} f(x) \leq \sup_{x\in\Omega} f(x); \tag{2.16}$$

if, in addition, $f(y) \leq \operatorname*{ess\,sup}_{x\in\Omega} f(x)$ for every $y \in \Omega$, then (2.16) holds true with the equality sign.

(ii) Prove that if $f = \widetilde{f}$ a.e. in Ω then

$$\operatorname*{ess\ sup}_{x\in\Omega} f(x) = \operatorname*{ess\ sup}_{x\in\Omega} \widetilde{f}.$$

(iii) Deduce from (i)-(ii) that if f and g are non negative functions defined in Ω then

$$\operatorname*{ess\ sup}_{x\in\Omega}[f(x)+g(x)] \leq \operatorname*{ess\ sup}_{x\in\Omega} f(x) + \operatorname*{ess\ sup}_{x\in\Omega} g(x).$$

(iv) Let $f : \mathbb{R} \to \mathbb{R}$ be defined as

$$f(x) = \begin{cases} x & \text{if } x \in \mathbb{Q} \\ 0 & \text{if } x \notin \mathbb{Q}\ . \end{cases}$$

Compute $\|f\|_{L^\infty}$ and $\sup_{x\in\mathbb{R}} f(x)$. □

Exercise 2.4.4. Prove that for every $1 \leq p \leq \infty$ the sets $L^p(\Omega)$ are vector spaces. To this purpose, for every $1 \leq p < \infty$ use the following convexity inequality

$$|a+b|^p \leq 2^{p-1}(|a|^p + |b|^p) \qquad \forall a, b \in \mathbb{R}\ ,$$

and for $p = \infty$ use the inequality (iii) in the previous exercise. □

Definition 2.4.5. *Let $1 \leq p \leq \infty$; we will say that p' ($1 \leq p' \leq \infty$) is the* **conjugate exponent** *of p if $\frac{1}{p} + \frac{1}{p'} = 1$, with the convention that $\frac{1}{\infty} = 0$.*

For example: $p = \infty \Rightarrow p' = 1, p = 1 \Rightarrow p' = \infty, p = 3 \Rightarrow p' = \frac{3}{2}, p = 2 \Rightarrow p' = 2$. The conjugate exponent plays a crucial role in the following inequality:

Theorem 2.4.6. (Hölder inequality). *Let $\Omega \subseteq \mathbb{R}^n$ be measurable and let $p, p' \in [1, \infty]$ be conjugate. If $f \in L^p(\Omega)$ and $g \in L^{p'}(\Omega)$, then $fg \in L^1(\Omega)$ and*

$$\int_\Omega |fg| = \|fg\|_{L^1} \leq \|f\|_{L^p} \cdot \|g\|_{L^{p'}}. \tag{2.17}$$

Proof. If $p = 1$ or $p = \infty$, the proof (the same in both cases) is trivial: indeed, if $f \in L^\infty(\Omega)$ and $g \in L^1(\Omega)$, we have that $|f| \leq C$ a.e. in Ω and that g is integrable; therefore $fg \in L^1(\Omega)$ and

$$\int_\Omega |f(x)g(x)|\ dx \leq \text{ess sup}|f| \cdot \int_\Omega |g(x)|\ dx = \|f\|_{L^\infty} \cdot \|g\|_{L^1}.$$

If $1 < p < \infty$, we need to exploit the **Young inequality**:

$$ab \leq \frac{1}{p}a^p + \frac{1}{p'}b^{p'} \qquad \forall a \geq 0, \forall b \geq 0. \tag{2.18}$$

In view of (2.18) we infer that for every $\lambda > 0$ there holds:

$$|f(x)g(x)| = |\lambda^{\frac{p-1}{p}} f(x)\lambda^{-\frac{p-1}{p}} g(x)| \leq \frac{\lambda^{p-1}|f(x)|^p}{p} + \frac{|g(x)|^{p'}}{\lambda p'} \qquad \text{for a.e. } x \in \Omega.$$

It follows that $fg \in L^1(\Omega)$ and

$$\int_\Omega |f(x)g(x)|\, dx \leq \frac{\lambda^{p-1}}{p} \|f\|_{L^p}^p + \frac{1}{\lambda p'} \|g\|_{L^{p'}}^{p'} . \tag{2.19}$$

By choosing $\lambda = \|f\|_{L^p}^{-1} \cdot \|g\|_{L^{p'}}^{p'/p}$ (in order to minimize the right hand side of (2.19)) we conclude the proof. □

An important consequence of the Hölder inequality is stated in the

Proposition 2.4.7. *Let $\Omega \subset \mathbb{R}^n$ be a set of finite measure.*
(i) If $1 \leq p < q \leq \infty$ then $L^q(\Omega) \subset L^p(\Omega)$, moreover

$$\|u\|_{L^p} \leq |\Omega|^{\frac{1}{p}\left(1-\frac{p}{q}\right)} \|u\|_{L^q} \qquad \forall u \in L^q(\Omega)$$

with the notation $p/q = 0$ if $q = \infty$.
(ii) For every measurable function u there holds

$$\lim_{p\to\infty} \|u\|_{L^p} = \|u\|_{L^\infty} .$$

Proof. Assertion (i) follows from the Hölder inequality (2.17). For the proof of (ii) see [75, Theorem 8.1]; notice that the limit makes sense even if some norms take the value $+\infty$. □

Remark 2.4.8. We notice that the assumption of finite measure is essential: indeed, if for example $u \equiv C$ in $\mathbb{R}$ with $C \neq 0$, we have $u \in L^\infty(\mathbb{R})$ but $u \notin L^p(\mathbb{R})$ for every $1 \leq p < \infty$.

Another example is the following: if $\Omega = (1, \infty)$ and $u(x) = 1/x$ then $u \in L^2(1, \infty)$ but $u \notin L^1(1, \infty)$. □

We show that L^p spaces are normed spaces:

Theorem 2.4.9. *Let $\Omega \subseteq \mathbb{R}^n$ be measurable. Then $L^p(\Omega)$ is a vector space and $\|\cdot\|_{L^p}$ is a norm for every $p \in [1, \infty]$, provided that we identify functions coinciding a.e.*

Proof. The fact that $L^p(\Omega)$ is a vector space for every $p \in [1, \infty]$ has been left to the reader in the Exercises 2.4.3-2.4.4. Let us then show that the function $\|\cdot\|_{L^p}$ is a norm over $L^p(\Omega)$ for every $p \in [1,\infty]$. The cases $p = 1$ and $p = \infty$ follow again from Exercises 2.4.3-2.4.4, therefore we may assume that $1 < p < \infty$. There holds

$$\|f+g\|_{L^p}^p = \int_\Omega |f+g|^{p-1} \cdot |f+g| \le \int_\Omega |f+g|^{p-1} \cdot |f| + \int_\Omega |f+g|^{p-1} \cdot |g|.$$

But $|f+g|^{p-1} \in L^{p'}(\Omega)$. Indeed, since $|f+g|^p \in L^1(\Omega)$, we obtain that $(|f+g|^{p-1})^{\frac{p}{p-1}} \in L^1(\Omega)$. Thus we deduce that $(|f+g|^{p-1})^{p'} \in L^1(\Omega)$ and hence $|f+g|^{p-1} \in L^{p'}(\Omega)$. In view of the Hölder inequality (2.17), it turns out that

$$\|f+g\|_{L^p}^p \le \|f+g\|_{L^p}^{p-1}(\|f\|_{L^p} + \|g\|_{L^p}),$$

that is $\|f+g\|_{L^p} \le \|f\|_{L^p} + \|g\|_{L^p}$. Hence the triangle inequality is proved. The other properties of the norm, introduced in Definition 1.1.9, are immediate. □

The next step consists in showing the completeness of the L^p spaces:

Theorem 2.4.10. *Let $\Omega \subseteq \mathbb{R}^n$ be measurable. Then $L^p(\Omega)$ is a Banach space for $p \in [1,\infty]$ and is separable for $p \in [1,\infty)$. If $p = 2$, then the bilinear form*

$$(f,g) = \int_\Omega fg \qquad \forall f,g \in L^2(\Omega) \tag{2.20}$$

defines a scalar product in $L^2(\Omega)$ which, as a consequence, is a separable Hilbert space.

Proof. The proof of the completeness of $L^p(\Omega)$ requires the usage of a further result of passage to the limit under the integral sign (besides Theorem 2.3.7), therefore we refer the reader to [10, Theorem IV.8].

Here we limit ourselves to prove the separability. Let $p \in [1,\infty)$. Given a set $A \subset \mathbb{R}^n$ we will denote by χ_A the function which equals 1 in A and 0 in $\mathbb{R}^n \setminus A$. Let us define the family of sets

$$\mathcal{R} := \left\{ R \subset \mathbb{R}^n : R = \prod_{k=1}^n (a_k, b_k),\ a_k, b_k \in \mathbb{Q} \right\}$$

and the vector space V of all finite linear combinations with coefficients in $\mathbb{Q}$ of functions χ_R with $R \in \mathcal{R}$. From the countability of $\mathbb{Q}$ follows the countability of $\mathcal{R}$, and hence of V. If we prove that the subspace of $L^p(\Omega)$ of the restrictions to Ω of the functions of V is dense in $L^p(\Omega)$, the separability of $L^p(\Omega)$ is straightforward.

Let $f \in L^p(\Omega)$. For $\varepsilon > 0$ fixed, we want to show the existence of a function $\varphi \in V$ such that

$$\|f - \varphi\|_{L^p(\Omega)} < \varepsilon.$$

Considering the trivial extension of f outside Ω, it suffices to prove that

$$\|f - \varphi\|_{L^p(\mathbb{R}^n)} < \varepsilon, \tag{2.21}$$

since $\|f-\varphi\|_{L^p(\Omega)} \le \|f-\varphi\|_{L^p(\mathbb{R}^n)}$. Exploiting the decomposition $f = f^+ - f^-$, with f^+ and f^- as in Definition 2.3.5, it is enough to prove (2.21) in the case in which f is non negative. Let us begin by showing the existence of a function $g \in L^\infty(\mathbb{R}^n)$ with $\{x \in \mathbb{R}^n : g(x) \neq 0\} \subset K$ for some compact K, such that

$$\|f - g\|_{L^p(\mathbb{R}^n)} < \frac{\varepsilon}{3}. \tag{2.22}$$

To this purpose, consider the sequence $g_m(x) = \chi_{B_m}(x)T_m(f(x))$, with B_m the ball of radius m centered at the origin and $T_m : [0,\infty) \to \mathbb{R}$ defined as $T_m(s) = s$ if $s \in [0,m)$ and $T_m(s) = m$ if $s \in [m,\infty)$. Applying Theorem 2.3.7 we deduce that $g_m \to f$ in $L^p(\mathbb{R}^n)$, therefore we may choose $g := g_{\overline{m}}$ in (2.22) with $\overline{m}$ large enough.

Let now $N \in \mathbb{N}$ be the smallest integer such that $N > \|g\|_{L^\infty(\mathbb{R}^n)}$ and define for every $k \in \mathbb{N}$ the sequence

$$\psi_k(x) = \sum_{i=0}^{kN-1} \frac{i}{k}\, \chi_{A_k^i}(x)$$

with $A_k^i = \left\{x \in \mathbb{R}^n : \frac{i}{k} \le g(x) < \frac{i+1}{k}\right\}$. Since for k fixed the sets A_k^i with $i = 0, ..., kN-1$ are pairwise disjoint, by Theorem 2.2.9 we derive

$$\begin{aligned}\int_{\mathbb{R}^n} |g(x) - \psi_k(x)|^p dx &= \sum_{i=0}^{kN-1} \int_{A_k^i \cap K} |g(x) - \psi_k(x)|^p dx \le \sum_{i=0}^{kN-1} \frac{1}{k^p}|A_k^i \cap K| \\ &= \frac{1}{k^p}\left|\bigcup_{i=0}^{kN-1} (A_k^i \cap K)\right| = \frac{1}{k^p}|K|.\end{aligned}$$

In particular, if we fix $k > \frac{3|K|^{1/p}}{\varepsilon}$, we obtain

$$\|g - \psi_k\|_{L^p(\mathbb{R}^n)} < \frac{\varepsilon}{3}. \tag{2.23}$$

Using Definitions 2.2.2 and 2.2.4 and Proposition 1.6.5, we deduce that for every $i = 1, ..., kN-1$ there exist an open set $F_k^i \supseteq A_k^i$ and a pluri-interval $R_k^i \subseteq F_k^i$ union of elements of $\mathcal{R}$ such that

$$\max\{|F_k^i \setminus A_k^i|, |F_k^i \setminus R_k^i|\} < \frac{1}{2}\left(\frac{2\varepsilon}{3N(kN-1)}\right)^p.$$

Once we set

$$\varphi_k(x) = \sum_{i=0}^{kN-1} \frac{i}{k}\chi_{R_k^i}(x),$$

we have

$$\|\psi_k - \varphi_k\|_{L^p(\mathbb{R}^n)} \le \sum_{i=0}^{kN-1} \frac{i}{k} \left(\int_{\mathbb{R}^n} |\chi_{A_k^i} - \chi_{R_k^i}|^p dx \right)^{1/p} \tag{2.24}$$
$$= \sum_{i=0}^{kN-1} \frac{i}{k} \left(|R_k^i \setminus A_k^i| + |A_k^i \setminus R_k^i| \right)^{1/p} < \frac{2\varepsilon}{3kN(kN-1)} \sum_{i=0}^{kN-1} i = \frac{\varepsilon}{3}.$$

Since $\varphi_k \in V$, by combining (2.22), (2.23) and (2.24), we infer that (2.21) holds true. Thus the separability of $L^p(\Omega)$ is proved.

Finally, we observe that (2.20) is well defined thanks to (2.17). □

The next result shows that strong convergence in L^p implies pointwise convergence, up to subsequences.

Theorem 2.4.11. *Let $\Omega \subseteq \mathbb{R}^n$ be measurable and let $1 \le p \le \infty$. Let $\{f_m\} \subset L^p(\Omega)$ and let $f \in L^p(\Omega)$ be such that $f_m \to f$ in $L^p(\Omega)$. Then there exists a subsequence $\{f_{m_k}\}$ such that $f_{m_k} \to f$ a.e. in Ω, namely $f_{m_k}(x) \to f(x)$ for a.e. $x \in \Omega$.*

Proof. See [10, Teorema IV.9]. □

Remark 2.4.12. Except for the case $p = \infty$, in Theorem 2.4.11 it is not possible to obtain the pointwise convergence a.e. without passing to subsequences. □

Let us now characterize the dual space of L^p:

Theorem 2.4.13. (Riesz representation Theorem). *Let $\Omega \subseteq \mathbb{R}^n$ be measurable, let $1 \le p < \infty$ and let $\varphi \in (L^p(\Omega))'$. Then there exists a unique $u \in L^{p'}(\Omega)$ such that*

$$\langle \varphi, f \rangle = \int_\Omega uf \qquad \forall f \in L^p(\Omega).$$

Moreover there holds $\|u\|_{L^{p'}} = \|\varphi\|_{(L^p)'}$.

Proof. See [10, Teorema IV.11 and IV.14]. □

This theorem is very important. It states that every linear continuous functional over $L^p(\Omega)$ (with $1 \le p < \infty$) can be represented (isometrically) by means of a suitable function belonging to $L^{p'}(\Omega)$ which turns out to be unique. In other words, we may say that

$$(L^p(\Omega))' = L^{p'}(\Omega) \qquad \forall p \in [1, \infty).$$

Remark 2.4.14. One may wonder what it is possible to say about the space $(L^\infty(\Omega))'$. It would be natural to think that such space can be identified with $L^1(\Omega)$, as 1 is formally the conjugate exponent of ∞. Nevertheless, this is not true! We refer to Example 2.6.4 for a linear continuous functional over $C^0(\overline{\Omega})$ which *is not* a function of $L^1(\Omega)$. □

Although for $p \neq 2$ L^p spaces are not Hilbert spaces, Theorem 2.4.13 suggests to define a weak convergence (see Definition 1.4.10) in L^p in the following way:

Definition 2.4.15. *Let $\Omega \subseteq \mathbb{R}^n$ be measurable and let $1 \leq p \leq \infty$. We will say that $\{f_m\} \subset L^p(\Omega)$ converges weakly to f in $L^p(\Omega)$ if*

$$\int_\Omega f_m v \to \int_\Omega f v \qquad \forall v \in L^{p'}(\Omega).$$

Remark 2.4.16. We point out that the definition of weak convergence just introduced is not completely rigorous in the case $p = \infty$, since the dual space of $L^\infty(\Omega)$ is not $L^1(\Omega)$. When $p = \infty$, the notion introduced in Definition 2.4.15 is commonly called *weak star convergence* and is denoted by $f_m \stackrel{*}{\rightharpoonup} f$ in $L^\infty(\Omega)$. □

Exploiting the definition of weak convergence introduced in the L^p spaces, a version of Theorem 1.4.13 holds true:

Theorem 2.4.17. *Let $\Omega \subseteq \mathbb{R}^n$ be measurable and let $1 \leq p \leq \infty$.*

(i) If $\{f_m\} \subset L^p(\Omega)$ and $f \in L^p(\Omega)$ satisfy $f_m \rightharpoonup f$, then $\{f_m\}$ is bounded in $L^p(\Omega)$ and there holds

$$\|f\|_{L^p} \leq \liminf_{m\to\infty} \|f_m\|_{L^p}.$$

(ii) If $1 < p \leq \infty$ and $\{f_m\}$ is bounded, then there exist a subsequence $\{f_{m_k}\}$ and a function $f \in L^p(\Omega)$ such that $f_m \rightharpoonup f$ in $L^p(\Omega)$.

Proof. See [10, Proposizione III.5], [10, Proposizione III.12], [10, Teorema III.15] and [10, Teorema III.27]. □

Let $\Omega \subset \mathbb{R}^n$ be open and let $f \in C^\infty(\Omega)$. We call **support** of f and we denote it by

$$\operatorname{supp} f = \overline{\{x \in \Omega; f(x) \neq 0\}} \tag{2.25}$$

the closure of the set in which $f \neq 0$. Now we define the space

$$\mathcal{D}(\Omega) = \{f \in C^\infty(\Omega); \quad \operatorname{supp} f \text{ compact in } \Omega\}.$$

Basically, $\mathcal{D}(\Omega)$ contains those regular functions in Ω which vanish before reaching the boundary $\partial\Omega$ or before reaching the infinity (if Ω is unbounded). A simple example of function belonging to $\mathcal{D}(\mathbb{R}^n)$ is the following:

$$u(x) = \begin{cases} e^{1/(|x|^2-1)} & \text{if } |x| < 1 \\ 0 & \text{if } |x| \geq 1\,. \end{cases}$$

Clearly, the functions that belong to $\mathcal{D}(\Omega)$ are "precious" and have great properties. There holds the following

Proposition 2.4.18. *Let $\Omega \subseteq \mathbb{R}^n$ be measurable and let $p \in [1, \infty)$. Then the space $\mathcal{D}(\Omega)$ is dense in $L^p(\Omega)$, that is*

$$\forall f \in L^p(\Omega),\ \forall \varepsilon > 0\ \ \exists f_\varepsilon \in \mathcal{D}(\Omega),\ \int_\Omega |f - f_\varepsilon|^p < \varepsilon.$$

Proof. See [10, Corollario IV.23]. □

This proposition states that it is possible to approximate an L^p function with a "very regular" function.

Definition 2.4.19. *Let $\Omega \subseteq \mathbb{R}^n$ be an open set and let $1 \leq p \leq \infty$. We denote by $L^p_{loc}(\Omega)$ the space*

$$L^p_{loc}(\Omega) = \{f \in L^p(K),\ \forall K \text{ compact} \subset \Omega\}.$$

The functions of the space $L^p_{loc}(\Omega)$ (in which the subscript "loc" stays for local) are the ones which have a "good behavior" inside Ω, but for which we do not have any information near $\partial\Omega$. For example, the function $f(x) = \frac{1}{x}$ is not integrable in 0 and $+\infty$, then it does not belong to $L^1(0, \infty)$ but it belongs to the space $L^1_{loc}(0, \infty)$.

Exercise 2.4.20. Let $1 \leq p < q \leq +\infty$ and let $\Omega \subseteq \mathbb{R}^n$ be measurable. Show that $L^q_{loc}(\Omega) \subset L^p_{loc}(\Omega)$. Hint: consider $f \in L^q_{loc}(\Omega)$ and split Ω into the sets in which $|f| > 1$ and $|f| \leq 1$. □

Even if it is not possible to define a norm over the space $L^p_{loc}(\Omega)$, however it is possible to introduce a notion of convergence.

Definition 2.4.21. *Let $\{u_k\}$ be a sequence in $L^p_{loc}(\Omega)$ with $p \geq 1$. We say that $\{u_k\}$ converges in $L^p_{loc}(\Omega)$ to $u \in L^p_{loc}(\Omega)$ if $u_k \to u$ in $L^p(K)$ for every compact set $K \subset \Omega$.*

We now state the following property:

Proposition 2.4.22. *Let $\Omega \subseteq \mathbb{R}^n$ be open and let $f \in L^1_{loc}(\Omega)$ satisfy*

$$\int_\Omega f\varphi = 0 \qquad \forall \varphi \in \mathcal{D}(\Omega).$$

Then $f = 0$ a.e. in Ω.

Proof. See [10, Corollario IV.24]. □

Proposition 2.4.22 should not be surprising. If we consider $f \in L^2(\Omega)$, the proposition says that f is orthogonal to the space $\mathcal{D}(\Omega)$, in the sense of the scalar product (2.20). But since $\mathcal{D}(\Omega)$ is dense in $L^2(\Omega)$ (cf. Proposition 2.4.18), this means that f is orthogonal to the whole space $L^2(\Omega)$ and then it cannot be other than a function vanishing a.e.

We conclude this section saying some words about complex L^p spaces. These spaces are the basis for defining the integral transforms that we will treat in Chapter 5.

Given a function f defined in a measurable set $A \subseteq \mathbb{R}^n$ with values in $\mathbb{C}$, we will denote by $\mathrm{Re}f$ and $\mathrm{Im}f$ the real and imaginary part of f respectively:

$$f = \mathrm{Re}f + i\mathrm{Im}f.$$

Moreover we will denote by $|\cdot|$ the usual modulus in $\mathbb{C}$.

Definition 2.4.23. *Let $1 \leq p \leq \infty$ and let $\Omega \subseteq \mathbb{R}^n$ be measurable. We denote by $L^p(\Omega, \mathbb{C})$ the space of functions $f : \Omega \to \mathbb{C}$ such that $\mathrm{Re}f$, $\mathrm{Im}f \in L^p(\Omega)$. Moreover, we define the norms*

$$\|f\|_{L^p} := \left(\int_\Omega |f(x)|^p dx \right)^{1/p} \qquad \text{if } 1 \leq p < \infty$$

and

$$\|f\|_{L^\infty} := \operatorname*{ess\,sup}_{x \in \Omega} |f(x)|.$$

The space $L^p(\Omega, \mathbb{C})$ equipped with this norm becomes a complex Banach space. Finally, given a function $f \in L^1(\Omega, \mathbb{C})$, the integral of such a function is simply defined as

$$\int_\Omega f(x)dx := \int_\Omega \mathrm{Re}f(x)dx + i \int_\Omega \mathrm{Im}f(x)dx.$$

We remark that (as it happens in the real case) a measurable function $f : \Omega \to \mathbb{C}$ belongs to $L^1(\Omega, \mathbb{C})$ if and only if $|f| \in L^1(\Omega)$.

Remark 2.4.24. It is possible to endow the space $L^2(\Omega, \mathbb{C})$ with the scalar product defined by

$$(f, g) = \int_\Omega f\overline{g} \qquad \forall f, g \in L^2(\Omega, \mathbb{C}).$$

This product is well defined in $L^2(\Omega, \mathbb{C})$ since $|f\overline{g}| \in L^1(\Omega)$; indeed there holds

$$|f\overline{g}| = |f| \cdot |g| \leq \frac{|f|^2}{2} + \frac{|g|^2}{2} \in L^1(\Omega). \qquad \square$$

2.5 Fourier series

Here we give a simple example of how the abstract results of Chapter 1 can be applied to the study of concrete problems. We will study the separable Hilbert space $L^2(0, 2\pi)$ with the scalar product defined by

$$(u, v)_{L^2} := \frac{1}{\pi} \int_0^{2\pi} uv \qquad \forall u, v \in L^2(0, 2\pi) \tag{2.26}$$

and with the induced norm

$$\|u\|_{L^2}^2 = \frac{1}{\pi}\int_0^{2\pi} |u|^2 \qquad \forall u \in L^2(0,2\pi);$$

we observe that this norm is equivalent (cf. Definition 1.2.10) to the standard one (2.15). Let us consider the (countable) set of functions

$$\mathcal{B} := \left\{ \frac{1}{\sqrt{2}},\ \cos x,\ \sin x,\ \cos 2x,\ \sin 2x,\ ..., \cos mx,\ \sin mx,\ ... \right\}. \tag{2.27}$$

This family of functions has the following property:

Theorem 2.5.1. *The family $\mathcal{B}$ defined in* (2.27) *is a Hilbertian basis in* $L^2(0,2\pi)$ *with respect to the scalar product defined in* (2.26).

Proof. We leave to the reader the easy exercise of showing that, considering the scalar product defined in (2.26) and the induced norm $\|\cdot\|_{L^2}$, one has:
- for every $e \in \mathcal{B}$ it holds $\|e\|_{L^2} = 1$;
- for every pair of distinct functions $e_1, e_2 \in \mathcal{B}$ it holds $(e_1, e_2)_{L^2} = 0$.

This proves that $\mathcal{B}$ is an orthonormal system. It would remain to show that $\mathcal{B}$ generates $L^2(0,2\pi)$; the proof of this fact is delicate and we refer the interested reader to [60, Paragrafo 4.24]. □

Theorem 2.5.1 states that, given any function $u \in L^2(0,2\pi)$, it is always possible to determine two real sequences $\{a_n\}$ and $\{b_n\}$ such that

$$u(x) = \frac{a_0}{2} + \sum_{n=1}^{\infty} \Big(a_n \cos nx + b_n \sin nx \Big), \tag{2.28}$$

meaning that u is the limit in $L^2(0,2\pi)$ of the sequence of partial sums

$$S_k(x) = \frac{a_0}{2} + \sum_{n=1}^{k} \Big(a_n \cos nx + b_n \sin nx \Big).$$

According to Definition 1.1.12, this means that

$$\lim_{k\to\infty} \int_0^{2\pi} |u(x) - S_k(x)|^2\, dx = 0\ . \tag{2.29}$$

Moreover, Theorem 2.5.1 allows to compute

$$\|S_k\|_{L^2}^2 = \frac{a_0^2}{2} + \sum_{n=1}^{k} (a_n^2 + b_n^2)\ .$$

Since $S_k \to u$ in $L^2(0, 2\pi)$, we infer that

$$\lim_{k\to\infty} \left[\frac{a_0^2}{2} + \sum_{n=1}^{k} (a_n^2 + b_n^2)\right] < \infty$$

that is $\{a_n\}$, $\{b_n\} \in \ell^2$. Since $L^2(0, 2\pi)$ is a separable Hilbert space (cf. Theorem 2.4.10), this explains the meaning of Theorem 1.6.9. In view of (2.29) and (1.7) we infer that also $S_k \rightharpoonup u$ in $L^2(0, 2\pi)$ and hence

$$(u(x), \cos nx)_{L^2} = \frac{1}{\pi}\int_0^{2\pi} u(x)\cos nx\, dx = \frac{1}{\pi}\lim_{k\to\infty}\int_0^{2\pi} S_k(x)\cos nx\, dx = a_n$$

for every $n \geq 0$, where we have used the orthonormality of $\mathcal{B}$. Proceeding in a similar way for the function $\sin nx$, we obtain

$$a_0 = \frac{1}{\pi}\int_0^{2\pi} u(x)\, dx, \quad a_n = \frac{1}{\pi}\int_0^{2\pi} u(x)\cos nx\, dx, \quad b_n = \frac{1}{\pi}\int_0^{2\pi} u(x)\sin nx\, dx \tag{2.30}$$

for every $n \geq 1$. Therefore, by combining Theorem 1.6.8 with the last lines, we have obtained the

Theorem 2.5.2. (Parseval Theorem). *Let $u \in L^2(0, 2\pi)$; then*

$$\frac{1}{\pi}\int_0^{2\pi} |u(x)|^2\, dx = \|u\|_{L^2}^2 = \frac{a_0^2}{2} + \sum_{n=1}^{\infty}(a_n^2 + b_n^2)$$

where a_n and b_n are the Fourier coefficients of u and are defined by (2.30).

Clearly, the construction described above can be done also for intervals of the form $(0, T)$, with a generic $T > 0$. In that case, a function belonging to $L^2(0, T)$ has Fourier coefficients

$$a_0 = \frac{2}{T}\int_0^{T} u(x)\, dx, \quad a_n = \frac{2}{T}\int_0^{T} u(x)\cos\frac{2\pi nx}{T}\, dx, \quad b_n = \frac{2}{T}\int_0^{T} u(x)\sin\frac{2\pi nx}{T}\, dx \tag{2.31}$$

for every $n \geq 1$ and satisfies

$$u(x) = \frac{a_0}{2} + \sum_{n=1}^{\infty}\left(a_n\cos\frac{2\pi nx}{T} + b_n\sin\frac{2\pi nx}{T}\right).$$

If instead we keep the size 2π fixed, we can also consider the interval $(-\pi, \pi)$. In this case, the formulas (2.30) remain the same (it suffices to change the bounds of integration) and, in addition, we have further interesting properties that we leave to the reared as an exercise:

Exercise 2.5.3. Let $u \in L^2(-\pi, \pi)$. Prove that if u is even then $b_n = 0$ for every n, while if u is odd then $a_n = 0$ for every n. Moreover, show that the subspaces of $L^2(-\pi, \pi)$ of even and odd functions are orthogonal, in the sense of Definition 1.5.2; finally, rewrite the statement of Theorem 1.5.3 in this setting. □

The well-known Euler representation of the exponential of an imaginary number, $e^{i\vartheta} = \cos\vartheta + i \sin\vartheta$, suggests to write Fourier series in terms of exponentials. For simplicity, we only deal with the interval $[0, 2\pi]$. We know that

$$\cos(nx) = \frac{e^{inx} + e^{-inx}}{2}, \qquad \sin(nx) = \frac{e^{inx} - e^{-inx}}{2i}.$$

For all $n \geq 0$, let

$$u_n = \frac{a_n - ib_n}{2}, \quad u_{-n} = \frac{a_n + ib_n}{2} = \overline{u_n}\ ; \tag{2.32}$$

in particular, $u_0 = \frac{a_0}{2}$. Then, by (2.30) and with a simple computation, we get

$$u_n = \frac{a_n - ib_n}{2} = \frac{1}{2\pi}\int_0^{2\pi} u(x)[\cos(nx) - i\sin(nx)]dx = \frac{1}{2\pi}\int_0^{2\pi} u(x)e^{-inx}\,dx,$$

$$u_{-n} = \frac{a_n + ib_n}{2} = \frac{1}{2\pi}\int_0^{2\pi} u(x)[\cos(nx) + i\sin(nx)]dx = \frac{1}{2\pi}\int_0^{2\pi} u(x)e^{inx}\,dx,$$

where both these formulas hold for all $n \geq 0$. From (2.32) we infer $a_n = u_n + u_{-n}$ and $b_n = i(u_n - u_{-n})$ so that, by (2.30) and (2.28),

$$u(x) = \frac{a_0}{2} + \sum_{n=1}^{\infty}\Big(a_n\cos(nx) + b_n\sin(nx)\Big) = \sum_{n=-\infty}^{\infty} u_n\, e^{inx}.$$

Moreover, the Parseval identity resembles even more to the Pythagoras Theorem:

$$\frac{1}{2\pi}\int_0^{2\pi} u(x)^2\,dx = \sum_{n=-\infty}^{\infty} |u_n|^2.$$

Let us now investigate different kinds of convergence of Foureir series. Concerning the total convergence (which implies the uniform convergence), note that

$$\sum_{n=1}^{\infty}(|a_n| + |b_n|) < \infty \implies \text{(2.28) converges totally on } [0, 2\pi]. \tag{2.33}$$

For instance, la trigonometric series

$$\sum_{n=1}^{\infty}\left(\frac{\cos(nx)}{n^2} + \frac{\sin(nx)}{n^2}\right)$$

converges totally. By formally differentiating this series we obtain

$$\sum_{n=1}^{\infty}\left(\frac{\cos(nx)}{n}-\frac{\sin(nx)}{n}\right)$$

and we see that the convergence has become more difficult: we conclude that, differently from power series, the trigonometric series may fail to converge after differentiation.

The condition in (2.33) appears too restrictive for the sole pointwise convergence: there is a weaker sufficient condition, known as the **Dirichlet criterion**:

$$a_n, b_n \downarrow 0 \implies \text{(2.28) converges pointwise in } (0,2\pi). \tag{2.34}$$

The symbol $\downarrow 0$ means that the sequences decrease monotonically to 0 and, hence, that they are positive. Moreover, the sine series obviously converges to 0 for $x = 0, 2\pi$ while the cosine series may not converge in these points (think about $a_n = \frac{1}{n}$ and $x = 0$).

So far we have seen some convergence results for the Fourier series. Let us recall the hierarchies between different kinds of convergences:

uniform convergence ⇒ squared mean convergence
⇒ a.e. convergence, up to a subsequence.

Clearly, also the total convergence implies the squared mean convergence while for the last implication, see Theorem 2.4.11.

But there are also different criteria which directly ensure the pointwise convergence without involving the squared mean convergence or conditions (2.33) and (2.34). Let us just see one of them.

Definition 2.5.4. *We say that $u : [0,2\pi] \to \mathbb{R}$ is* **piecewise regular** *if it is bounded in $[0,2\pi]$ and if the interval $[0,2\pi]$ may be decomposed in a finite number of subintervals in which u is continuous and differentiable; moreover, at the endpoints of each subinterval, the limits of both u and u' exist and are finite.*

Obviously, if $u \in C^1[0,2\pi]$ then u is piecewise regular but the latter class is much wider. There may be angle points and even discontinuities provided that both u and u' have finite limits on both sides of the discontinuities. But vertical asymptotes or points with vertical tangent are not allowed. It is clear that any piecewise regular function u belongs to $L^2(0,2\pi)$ and hence its Fourier series converges in squared mean value to u. But then we know that, up to a subsequence and to a set of points of finite measure, the convergence is also pointwise. These particular points deserve some attention.

Given a piecewise regular function u and given a point $x_0 \in [0,2\pi]$ we denote, respectively, by $u(x_0^-)$ and $u(x_0^+)$ the left and right limits of u at x_0: these values are finite. By periodicity of u we know that $u(0^\pm) = u(2\pi^\pm)$ and therefore we may define these limits also at the endpoints of the interval. If u is continuous at x_0 we clearly have $u(x_0) = u(x_0^\pm)$. We may now state

Theorem 2.5.5. *Let $u : [0, 2\pi] \to \mathbb{R}$ be piecewise regular. Then its Fourier series (2.28) converges in each point $x_0 \in [0, 2\pi]$ to the average of the two limits $u(x_0^\pm)$:*

$$\frac{a_0}{2} + \sum_{n=1}^{\infty} \Big(a_n \cos(nx_0) + b_n \sin(nx_0) \Big) = \frac{u(x_0^+) + u(x_0^-)}{2}$$

with the convention that $u(0^\pm) = u(2\pi^\pm)$. In particular, if u is continuous at x_0, then the series converges to $u(x_0)$.

2.6 Distributions

We start with the following notation: for every multi-index $\alpha = (\alpha_1, ..., \alpha_n)$ with $\alpha_1, ..., \alpha_n$ non negative integers, we define $|\alpha| := \sum_{i=1}^n \alpha_i$ and

$$D^\alpha \varphi = \frac{\partial^{|\alpha|} \varphi}{\partial x_1^{\alpha_1} ... \partial x_n^{\alpha_n}} \qquad \forall \varphi \in \mathcal{D}(\Omega). \tag{2.35}$$

Over the space $\mathcal{D}(\Omega)$ we introduce the following notion of convergence:

Definition 2.6.1. *Let $\{\varphi_k\} \subset \mathcal{D}(\Omega)$ be a sequence and let $\varphi \in \mathcal{D}(\Omega)$. We say that $\varphi_k \to \varphi$ in $\mathcal{D}(\Omega)$ if the following conditions hold true:*
(i) there exists a compact set $K \subset \Omega$ such that supp $\varphi_k \subset K$ *for every $k \in \mathbb{N}$,*
(ii) $D^\alpha \varphi_k \to D^\alpha \varphi$ uniformly in Ω for every multi-index α.

At this point we can define the space of distributions.

Definition 2.6.2. *Let $\Omega \subseteq \mathbb{R}^n$ be an open set. We will call* **distribution** *any functional $\Lambda : \mathcal{D}(\Omega) \to \mathbb{R}$ which is linear and continuous, namely which satisfies*

$$\Lambda(\varphi_k) \to \Lambda(\varphi)$$

for every sequence such that $\varphi_k \to \varphi$ in $\mathcal{D}(\Omega)$. We shall denote by $\mathcal{D}'(\Omega)$ the space of distributions. Moreover, given a generic distribution $\Lambda \in \mathcal{D}'(\Omega)$ and a generic test function $\varphi \in \mathcal{D}(\Omega)$, the number $\Lambda(\varphi)$ will be denoted by $\langle \Lambda, \varphi \rangle$.

Example 2.6.3. It is possible to associate with each function $u \in L^1_{loc}(\Omega)$ the linear continuous functional Λ_u as follows: setting

$$\langle \Lambda_u, \varphi \rangle := \int_\Omega u\varphi \qquad \forall \varphi \in \mathcal{D}(\Omega) ,$$

it holds $\Lambda_u \in \mathcal{D}'(\Omega)$. Indeed, if $\varphi_k \to \varphi$ in $\mathcal{D}(\Omega)$, letting $K \subset \Omega$ be a compact set containing suppφ and suppφ_k for every k, thanks to the Hölder inequality (2.17), we have

$$|\Lambda_u(\varphi_k - \varphi)| = \left| \int_\Omega u(\varphi_k - \varphi) \right| = \left| \int_K u(\varphi_k - \varphi) \right| \leq \|u\|_{L^1(K)} \cdot \|\varphi_k - \varphi\|_{L^\infty(K)} \to 0$$

as $k \to \infty$. The function which associates with every function $u \in L^1_{loc}(\Omega)$ the distribution Λ_u is injective. In fact, if $u, v \in L^1_{loc}(\Omega)$ are such that $\Lambda_u = \Lambda_v$ then

$$\int_\Omega (u - v)\varphi = 0 \qquad \forall \varphi \in \mathcal{D}(\Omega)$$

and hence, in view of Proposition 2.4.22, we conclude that $u = v$ a.e. in Ω, that is $u = v$ in $L^1_{loc}(\Omega)$. Thus we may identify every function $u \in L^1_{loc}(\Omega)$ with the corresponding distribution Λ_u and consider $L^1_{loc}(\Omega)$ as a subspace of $\mathcal{D}'(\Omega)$. The duality product $\langle \Lambda_u, \varphi \rangle$ will be simply denoted by $\langle u, \varphi \rangle$ and will coincide with $\int_\Omega u\varphi$. □

Another example of distribution is the following:

Example 2.6.4. Given $y \in \Omega$, consider the linear functional $\delta_y : \mathcal{D}(\Omega) \to \mathbb{R}$ defined by

$$\delta_y(\varphi) := \varphi(y).$$

It is immediate to check that this functional is a distribution. Such distribution is called **Dirac delta function** centered at y. It is natural to ask whether such distribution can be represented by an $L^1_{loc}(\Omega)$ function, namely whether there exists a function $u \in L^1_{loc}(\Omega)$ such that

$$\langle \delta_y, \varphi \rangle = \int_\Omega u\varphi \qquad \forall \varphi \in \mathcal{D}(\Omega).$$

The answer is negative: by contradiction, suppose that such a function $u \in L^1_{loc}(\Omega)$ exists. Then, for every $\varphi \in \mathcal{D}(\Omega)$ such that $\text{supp}\varphi \subset \Omega\backslash\{y\}$ it holds

$$\int_\Omega u\varphi = \langle \delta_y, \varphi \rangle = \varphi(y) = 0.$$

In view of Proposition 2.4.22 applied to the open set $\Omega\backslash\{y\}$, we conclude that $u = 0$ a.e. in $\Omega\backslash\{y\}$ and hence $u = 0$ a.e. in Ω. Consequently, we have

$$\varphi(y) = \langle \delta_y, \varphi \rangle = \int_\Omega u\varphi = 0 \quad \forall \varphi \in \mathcal{D}(\Omega)$$

and this is obviously absurd, since the elements of $\mathcal{D}(\Omega)$ do not necessarily vanish at y. □

It is possible to generalize the notion of support, previously introduced for functions, also to distributions.

Definition 2.6.5. *Let $\Omega \subseteq \mathbb{R}^n$ be an open set and let $\Lambda \in \mathcal{D}'(\Omega)$. If ω is an open subset of Ω we will say that the distribution Λ vanishes on ω if $\langle \Lambda, \varphi \rangle = 0$ for every $\varphi \in \mathcal{D}(\omega)$. Let W be the union of all the open sets $\omega \subseteq \Omega$ on which Λ vanishes. Then we define the* **support** *of Λ as the complement of the set W, and we denote it by* supp Λ.

We underline that such definition is perfectly compatible with (2.25), in the sense that if $u \in C^0(\Omega) \cap L^1_{loc}(\Omega)$ then the set

$$\text{supp } u = \overline{\{x \in \Omega;\ u(x) \neq 0\}}$$

coincides with the support of the distribution associated to u. Moreover, this definition allows to give a meaning to the support of L^p functions. Recall that the elements of L^p are not properly functions, but rather equivalence classes of functions (equal a.e.) and if the support of an element of L^p was defined as in (2.25), then it would depend on the choice of the representative of the equivalence class. Essentially, when dealing with the support of an L^p function, we will always mean the distributional sense.

Example 2.6.6. The functions of $L^1(\mathbb{R}^n)$ with compact support and the Dirac delta functions are distributions with compact support. □

Also in the space $\mathcal{D}'(\Omega)$ it is possible to introduce a convergence.

Definition 2.6.7. *Let $\{u_k\}$ be a sequence in $\mathcal{D}'(\Omega)$. We say that u_k converges to $u \in \mathcal{D}'(\Omega)$* **in the sense of distributions** *if*

$$\langle u_k, \varphi \rangle \to \langle u, \varphi \rangle \qquad \forall \varphi \in \mathcal{D}(\Omega).$$

Example 2.6.8. Let $\{u_k\} \subset L^1_{loc}(\Omega)$ and $u \in L^1_{loc}(\Omega)$ such that $u_k \to u$ in $L^1_{loc}(\Omega)$, see Definition 2.4.21. Then $u_k \to u$ in $\mathcal{D}'(\Omega)$. Indeed, for every fixed $\varphi \in \mathcal{D}(\Omega)$, let $K \subset \Omega$ be a compact set such that supp $\varphi \subset K$. Since $u_k \to u$ in $L^1(K)$, it holds

$$\lim_{k\to\infty} \langle u_k, \varphi \rangle = \lim_{k\to\infty} \int_\Omega u_k \varphi = \lim_{k\to\infty} \int_K u_k \varphi = \int_K u\varphi = \int_\Omega u\varphi = \langle u, \varphi \rangle. \qquad \square$$

Example 2.6.9. We construct a sequence of functions of $L^1(\mathbb{R}^n)$ converging to the Dirac delta function in the sense of distributions. Let $u \in L^1(\mathbb{R}^n)$ be an arbitrary non negative function such that $\|u\|_{L^1} = 1$ and let $\{u_k\}$ be the sequence defined as follows: $u_k(x) = k^n u(kx)$. Using the change of variables $x = y/k$, we obtain

$$\|u_k\|_{L^1} = k^n \int_{\mathbb{R}^n} |u(kx)|dx = k^n \int_{\mathbb{R}^n} |u(y)|k^{-n}dy = \|u\|_{L^1} = 1.$$

Since $\int_{\mathbb{R}^n} u_k = \|u_k\|_{L^1} = 1$, using again the change of variables $x = y/k$, we have

$$\left| \int_{\mathbb{R}^n} u_k(x)\varphi(x)dx - \varphi(0) \right| = \left| \int_{\mathbb{R}^n} u_k(x)[\varphi(x) - \varphi(0)]dx \right|$$

$$\leq k^n \int_{\mathbb{R}^n} u(kx)|\varphi(x) - \varphi(0)|dx = \int_{\mathbb{R}^n} u(y) \left| \varphi\left(\frac{y}{k}\right) - \varphi(0) \right| dy. \tag{2.36}$$

On the other hand, being φ a continuous function, we infer

$$\lim_{k\to\infty} u(y)\left|\varphi\left(\frac{y}{k}\right)-\varphi(0)\right| = 0 \qquad \text{a.e. in } \mathbb{R}^n.$$

Since φ is bounded in $\mathbb{R}^n$, we also have that

$$|u(y)|\left|\varphi\left(\frac{y}{k}\right)-\varphi(0)\right| \le 2\left(\sup_{\mathbb{R}^n}|\varphi|\right)\cdot|u(y)| \in L^1(\mathbb{R}^n) \qquad \text{a.e. in } \mathbb{R}^n \qquad \forall k\in\mathbb{N}.$$

Thus, exploiting Theorem 2.3.7, we get

$$\int_{\mathbb{R}^n} u_k(x)\varphi(x)dx \to \varphi(0) \qquad \forall\varphi\in\mathcal{D}(\mathbb{R}^n),$$

which is a reformulation of

$$\lim_{k\to\infty}\langle u_k,\varphi\rangle = \langle\delta_0,\varphi\rangle \qquad \forall\varphi\in\mathcal{D}(\mathbb{R}^n)$$

as $k\to\infty$. In conclusion, it holds $u_k\to\delta_0$ in $\mathcal{D}'(\mathbb{R}^n)$ as $k\to\infty$. □

The previous example shows how the Dirac delta function can be interpreted in a mathematical way: it is the limit (in the sense of distributions) of a sequence of L^1 functions which "tend to concentrate their mass" at the center of the delta. Now we present another significant example of distribution which is defined as the limit, in the sense of distributions, of a sequence of functions.

Example 2.6.10. Consider the following measurable function defined a.e. in $\mathbb{R}$: $u(x) = 1/x$ for every $x\neq 0$. This function does not belong to $L^1_{loc}(\mathbb{R})$, since it is not integrable in a neighborhood of $x=0$. However, it is possible to associate with this function a distribution that will be called **principal value** of $1/x$ and will be denoted by p.v.$(1/x)$. For every $\varepsilon>0$ we denote by χ_ε the function

$$\chi_\varepsilon(x) = \begin{cases} 1 & \text{if } x\in(-\infty,-\varepsilon)\cup(\varepsilon,\infty) \\ 0 & \text{if } x\in[-\varepsilon,\varepsilon]. \end{cases}$$

The distribution principal value is defined as

$$\text{p.v.}\left(\frac{1}{x}\right) := \lim_{\varepsilon\to 0^+}\frac{\chi_\varepsilon(x)}{x},$$

where the limit is meant in the sense of distributions. Let us check that such definition makes sense. Let $\varphi\in\mathcal{D}(\mathbb{R})$ and let $R>0$ be such that $\operatorname{supp}\varphi\subset(-R,R)$ (thus R depends on φ). Since $1/x$ is odd, for $\varepsilon<R$ it holds

$$\int_{\mathbb{R}}\frac{\chi_\varepsilon(x)}{x}\varphi(x)dx = \int_{\{\varepsilon<|x|<R\}}\frac{\varphi(x)}{x}dx = \int_{\{\varepsilon<|x|<R\}}\frac{\varphi(x)-\varphi(0)}{x}dx. \tag{2.37}$$

Since φ is differentiable in all $\mathbb{R}$, the function $(\varphi(x) - \varphi(0))/x$ is bounded in a neighborhood of zero and hence we obtain

$$\lim_{\varepsilon \to 0^+} \int_{\mathbb{R}} \frac{\chi_\varepsilon(x)}{x} \varphi(x) dx = \int_{-R}^{R} \frac{\varphi(x) - \varphi(0)}{x} \, dx.$$

Thus we can set

$$\left\langle \text{p.v.} \left(\frac{1}{x} \right), \varphi \right\rangle = \int_{-R}^{R} \frac{\varphi(x) - \varphi(0)}{x} \, dx. \tag{2.38}$$

We leave as an exercise the check that p.v.$(1/x)$ is actually a distribution. □

Let us define the derivative of a distribution.

Definition 2.6.11. *Let $u \in \mathcal{D}'(\Omega)$ and let α be a multi-index. We define $D^\alpha u$ the distribution given by*

$$\langle D^\alpha u, \varphi \rangle := (-1)^{|\alpha|} \langle u, D^\alpha \varphi \rangle \qquad \forall \varphi \in \mathcal{D}(\Omega). \tag{2.39}$$

Let us verify that the linear functional $D^\alpha u : \mathcal{D}(\Omega) \to \mathbb{R}$ is continuous, namely it defines a distribution. Let then $\{\varphi_k\} \subset \mathcal{D}(\Omega)$ and $\varphi \in \mathcal{D}(\Omega)$ be such that $\varphi_k \to \varphi$ in $\mathcal{D}(\Omega)$. It is easy to see that $D^\alpha \varphi_k \to D^\alpha \varphi$ in $\mathcal{D}(\Omega)$. Consequently, since u is a continuous functional over $\mathcal{D}(\Omega)$, we have

$$\lim_{k \to \infty} \langle D^\alpha u, \varphi_k \rangle = \lim_{k \to \infty} (-1)^{|\alpha|} \langle u, D^\alpha \varphi_k \rangle = (-1)^{|\alpha|} \langle u, D^\alpha \varphi \rangle = \langle D^\alpha u, \varphi \rangle.$$

This definition of distributional derivative is justified by what happens when the distribution is a regular function that admits classical derivatives. We start considering the particular case of a function $u \in C^1(\overline{\Omega})$, with Ω an open bounded set with C^1 boundary (see Definition 3.1.2). For every $\varphi \in \mathcal{D}(\Omega)$, an integration by parts leads to

$$\int_\Omega \frac{\partial u}{\partial x_i} \varphi = - \int_\Omega u \frac{\partial \varphi}{\partial x_i}$$

being the integral over $\partial\Omega$ zero since φ vanishes on the boundary. Iterating this procedure, it is easy to verify that if $u \in C^k(\overline{\Omega})$ then for every multi-index α with $|\alpha| = k$ we obtain

$$\int_\Omega (D^\alpha u) \varphi = (-1)^{|\alpha|} \int_\Omega u D^\alpha \varphi \quad \forall \varphi \in \mathcal{D}(\Omega).$$

This shows that if $u \in C^k(\overline{\Omega})$ then the distributional derivatives of u up to order k coincide with the classical ones; in short, (2.39) is the generalization of k integrations by parts.

Example 2.6.12. Consider on $(-1,1)$ the function $u(x) = x^+ = \max\{x, 0\}$ which displays a corner at zero. Let us try to compute its distributional derivative. Taking an arbitrary $\varphi \in \mathcal{D}(-1,1)$, integrating by parts and exploiting the fact that φ vanishes on the boundary of $(-1,1)$, we obtain

$$\int_{-1}^{1} u(x)\varphi'(x)dx = \int_{-1}^{0} u(x)\varphi'(x)dx + \int_{0}^{1} u(x)\varphi'(x)dx = \int_{0}^{1} x\varphi'(x)dx$$

$$= [x\varphi(x)]_0^1 - \int_0^1 \varphi(x)dx = -\int_0^1 \varphi(x)dx = -\int_{-1}^{1} H(x)\varphi(x)dx$$

where $H(x)$ is called **Heaviside function** and is defined by

$$H(x) = \begin{cases} 0 & \text{if } x < 0 \\ 1 & \text{if } x \geq 0. \end{cases} \tag{2.40}$$

Therefore, even if it is not differentiable in the classical sense, the function u admits a distributional derivative, which coincides with the Heaviside function. At this point, it is natural to wonder if also the function H (which belongs to $L^1_{loc}(\mathbb{R}^n)$) admits a derivative in the sense of distributions. For every $\varphi \in \mathcal{D}(-1,1)$ it holds

$$\int_{-1}^{1} H(x)\varphi'(x)dx = \int_0^1 \varphi'(x)dx = \varphi(1) - \varphi(0) = -\varphi(0) = -\langle \delta_0, \varphi \rangle.$$

Hence, the distributional derivative of H is the Dirac delta function centered at the origin. This happens since the function H has a jump discontinuity of size 1 at the origin. □

We conclude this section by introducing the notion of tempered distribution. We begin by defining the space of functions rapidly vanishing at infinity.

Definition 2.6.13. *We shall say that a function $\varphi \in C^\infty(\mathbb{R}^n)$ is* **rapidly vanishing at infinity** *if for every multi-index α and for every non negative integer N there holds $|x|^N D^\alpha \varphi \in L^\infty(\mathbb{R}^n)$. We shall denote by $\mathcal{S}(\mathbb{R}^n)$ the space of functions rapidly vanishing at infinity. Moreover, we will say that a sequence $\{\varphi_k\} \subset \mathcal{S}(\mathbb{R}^n)$ converges to $\varphi \in \mathcal{S}(\mathbb{R}^n)$ if for every non negative integer N and for every multi-index α it holds*

$$|x|^N D^\alpha \varphi_k \to |x|^N D^\alpha \varphi \quad \text{in } L^\infty(\mathbb{R}^n)$$

as $k \to \infty$.

Basically, a rapidly vanishing function has an "infinite rate of vanishing at infinity", namely it is "almost with compact support". The easiest example of such a function is $\varphi(x) = e^{-|x|^2}$.

Clearly, $\mathcal{D}(\mathbb{R}^n)$ is a subspace of $\mathcal{S}(\mathbb{R}^n)$ and if a sequence converges in $\mathcal{D}(\mathbb{R}^n)$ then it converges in $\mathcal{S}(\mathbb{R}^n)$ as well: hence, the inclusion $\mathcal{D}(\mathbb{R}^n) \subset \mathcal{S}(\mathbb{R}^n)$ is a continuous function.

We are now in position to define the space of tempered distributions.

Definition 2.6.14. *We will call* **tempered distribution** *any linear functional* $\Lambda : \mathcal{S}(\mathbb{R}^n) \to \mathbb{R}$ *which is continuous, in the sense that*

$$\Lambda(\varphi_k) \to \Lambda(\varphi)$$

for every sequence $\{\varphi_k\} \subset \mathcal{S}(\mathbb{R}^n)$ *such that* $\varphi_k \to \varphi$ *in* $\mathcal{S}(\mathbb{R}^n)$. *We shall denote the space of tempered distributions by* $\mathcal{S}'(\mathbb{R}^n)$. *Then we will say that a sequence* $\{u_k\} \subset \mathcal{S}'(\mathbb{R}^n)$ *converges to* $u \in \mathcal{S}'(\mathbb{R}^n)$ **in the sense of tempered distributions** *if* $\langle u_k, \varphi\rangle \to \langle u, \varphi\rangle$ *for every* $\varphi \in \mathcal{S}(\mathbb{R}^n)$.

In view of the continuity of the embedding $\mathcal{D}(\mathbb{R}^n) \subset \mathcal{S}(\mathbb{R}^n)$, we may consider $\mathcal{S}'(\mathbb{R}^n)$ as a subspace of $\mathcal{D}'(\mathbb{R}^n)$. In fact, if $\Lambda : \mathcal{S}(\mathbb{R}^n) \to \mathbb{R}$ is a tempered distribution then its restriction to $\mathcal{D}(\mathbb{R}^n)$ is actually a distribution. We remark that $\mathcal{S}'(\mathbb{R}^n) \subset \mathcal{D}'(\mathbb{R}^n)$ is a strict inclusion, as we can notice in the following example.

Example 2.6.15. In dimension $n = 1$ let us consider the function $u(x) = e^x$. Since $u \in L^1_{loc}(\mathbb{R})$, there hold $u \in \mathcal{D}'(\mathbb{R})$ and

$$\langle u, \varphi\rangle = \int_{\mathbb{R}} e^x \varphi(x)dx \qquad \forall \varphi \in \mathcal{D}(\mathbb{R}).$$

Choosing $\varphi(x) = e^{-\sqrt{1+|x|^2}} \in \mathcal{S}(\mathbb{R})$, we observe that

$$\langle u, \varphi\rangle = \int_{\mathbb{R}} e^x e^{-\sqrt{1+|x|^2}}\, dx = +\infty.$$

Thus it is not possible to extend the functional u to the whole space $\mathcal{S}'(\mathbb{R})$. □

The previous example suggests that, in order to be a tempered distribution, a function $u \in L^1_{loc}(\mathbb{R}^n)$ should not go to infinity "too rapidly". In the next example we show how a restriction on the growth at infinity is sufficient to guarantee that a function of $L^1_{loc}(\mathbb{R}^n)$ belongs to $\mathcal{S}'(\mathbb{R}^n)$.

Example 2.6.16. Let $u \in L^1_{loc}(\mathbb{R}^n)$ satisfy

$$\int_{\mathbb{R}^n} (1 + |x|)^{-N} |u(x)|dx < +\infty \tag{2.41}$$

for some non negative integer N.

We prove that $u \in \mathcal{S}'(\mathbb{R}^n)$. Let $\{\varphi_k\} \subset \mathcal{S}(\mathbb{R}^n)$ and let $\varphi \in \mathcal{S}(\mathbb{R}^n)$ be such that $\varphi_k \to \varphi$ in $\mathcal{S}(\mathbb{R}^n)$ as $k \to \infty$. Then

$$\lim_{k\to\infty} \int_{\mathbb{R}^n} u(x)\varphi_k(x)dx = \lim_{k\to\infty} \int_{\mathbb{R}^n} (1 + |x|)^{-N} u(x)(1 + |x|)^N \varphi_k(x)dx. \tag{2.42}$$

But $(1 + |x|)^{-N} u(x) \in L^1(\mathbb{R}^n)$, while

$$(1 + |x|)^N \varphi_k(x) \to (1 + |x|)^N \varphi(x) \quad \text{in } L^\infty(\mathbb{R}^n).$$

This allows to pass to the limit in (2.42), obtaining

$$\lim_{k\to\infty}\int_{\mathbb{R}^n} u(x)\varphi_k(x)dx = \int_{\mathbb{R}^n} u(x)\varphi(x)dx,$$

thus $u \in \mathcal{S}'(\mathbb{R}^n)$.

A particular case of functions u that verify condition (2.41) are the functions $u \in L^p(\mathbb{R}^n)$ for some $p \in [1,\infty]$. If $p=1$, it suffices to choose $N=0$. If $1<p\le\infty$, in view of the Hölder inequality (2.17), we have to impose $(1+|x|)^{-N} \in L^{p'}(\mathbb{R}^n)$, with $p'=p/(p-1)$ if $p<\infty$ and $p'=1$ if $p=\infty$. Passing to radial coordinates, the condition becomes

$$\int_{\mathbb{R}^n}(1+|x|)^{-Np'}dx<\infty \Longleftrightarrow \int_0^\infty (1+r)^{-Np'}r^{n-1}dr<\infty \Longleftrightarrow N>n/p'. \qquad \square$$

We conclude this section with other significant examples of tempered distributions.

Example 2.6.17. If $\Lambda \in \mathcal{D}'(\mathbb{R}^n)$ has compact support, then $\Lambda \in \mathcal{S}'(\mathbb{R}^n)$: in particular, $\delta_0 \in \mathcal{S}'(\mathbb{R}^n)$. We have to show that the linear continuous functional $\Lambda : \mathcal{D}(\mathbb{R}^n) \to \mathbb{R}$ admits a linear continuous extension $\widetilde{\Lambda} : \mathcal{S}(\mathbb{R}^n) \to \mathbb{R}$. Let us denote by K the support of Λ and let us consider a function $\psi \in \mathcal{D}(\mathbb{R}^n)$ such that $\psi = 1$ on some open set U containing K. We set

$$\widetilde{\Lambda}(\varphi) := \Lambda(\psi\varphi) \qquad \forall\varphi\in\mathcal{S}(\mathbb{R}^n).$$

If $\{\varphi_k\}$ is a sequence such that $\varphi_k \to \varphi$ in $\mathcal{S}(\mathbb{R}^n)$, then (for every multi-index α) $D^\alpha\varphi_k \to D^\alpha\varphi$ uniformly in $\mathbb{R}^n$ and hence $D^\alpha(\psi\varphi_k) \to D^\alpha(\psi\varphi)$ in $\mathcal{D}(\mathbb{R}^n)$. Since Λ is continuous on $\mathcal{D}(\mathbb{R}^n)$ we immediately infer $\widetilde{\Lambda}(\varphi_k) \to \widetilde{\Lambda}(\varphi)$. This proves that $\widetilde{\Lambda}$ is a tempered distribution. Furthermore, $\widetilde{\Lambda}$ is an extension of Λ: since $\varphi \in \mathcal{D}(\mathbb{R}^n)$ it holds

$$\widetilde{\Lambda}(\varphi)-\Lambda(\varphi)=\Lambda(\psi\varphi)-\Lambda(\varphi)=\Lambda((\psi-1)\varphi)=0,$$

where the last equality follows from the definition of support of a distribution and from the fact that $\operatorname{supp}((\psi-1)\varphi)\subseteq K^c$. $\square$

Example 2.6.18. The distribution $\text{p.v.}(\frac{1}{x})$ introduced in Example 2.6.10 is tempered. Indeed, if $\varphi\in\mathcal{S}(\mathbb{R})$ and we fix $R>0$ arbitrarily, instead of (2.37) we obtain

$$\int_{\mathbb{R}}\frac{\chi_\varepsilon(x)}{x}\varphi(x)dx=\int_{\{|x|>\varepsilon\}}\frac{\varphi(x)}{x}dx=\int_{\{\varepsilon<|x|<R\}}\frac{\varphi(x)-\varphi(0)}{x}dx+\int_{\{|x|>R\}}\frac{\varphi(x)}{x}dx$$

and, passing to the limit as $\varepsilon\to 0$, we deduce that

$$\left\langle \text{p.v.}\left(\frac{1}{x}\right),\varphi\right\rangle=\int_{-R}^{R}\frac{\varphi(x)-\varphi(0)}{x}\,dx+\int_{\{|x|>R\}}\frac{\varphi(x)}{x}dx,$$

which replaces (2.38). Note that the definition above does not depend on $R>0$. At this point, it is immediate to verify that such linear functional is continuous with respect to the convergence in $\mathcal{S}(\mathbb{R})$. $\square$

2.7 Advantages of the Lebesgue integral

In this section we summarize the main differences between Riemann and Lebesgue integral, trying to emphasize the advantages of the latter.

In the construction of the Riemann integral the procedure consists in slicing the *domain* of the function, while in the Lebesgue integral the procedure consists in slicing its *codomain*; in both cases the further step is to pass to the limit as the partitions become thicker and thicker. For the Riemann integral, after having fixed the partition, one selects the ordinate in the codomain. This choice risks not to be representative of f, unless f is continuous or at least it has not too many discontinuities. Thus we may ask: how discontinuous can the function be, so that it is still Riemann integrable? The answer is given by the following

Theorem 2.7.1. (Lebesgue Theorem about Riemann integrable functions). *Let $\Omega \subset \mathbb{R}^n$ be an open set. A bounded function $f : \Omega \to \mathbb{R}$ is Riemann integrable if and only if it is continuous a.e.*

The reader should pay attention not to confuse functions which are continuous a.e. with functions which coincide a.e. with a continuous function. The function $[x]$ (integral part of x) is continuous a.e. but it does not coincide a.e. with a continuous function. The Dirichlet function (2.3) coincides a.e. with the zero function but is it not continuous a.e. (actually, it is nowhere continuous!).

Theorem 2.7.1 shows the first convenience of Lebesgue integral: it allows to integrate a larger class of bounded functions, the class of measurable functions.

In order to describe the second advantage, let us consider again the sequence (2.4). Every function f_k is Riemann integrable and has zero (Riemann) integral. Hence, $\{f_k\}$ is a Cauchy sequence (see Definition 1.2.1) with respect to the norm defined by the Riemann integral. However, we well know that its pointwise limit (2.3) is not Riemann integrable. Thus, the space of Riemann integrable functions is not closed under pointwise convergence. In Sections 2.4 and 2.5 we have enlightened the importance of completeness in the spaces of integrable functions. This property allows to consider a Hilbert space (such as L^2) just like $\mathbb{R}^\infty$. In order to better understand the great advantage that completeness (and the Hilbertian structure) offers, we refer the reader to Chapter 7.

Finally, thanks to the Lebesgue integral, we can prove some good properties of the space of distributions. As an example, let us consider the sequence introduced in Example 2.6.9; the main difficulty in determining its limit has been overcome by applying Theorem 2.3.7 that, as we know, is not valid for the Riemann integral (cf. Exercise 2.1.1).

Chapter 3

Classical solutions of linear partial differential equations

"To draw quantitative conclusions we must use the language of Mathematics. Most of the fundamental ideas of science are essentially simple, and may, as a rule, be expressed in a language comprehensible to everyone. To follow up these ideas demands the knowledge of highly refined technique of investigation. Mathematics as a tool of reasoning is necessary if we wish to draw conclusions which may be compared with experiment. [...] The price which must be paid for abandoning the language of Mathematics is a loss in precision, and the necessity of sometimes quoting results without showing how they were reached."

Albert Einstein

[The evolution of Physics (1938)]

3.1 Introduction

Mathematical modelling occupies an important position in the scientific research and in several technical and industrial activities. Indeed, exploiting suitable mathematical models, it is possible to describe, predict and control the development of most of phenomena which characterize the real world. The mathematical modelling process begins necessarily with a preliminary analysis of the observed phenomenon which leads to a suitable schematization of the problem; then it follows the identification of a set of equations and/or other mathematical relations which can describe and predict the behavior of the system under study. Having at our disposal a mathematical model, we can proceed, for instance, with a numerical simulation followed by an experimental verification in order to validate the obtained results: this method proved successful in the scientific and industrial fields, also thanks to the high performances of the computers of last generation. As an alternative to the numerical simulation, it can be set a non deterministic analysis with the tools of probability and statistics. In general, a mathematical model of a phenomenon is built starting from the *general laws* and from the *constitutive relations*. The former are laws valid for all the phenomena of the same kind: in Mechanics, for instance, the general laws are conservation or balance laws (mass, energy, linear momentum,...). The constitutive laws, instead, concern only the phenomenon under examination and are often of experimental nature.

Irrespective to the nature of the phenomenon under study, by applying the general laws and the constitutive laws for the description of a suitable mathematical model, one often arrives to a partial differential equation or a system of partial differential equations. It is defined **partial differential equation** of order m an equation of the kind

$$F\left(x_1,\dots,x_n,u,\nabla u,D^2u,...,D^\alpha u\right)=0 \qquad \forall|\alpha|\le m \tag{3.1}$$

which relates the n independent variables $x_1,\dots,x_n$ $(n>1)$ to the unknown function $u=u(x_1,\dots,x_n)$ and with its partial derivatives up to order m included. In (3.1), D^2u represents the Hessian matrix of u, while $D^\alpha u$ is defined according to the multi-index representation introduced in (2.35). We shall assume that F is a continuous function with respect to all its arguments. We consider functions $u=u(x_1,\dots,x_n)$ which, together with their derivatives up to order m, are continuous over an open subset Ω of $\mathbb{R}^n$ and, possibly, also on the boundary $\partial\Omega$. If the condition expressed in (3.1) is satisfied by inserting the function $u(x_1,\dots,x_n)$ and its derivatives into the left hand side of (3.1), we will say that u is a *solution* to (3.1). According to the circumstances, we will use different notations to denote the partial derivatives; for instance, we will indistinctly denote the first and second order partial derivatives of u with respect to t with

$$u_t=\partial_t u=\frac{\partial u}{\partial t}\,, \qquad u_{tt}=\partial_{tt}u=\frac{\partial^2 u}{\partial t^2}.$$

Equation (3.1) is called **linear** if F is linear with respect to u and to all the derivatives of u; otherwise it is called **nonlinear**. Nonlinear equations are further divided into **semilinear** (if F is linear with respect to all the derivatives of u), **quasilinear** (if F is linear with respect to the derivatives of u of highest order), **fully nonlinear** (if F is nonlinear with respect to the derivatives of u of highest order).

We recall that the solutions of an *ordinary* differential equation of order m (in normal form) depend on m arbitrary *costants*. In turn, the solutions of a partial differential equation of order m depend on m arbitrary functions. Thus, in order to obtain the existence of a unique solution to (3.1), it is necessary to specify some properties of the solution, by adding some constraints which define new problems. Let us list the main ones:

- the *Cauchy problem*, for which the values of the unknown u and of all its normal derivatives up to order $m-1$ are prescribed on an $(n-1)$-dimensional surface $\Sigma \subset \Omega$;

- the *boundary value problem*, for which the values of "some" derivatives of u are prescribed on the whole boundary $\partial\Omega$.

For the boundary value problems, the equation must be of order $m \geq 2$ (think of what happens in the case $n = 1$!). When $m = 2$, the most studied boundary value problems are the *Dirichlet problem* and the *Neumann problem*, which prescribe respectively the value of u and the value of the normal derivative $\frac{\partial u}{\partial \nu}$ at every point of the boundary.

Example 3.1.1. Let us consider the following third order equation:

$$\frac{\partial^3 u(x,y,t)}{\partial t^3} = 0$$

being $u = u(x,y,t)$ the unknown function. At $t = \bar{t}$, let the initial conditions be (Cauchy problem):

$$u = u(x,y,\bar{t}) = u_0(x,y) \qquad \frac{\partial u(x,y,\bar{t})}{\partial t} = u_1(x,y) \qquad \frac{\partial^2 u(x,y,\bar{t})}{\partial t^2} = u_2(x,y)$$

where $u_0(x,y)$, $u_1(x,y)$, $u_2(x,y)$ are three given functions. With three consecutive integrations with respect to t, we find the solution

$$u(x,y,t) = u_0(x,y) + (t-\bar{t})u_1(x,y) + \frac{(t-\bar{t})^2}{2}u_2(x,y)$$

which, therefore, turns out to be uniquely determined. □

If we have further information, in the form of initial conditions and/or boundary conditions or something else, it is then necessary to establish whether the problem:

- admits at least one solution;

- admits at most one solution;

- the (unique) solution depends continuously on the initial data.

The last property, also called *local stability of the solution with respect to the data*, means that a small variation of the data generates a small variation of the solution; if this is not the case, as the data vary one may find solutions very different from the original one, characterized by non negligible errors. A problem is said to be **well posed** if it is endowed with the three afore-mentioned properties. Therefore, a well posed problem models a phenomenon that, for every choice of the data, gives a unique solution; moreover, since the data can be found experimentally and then they might be inaccurate, we require also the "low sensitivity" of the solution with respect to the data themselves.

In the following sections of this chapter we deal with linear partial differential equations, postponing to the last chapter the analysis of the nonlinear cases. Note that, while the theory concerning linear equations is sufficiently developed and funded with principles with general validity, nonlinear equations are not set within a general unifying theory.

We conclude by observing that, unlike what happens in dimension $n = 1$, if $n \geq 2$ the subsets of $\mathbb{R}^n$ can have very complicated geometries and their boundaries may play an important role, especially in the regularity theory. Thus it is necessary to characterize quantitatively the "goodness" of a boundary.

Given $x = (x_1, ..., x_n) \in \mathbb{R}^n$, set $x' = (x_1, ..., x_{n-1}) \in \mathbb{R}^{n-1}$ and define the following sets

$$B := \{x \in \mathbb{R}^n;\ |x| < 1\}, \quad Q_+ := \{x \in B;\ x_n > 0\}, \quad Q_0 := \{x \in B;\ x_n = 0\}.$$

Definition 3.1.2. *We say that the boundary of an open set $\Omega \subset \mathbb{R}^n$ is of class C^k with $1 \leq k \leq \infty$, if for every $x \in \partial\Omega$ there exist an open neighborhood U of x and an invertible function $\Lambda : B \to U$ with $\Lambda \in C^k(\overline{B})$ and $\Lambda^{-1} \in C^k(\overline{U})$ such that*

$$\Lambda(Q_+) = \Omega \cap U \qquad \text{and} \qquad \Lambda(Q_0) = \partial\Omega \cap U.$$

If the function Λ is only Lipschitz then we shall say that Ω is a domain with Lipschitz boundary.

Example 3.1.3. Consider a disk and a square in the plane:

$$\Omega_1 = \{x \in \mathbb{R}^2;\ |x| < 1\}, \qquad \Omega_2 = (0,1) \times (0,1).$$

The boundary of Ω_1 is of class C^∞ while the boundary of Ω_2 is just Lipschitz because of the corners. □

3.2 Transport equation

3.2.1 Description of the physical model

The simple model that we are about to introduce describes how a certain substance contained into a fluid is transported by the motion of the fluid itself. We want to study the evolution of its volumic mass, in space and time. Thus we are led to consider a model in which the volumic mass (density) ρ of the substance is a function depending on the position $x \in \mathbb{R}^n$ and on the time t. If we consider a generic portion of volume of fluid $\Omega \subset \mathbb{R}^n$, then the mass of the substance contained into the region Ω is given by

$$\int_\Omega \rho \, dx.$$

The variation in time of the mass of the substance contained into Ω is due to the flow of the substance through the boundary of Ω and to the possible presence of sources of mass $f = f(x,t)$ which supply the substance itself directly into the region Ω. We adopt the following convention: if some mass of substance is injected into the control region Ω, we set $f = f(x,t) > 0$. In order to characterize the passage of substance through $\partial\Omega$, we introduce the surface flux density of the substance through the boundary as the vector quantity $\mathbf{h} = \mathbf{h}(x,t)$. Therefore, the total flow of the substance through the boundary is given by

$$\int_{\partial\Omega} \mathbf{h} \cdot \nu \, dS$$

with ν the outer unit normal vector to $\partial\Omega$.

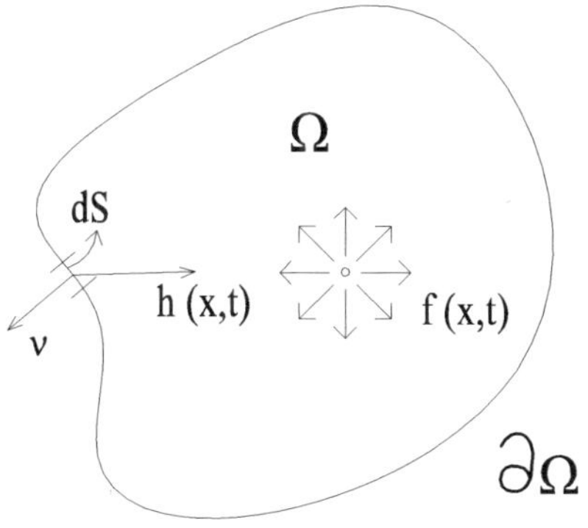

Control region Ω of a fluid

Then, thanks to the law of conservation of mass, we have

$$\frac{d}{dt}\int_\Omega \rho \, dx = -\int_{\partial\Omega} \mathbf{h} \cdot \nu \, dS + \int_\Omega f \, dx \, .$$

Differentiating under the integral sign and exploiting the divergence Theorem, we obtain

$$\int_\Omega (\rho_t + \mathrm{div}\mathbf{h} - f) \, dx = 0$$

and, by the arbitrariness of the volume amount Ω, we arrive to the equation

$$\rho_t + \mathrm{div}\mathbf{h} = f.$$

It is reasonable to assume that the surface flux density $\mathbf{h}$ is of the form $\mathbf{h} = \rho\mathbf{v}$ where $\mathbf{v}$ represents the velocity of the fluid. Hence, the differential equation becomes

$$\rho_t + \mathbf{v} \cdot \nabla\rho + (\mathrm{div}\mathbf{v})\rho = f \tag{3.2}$$

and is called transport equation. Let us examine equation (3.2) in the case in which the velocity $\mathbf{v}$ is constant $\mathbf{v} = b \in \mathbb{R}^n$.

3.2.2 Solution of the transport equation

Let $x = (x_1, ..., x_n) \in \mathbb{R}^n$ and $t \in (0, \infty)$. For every differentiable function $u = u(x,t)$, we adopt the notations

$$\nabla_x u = (u_{x_1}, ..., u_{x_n}) \qquad \nabla_{x,t} u = (u_{x_1}, ..., u_{x_n}, u_t) \ .$$

If in the equation (3.2) we assume that the velocity $\mathbf{v}$ is constant $\mathbf{v} = b \in \mathbb{R}^n$ and we set $u = \rho$, we arrive to the Cauchy problem

$$\begin{cases} u_t(x,t) + b \cdot \nabla_x u(x,t) = f(x,t) & \text{in } \mathbb{R}^n \times (0,\infty) \\ u(x,0) = g(x) & \text{in } \mathbb{R}^n. \end{cases} \tag{3.3}$$

Let us begin by considering the simplest case in which $f = 0$ (no mass source):

$$\begin{cases} u_t(x,t) + b \cdot \nabla_x u(x,t) = 0 & \text{in } \mathbb{R}^n \times (0,\infty) \\ u(x,0) = g(x) & \text{in } \mathbb{R}^n \ . \end{cases} \tag{3.4}$$

Setting $v = (b, 1) \in \mathbb{R}^{n+1}$, in view of the gradient rule, the differential equation in (3.4) can be rewritten as

$$\frac{\partial u}{\partial v} = \nabla_{x,t} u \cdot v = 0$$

showing the fact that u is constant along the straight lines parallel to v. Such lines are called **characteristic curves** and their equations, in parametric form, are

$$\begin{cases} x(s) = x_0 + sb \\ t(s) = t_0 + s \end{cases} \qquad (s \in \mathbb{R}) \qquad \forall (x_0, t_0) \in \mathbb{R}^{n+1} \ .$$

But then (3.4) has solution

$$u(x,t) = u(x - tb, 0) = g(x - tb) \ . \tag{3.5}$$

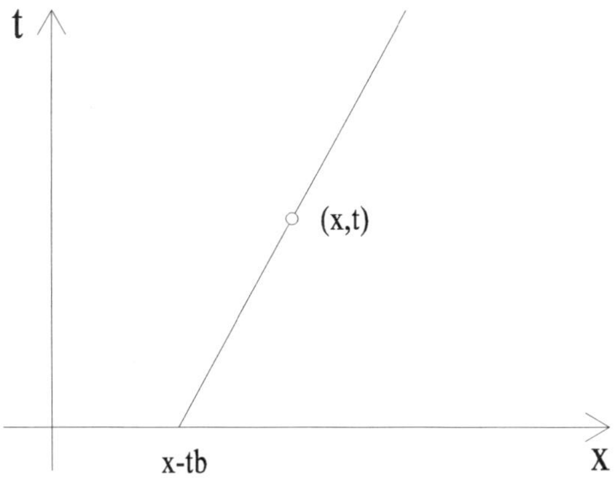

Transport of the datum along the characteristics

Basically, the initial datum is "transported" along the characteristic lines.

Let us now pass to the complete equation (3.3). We define the function $z : \mathbb{R} \to \mathbb{R}$ as

$$z(s) = u(x + sb, t + s) \ , \quad s \in \mathbb{R} \ .$$

Differentiating, we obtain $z'(s) = \nabla_x u \cdot b + u_t = f(x + sb, t + s)$. Hence, integrating between $-t$ and 0 we infer

$$u(x,t) - g(x - tb) = z(0) - z(-t) = \int_{-t}^{0} f(x + sb, t + s) ds$$

$$= [\sigma = t + s] = \int_0^t f(x + (\sigma - t)b, \sigma) d\sigma \ .$$

In conclusion, the solution of (3.3) is given by

$$u(x,t) = g(x - tb) + \int_0^t f(x + (\sigma - t)b, \sigma) d\sigma \ . \tag{3.6}$$

This formula shows that the solution does not only depend on the initial datum transported along the characteristic but also on the contribution of the forcing term f on the part of characteristic which "precedes" (x,t) (namely the segment joining the points $(x - tb, 0)$ and (x,t)). We have then proved the following

Theorem 3.2.1. *Let $g \in C^1(\mathbb{R}^n)$ and $f \in C^1(\mathbb{R}^n \times [0,\infty))$. Then the unique solution $u \in C^1(\mathbb{R}^n \times [0,\infty))$ of problem (3.3) is given by formula (3.6).*

A possible variant of (3.4) consists in adding a damping or dilation term. Let $c \neq 0$ and let us consider the problem

$$\begin{cases} u_t(x,t) + b \cdot \nabla_x u(x,t) + cu = 0 & \text{in } \mathbb{R}^n \times (0,\infty) \\ u(x,0) = g(x) & \text{in } \mathbb{R}^n \ . \end{cases} \tag{3.7}$$

We set $w(x,t) = e^{ct}u(x,t)$; then $u_t = -ce^{-ct}w + e^{-ct}w_t$ and thus w solves (3.4). Hence, by (3.5) we infer that the solution of (3.7) is given by

$$u(x,t) = e^{-ct}g(x - tb) ,$$

showing that the initial datum propagates along the characteristics with a damping coefficient (if $c > 0$) or a dilation coefficient (if $c < 0$) which intensifies over time.

Exercise 3.2.2. Determine the solution of the equation

$$\begin{cases} u_t(x,t) + b \cdot \nabla_x u(x,t) + cu = f(x,t) & \text{in } \mathbb{R}^n \times (0,\infty) \\ u(x,0) = g(x) & \text{in } \mathbb{R}^n . \end{cases}$$

Hint: combine (3.6) with the strategy adopted to solve (3.7). □

3.3 More general first order equations

The transport equation (3.3) is a particular case of first order linear equation. The most general form of such equation reads

$$F(\nabla u(x), u(x), x) = 0 , \qquad x \in \Omega \subset \mathbb{R}^n , \tag{3.8}$$

where $F : \mathbb{R}^{2n+1} \to \mathbb{R}$ is a continuous function, linear with respect to the first $n+1$ components. Equation (3.8) is associated to an "initial condition" of the kind

$$u(x) = g(x) \qquad x \in \Gamma , \tag{3.9}$$

where Γ is an $(n-1)$-dimensional surface contained into Ω (for instance, if $n = 2$, Γ is a curve). We look for a local solution to the Cauchy problem (3.8)-(3.9), namely a solution which is defined in a neighborhood of Γ. Unfortunately, this is not always a "well posed" problem: indeed, it may happen that both existence and uniqueness are not guaranteed.

Let us see an example with an elementary equation.

Example 3.3.1. Let us consider in $\mathbb{R}^2$ the equation

$$u_x = 0 . \tag{3.10}$$

Clearly, since u does not depend on x, the solutions to (3.10) are of the form $u(x,y) = h(y)$. As curve Γ we choose the straight line $y = 0$ and we impose the initial condition

$$u(x,0) = g(x) . \tag{3.11}$$

If g is constant ($g(x) \equiv k$), then there exist infinitely many solutions to (3.10)-(3.11): it is enough to take $u(x,y) = h(y)$ with an arbitrary h, provided that $h(0) = k$. If instead g is not constant, the equation (3.10) is not satisfied for $y = 0$ and hence problem (3.10)-(3.11) has no solution. If in place of $y = 0$ we choose the curve Γ given by the equation $y = x^2$, a problem of extension occurs at the origin, namely where the curve Γ is tangent to the line $y = 0$. At the points (x, y) with $y > 0$ we have to implement a method which makes the solution u recover the datum on the parabola, following the characteristics. Thus, in order to avoid any ambiguity, it is necessary to separate by a curve the points interior to the parabola which recover the datum on the different branches of the parabola (with $x > 0$ and $x < 0$ respectively). Clearly, this curve must pass through the origin. For simplicity, we take the half-line of points (x, y) with $x = 0$ and $y > 0$.

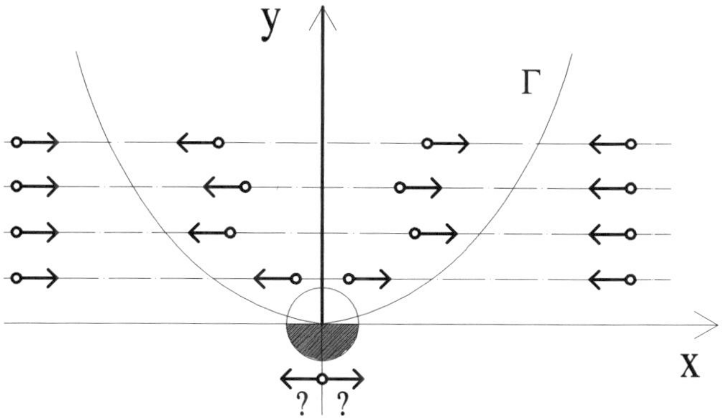

Recovery of the datum along the characteristics

In this way, we know where all the points with $y > 0$ and $x \neq 0$ recover the datum. But every neighborhood of the origin contains some points of the half-line (or an equivalent curve) and some points of the half-plane $y < 0$. For all these points, we do not know where to recover the datum. □

Let us now state, still for linear equations, a result of local existence and uniqueness for the following Cauchy problem in dimension $n = 2$:

$$\begin{cases} a(x,y)u_x + b(x,y)u_y = c(x,y) & \text{in } \Omega \\ u = g & \text{on } \Gamma \end{cases} \tag{3.12}$$

where Γ is the curve defined by the parametric equations

$$\begin{cases} x = \varphi(s) \\ y = \psi(s) \end{cases} \qquad (s \in I)$$

and I is an interval of $\mathbb{R}$. The condition $u = g$ on Γ is to be read as

$$u(\varphi(s), \psi(s)) = h(s) := g(\varphi(s), \psi(s)) \qquad \forall s \in I. \tag{3.13}$$

Let us assume that the functions a, b, c and φ, ψ, h are of class C^1 on the respective domains and let us introduce a concept of characteristic curves slightly different from the one presented in the previous section. We will call **characteristic curves** the curves in $\mathbb{R}^3$ that solve the system of ordinary equations

$$\begin{cases} x' = a(x, y) \\ y' = b(x, y) \\ z' = c(x, y). \end{cases} \tag{3.14}$$

Let us now give a geometric interpretation to the characteristic curves. For $(x_0, y_0) \in \Omega$ fixed arbitrarily, the vector $\nu_0 = (-u_x(x_0, y_0), -u_y(x_0, y_0), 1)$ represents the normal vector to the graph of u, while the vector $v_0 = (a(x_0, y_0), b(x_0, y_0), c(x_0, y_0))$ represents the tangent vector to the characteristic curve at the point $(x_0, y_0, u(x_0, y_0))$. Thus, equation (3.12) evaluated at (x_0, y_0) shows that *the characteristic curves are tangent at every point to the graph of* u.

Remark 3.3.2. If we consider the equation (3.3) and we set $y = t$, then the solutions to the system (3.14) reduce to

$$\begin{cases} x(s) = x(0) + bs \\ y(s) = y(0) + s \\ z(s) = z(0) + \int_0^s f(x(t), y(t))dt. \end{cases}$$

Thus, the characteristics defined in Section 3.2 are nothing else than the projections on the plane $z = 0$ of the characteristics defined above. □

Let us assume that $0 \in I$ and let us set $(x_0, y_0, z_0) = (\varphi(0), \psi(0), h(0))$; there holds the

Theorem 3.3.3. *Let a, b, c be functions of class C^1 in a neighborhood of the point (x_0, y_0) and let φ, ψ, h be functions of class C^1 in a neighborhood of $s = 0$. If*

$$J := b(x_0, y_0)\varphi'(0) - a(x_0, y_0)\psi'(0) \neq 0 \tag{3.15}$$

then there exists a neighborhood U of (x_0, y_0) such that problem (3.12)–(3.13) *admits a unique solution of class C^1 defined in U.*

Proof. See [62, Teorema 5.2, Capitolo 4]. □

The hypothesis (3.15) is a sufficient condition for the local existence and uniqueness of a solution for the Cauchy problem: to ensure the existence and uniqueness of a solution to (3.12) in a neighborhood of Γ one may assume that $J \neq 0$ at each point of Γ. However, the absence of this assumption does not prevent the local existence and uniqueness. Let us now discuss the meaning of condition (3.15). Let us assume that $J = 0$ and that (3.12) admits a local solution of class C^1. Then it holds

$$b(x_0, y_0)\varphi'(0) = a(x_0, y_0)\psi'(0). \tag{3.16}$$

On the other hand, in view of (3.12) and (3.13), we deduce that

$$au_x + bu_y = c \quad \text{and} \quad h'(s) = u_x(\varphi(s), \psi(s))\varphi'(s) + u_y(\varphi(s), \psi(s))\psi'(s). \tag{3.17}$$

By combining (3.16) and (3.17), we obtain, for $(x, y, z) = (x_0, y_0, z_0)$ and $s = 0$,

$$\begin{aligned} ah' &= au_x\varphi' + au_y\psi' = au_x\varphi' + bu_y\varphi' = c\varphi' \\ bh' &= bu_x\varphi' + bu_y\psi' = au_x\psi' + bu_y\psi' = c\psi'. \end{aligned} \tag{3.18}$$

In order to better understand the geometric meaning of (3.16), we introduce a curve γ in $\mathbb{R}^3$, characterized by the parametric equations

$$\gamma \begin{cases} x = \varphi(s) \\ y = \psi(s) \\ z = h(s) \end{cases} \qquad (s \in I).$$

By (3.16) and (3.18) we deduce that the vectors (φ', ψ', h') and (a, b, c), computed respectively at $s = 0$ and (x_0, y_0), are proportional to each other. In case both the vectors are non zero, we can affirm that the tangent to the curve γ at (x_0, y_0, z_0) is parallel to the tangent to the characteristic curve passing through the point (x_0, y_0, z_0) itself. Then, the point (x_0, y_0, z_0) will be called **characteristic** and the curve γ will be called **characteristic** at (x_0, y_0, z_0).

We can summarize the above results in the following statement:

Proposition 3.3.4. *If J defined in (3.15) is zero, then a necessary condition for the local existence and uniqueness of a solution of class C^1 for problem (3.12) – (3.13) in a neighborhood of (x_0, y_0) is that the curve γ is characteristic at (x_0, y_0, z_0).*

Let us go back to Example 3.3.1 and let us rephrase the conclusions in terms of Theorem 3.3.3 and of Proposition 3.3.4. If Γ was parametrized by $\varphi(s) = s$ and $\psi(s) = s^2$ and if the initial condition was $h(s) = s$ then it would not be possible to obtain the existence of a solution in a neighborhood of $(0, 0)$. Adopting the notations of Theorem 3.3.3 we infer that $a \equiv 1$, $b \equiv 0$, $c \equiv 0$, $(x_0, y_0, z_0) = (\varphi(0), \psi(0), h(0)) = (0, 0, 0)$ and $(\varphi'(s), \psi'(s), h'(s)) = (1, 2s, 1)$, so that

$$J = b(0, 0)\varphi'(0) - a(0, 0)\psi'(0) = 0.$$

The sufficient condition introduced in Theorem 3.3.3 is not satisfied. On the other hand, it holds $(\varphi'(0), \psi'(0), h'(0)) = (1, 0, 1)$ while $(a, b, c) = (1, 0, 0)$ and consequently not even the necessary condition of Proposition 3.3.4 is satisfied. This justifies the non existence of a local solution in a neighborhood of $(0, 0)$.

Let us consider again the equation (3.10), but now we choose Γ parametrized by $\varphi(s) = s$ and $\psi(s) = s^3$, with initial condition $h(s) = s$. Also in this case we obtain $J = 0$, $(\varphi'(0), \psi'(0), h'(0)) = (1, 0, 1)$ and $(a, b, c) = (1, 0, 0)$. Nevertheless, the function $u(x,y) = \sqrt[3]{y}$ is a global solution to the Cauchy problem, even if it does not fulfill the necessary condition stated in Proposition 3.3.4. This can be explained observing that $u(x,y) = \sqrt[3]{y}$ is a solution to the Cauchy problem, but it is not of class C^1 in a neighborhood of $(0,0)$.

Exercise 3.3.5. Consider the equation (3.10) and study the case in which Γ is parametrized by $\varphi(s) = s$ and $\psi(s) = s^3$, and the initial condition is $h(s) = s^3$.

3.4 Wave equation

The wave equation in spatial dimension $n \geq 1$ is a second order equation of the form

$$u_{tt} - c^2 \Delta u = f(x,t) \tag{3.19}$$

where $c > 0$ and Δ denotes the Laplace operator or Laplacian.

Several phenomena can be modeled by (3.19). In this section we will present some physical models based on the theory of elasticity, the wave propagation on the surface of a fluid, the sound wave propagation in a gas and the electromagnetic wave propagation.

The solution of the equation (3.19) depends significantly on the dimension n. Just think of the difference between the propagation of the waves across the water surface (case $n = 2$) and the propagation of sound in the space (case $n = 3$). For this reason, the discussion of the wave equation will be divided into three parts, separating the cases $n = 1$, $n = 2$ and $n = 3$. We will follow, with some substantial modifications, the procedure adopted in [41].

3.4.1 Wave equation in dimension $n = 1$

Vibrating string equation

Let us justify from a physical point of view the mathematical model that we will examine in this section. The goal is to study the transverse oscillations of a string which is subject to its weight and, possibly, to other external loads. We will assume that [16]:

– the string is so thin that it can be considered one-dimensional;

– the displacement of every point of the string takes place in a vertical plane Oxy which contains the axis (centroid) of the string, and it depends only on the time and the position of the point on the string itself;

– the string, in its equilibrium configuration, is straight and lies along the x axis. We will denote by $u = u(x,t)$ the vertical displacement at time t of the point of coordinate x, with respect to the equilibrium configuration;

– the inner force in the string at each point is tangent to the string itself in such a point. Such force, obtained applying axially external forces to the string, represents the tension of the string and will be denoted by $T(x) \geq 0$.

Calling $\alpha(x) \in [-\pi/2, \pi/2]$ the angle formed by the tangent to the string and the x axis, we shall denote by $\mathbf{t}$ the unit vector given by $\mathbf{t}(x) = (\cos(\alpha(x)), \sin(\alpha(x)))$. Therefore, on the regions in which the function u is increasing with respect to x it holds $\alpha \in [0, \pi/2]$, while where u is decreasing with respect to x it holds $\alpha \in [-\pi/2, 0]$.

Let $x_1 < x_2$ be arbitrary, then the tension acting on the portion of string between $u(x_1, t)$ and $u(x_2, t)$ reads

$$T(x_2)\mathbf{t}(x_2) - T(x_1)\mathbf{t}(x_1) = \Big[T\mathbf{t}(x)\Big]_{x_1}^{x_2}. \tag{3.20}$$

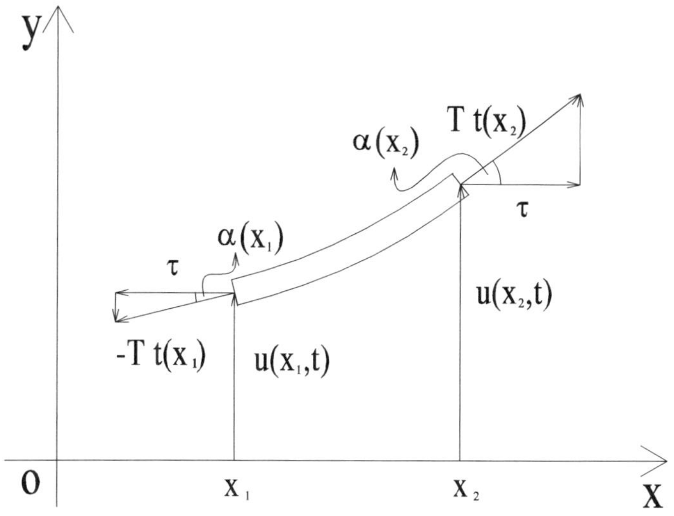

Element of a tense string

Since the displacement is purely vertical, it is reasonable to assume that, given two arbitrary points $x_1 < x_2$, the horizontal component of the tension acting on the portion of string between $u(x_1, t)$ and $u(x_2, t)$ vanishes. Thus, in view of (3.20), we obtain

$$T(x_2)\cos(\alpha(x_2)) - T(x_1)\cos(\alpha(x_1)) = 0$$

that is, for every x it holds

$$T(x)\cos(\alpha(x)) = \tau \geq 0 \tag{3.21}$$

with τ independent of x but possibly depending on t. On the other hand, by (3.20) and (3.21) we deduce that the vertical component of the tension acting on the portion of string between $u(x_1, t)$ and $u(x_2, t)$ is

$$T(x_2)\sin(\alpha(x_2)) - T(x_1)\sin(\alpha(x_1))$$

$$= \tau\left(\tan(\alpha(x_2)) - \tan(\alpha(x_1))\right) = \tau\left(u_x(x_2,t) - u_x(x_1,t)\right). \tag{3.22}$$

Let $\rho_0 = \rho_0(x)$ be the mass of the string per unit length in the condition of static equilibrium and let $\rho = \rho(x,t)$ be the mass per unit length at time t. Thanks to the mass conservation law, we can affirm that for every $x_1 < x_2$ there holds

$$\int_{x_1}^{x_2} \rho_0(x)\,dx = \int_{u(x_1,t)}^{u(x_2,t)} \rho(x,t)\,ds = \int_{x_1}^{x_2} \rho(x,t)\sqrt{1+u_x^2(x,t)}\,dx \qquad \forall t$$

and by the arbitrariness of the interval, we obtain

$$\rho_0(x) = \rho(x,t)\sqrt{1+u_x^2(x,t)}. \tag{3.23}$$

If g represents the gravity acceleration and if $F(x,t)$ is a vertical external load per unit mass, then from (3.23) we deduce that the total external load which acts on the portion of string between $u(x_1,t)$ and $u(x_2,t)$ is

$$\int_{u(x_1,t)}^{u(x_2,t)} \rho(F-g)\,ds = \int_{x_1}^{x_2} \rho(-g+F)\sqrt{1+u_x^2}\,dx = \int_{x_1}^{x_2} \rho_0 f\,dx\,, \tag{3.24}$$

where we have set $f = F - g$. Exploiting again (3.23), the total linear momentum of the string element with endpoints $u(x_1,t)$ and $u(x_2,t)$ turns out to be

$$\int_{u(x_1,t)}^{u(x_2,t)} \rho u_t\,ds = \int_{x_1}^{x_2} \rho u_t\sqrt{1+u_x^2}\,dx = \int_{x_1}^{x_2} \rho_0 u_t\,dx. \tag{3.25}$$

By combining (3.22), (3.24), (3.25) and the balance of the linear momentum, we get

$$\frac{d}{dt}\int_{x_1}^{x_2} \rho_0 u_t\,dx = \tau\left(u_x(x_2,t) - u_x(x_1,t)\right) + \int_{x_1}^{x_2} \rho_0 f\,dx$$

$$= \int_{x_1}^{x_2} [\tau u_{xx}(x,t) + \rho_0 f]dx$$

and, differentiating under the integral sign , we deduce

$$\int_{x_1}^{x_2} [\rho_0 u_{tt} - \tau u_{xx}(x,t) - \rho_0 f]dx = 0.$$

Finally, by the arbitrariness of x_1 and x_2, we obtain

$$\rho_0 u_{tt} - \tau u_{xx} = \rho_0 f.$$

If we set $c^2 = \tau/\rho_0$ (where c represents the longitudinal propagation velocity of the deformation wave), we are led to the equation

$$u_{tt} - c^2 u_{xx} = f. \tag{3.26}$$

Solution of the wave equation in dimension $n = 1$

For the sake of simplicity, in the following we will assume that the quantity $c > 0$ in (3.26) is constant. Let us consider the equation of a vibrating string with infinite length:

$$\begin{cases} u_{tt} - c^2 u_{xx} = f(x,t) & (x,t) \in \mathbb{R} \times (0,\infty) \\ u(x,0) = g(x) & x \in \mathbb{R} \\ u_t(x,0) = h(x) & x \in \mathbb{R} \,. \end{cases} \tag{3.27}$$

We want to study the evolution of the vibrating string starting from the initial conditions g and h that represent respectively the position and the velocity of the vertical displacement of the string at time $t = 0$.

Even if it is possible to proceed directly, we split (3.27) into the two following problems:

$$\begin{cases} u_{tt} - c^2 u_{xx} = 0 & (x,t)\in\mathbb{R}\times(0,\infty) \\ u(x,0) = g(x) & x \in \mathbb{R} \\ u_t(x,0) = h(x) & x \in \mathbb{R} \end{cases} \quad \text{and} \quad \begin{cases} u_{tt} - c^2 u_{xx} = f(x,t) & (x,t)\in\mathbb{R}\times(0,\infty) \\ u(x,0) = 0 & x \in \mathbb{R} \\ u_t(x,0) = 0 & x \in \mathbb{R} \,. \end{cases} \tag{3.28}$$

If $u_1 = u_1(x,t)$ and $u_2 = u_2(x,t)$ solve $(3.28)_1$ and $(3.28)_2$, the function $u = u_1 + u_2$ solves (3.27). Let us begin with problem $(3.28)_1$. We reduce the equation to a system, by setting

$$v_t(x,t) + c v_x(x,t) = 0 \,, \quad u_t(x,t) - c u_x(x,t) = v(x,t) \,. \tag{3.29}$$

The first equation of (3.29) is a (homogeneous) transport equation which, according to (3.5), has solutions of the type $v(x,t) = \varphi(x - ct)$, for some function φ. Once $v = v(x,t)$ is determined, the second equation of (3.29) becomes the (non homogeneous) transport equation $u_t(x,t) - c u_x(x,t) = \varphi(x - ct)$. According to (3.6), its solution is given by

$$\begin{aligned} u(x,t) &= g(x+ct) + \int_0^t \varphi(x + c(t-\sigma) - c\sigma) d\sigma \\ &= [y = x + ct - 2c\sigma] = g(x+ct) + \frac{1}{2c} \int_{x-ct}^{x+ct} \varphi(y) dy \,. \end{aligned} \tag{3.30}$$

Imposing the initial conditions, in view of [55, Capitolo 8, Corollario 1.17], we have

$$u_t(x,0) = h(x) = \varphi(x) + c g'(x) \;\Rightarrow\; \varphi(x) = h(x) - c g'(x) \,.$$

Thus, by (3.30), we get

$$u(x,t) = \frac{1}{2} \int_{x-ct}^{x+ct} \left[\frac{h(y)}{c} - g'(y) \right] dy + g(x+ct) \,,$$

from which we deduce the **D'Alembert formula**:

$$u(x,t) = \frac{1}{2}[g(x+ct) + g(x-ct)] + \frac{1}{2c}\int_{x-ct}^{x+ct} h(y)dy \ , \tag{3.31}$$

which gives the solution of problem $(3.28)_1$. The D'Alembert formula (3.31) shows that the solution is the sum of two terms of the form

$$u(x,t) = F(x+ct) + G(x-ct) \ , \tag{3.32}$$

that is, the sum of two progressive waves which move in opposite directions with the same velocity c. In particular, from (3.32) we deduce that if A, B, C, D are the vertices (clockwise) of a *characteristic quadrilateral* (namely a parallelogram lying in the plane (x,t), with sides having angular coefficients $\pm\frac{1}{c}$), then we have $u(A) + u(C) = u(B) + u(D)$. This way of constructing the solution is called *characteristic quadrilaterals method* and is very useful for the *Cauchy-Dirichlet problem.*

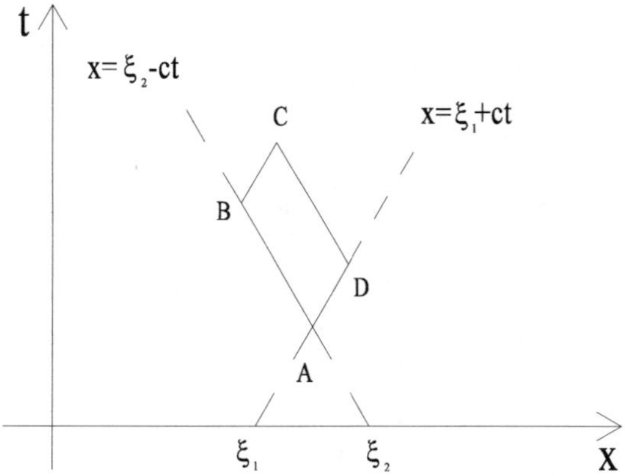

Characteristic quadrilateral

The interval $[x-ct, x+ct]$ is called **domain of dependence** because when the external force f is identically zero, the value $u(x,t)$ only depends on the behavior of the initial conditions g and h in $[x-ct, x+ct]$.

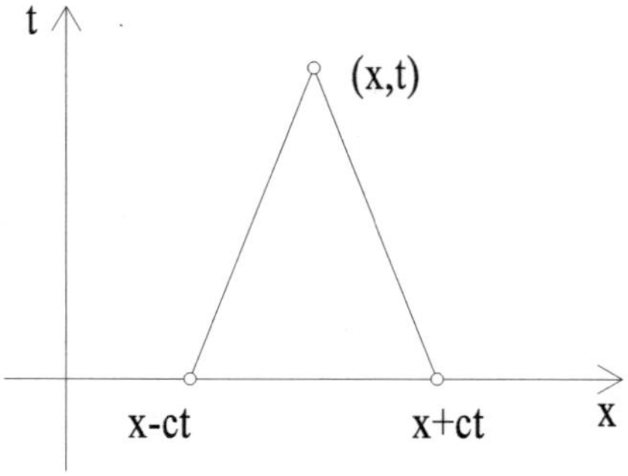

Domain of dependence

On the other hand, for $\xi \in \mathbb{R}$ fixed, the values of the initial conditions g and h at ξ influence the values of u in the whole region

$$\{(x,t) \in \mathbb{R} \times (0,\infty);\ \xi - ct \leq x \leq \xi + ct\}.$$

For this reason such a region of the plane is called **range of influence**.

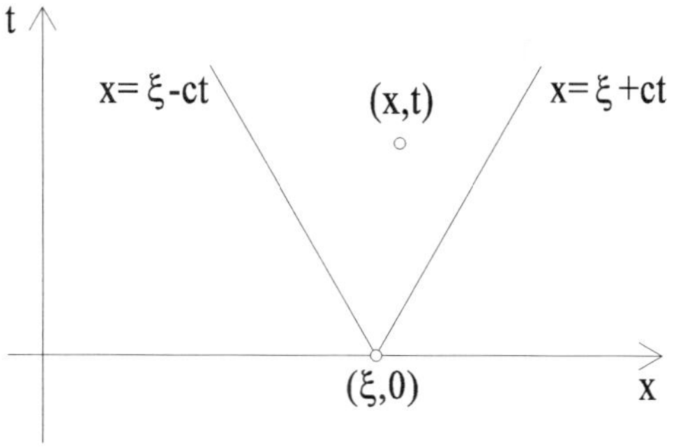

Range of influence

We now give the following

Definition 3.4.1. *Calling H the Heaviside function defined in* (2.40), *set*

$$K(x,z,t) = \frac{1}{2c}[H(x-z+ct) - H(x-z-ct)]. \tag{3.33}$$

The function K is called **fundamental solution of the wave equation** *in dimension $n = 1$ with pole at $z \in \mathbb{R}$.*

By (3.31), the solution of $(3.28)_1$ with $g = 0$ reads

$$u(x,t) = \frac{1}{2c}\int_{x-ct}^{x+ct} h(y)\,dy = \int_{\mathbb{R}} K(x,y,t)h(y)\,dy\ .$$

Hence, provided that we replace the integral with two brackets, for every fixed $z \in \mathbb{R}$ we can interpret the function K defined in (3.33) as the solution to the Cauchy problem

$$\begin{cases} u_{tt} - c^2 u_{xx} = 0 & \text{in } \mathbb{R} \times (0,\infty) \\ u(x,0) = 0 & \text{in } \mathbb{R} \\ u_t(x,0) = \delta_z & \text{in } \mathbb{R} \end{cases} \tag{3.34}$$

where δ_z is the Dirac delta function centered at z (see Example 2.6.4).

Let us now pass to problem $(3.28)_2$. In order to solve it, we will use the Duhamel *retard principle.* Let us consider $s \geq 0$ as a parameter and let us define $w = w(x,t;s)$ as the solution to the problem

$$\begin{cases} w_{tt}(x,t;s) - c^2 w_{xx}(x,t;s) = 0 & x \in \mathbb{R},\ t > s \\ w(x,s;s) = 0 & x \in \mathbb{R} \\ w_t(x,s;s) = f(x,s) & x \in \mathbb{R}\ . \end{cases}$$

By the D'Alembert formula (3.31), its solution is of the form

$$w(x,t;s) = \frac{1}{2c}\int_{x-c(t-s)}^{x+c(t-s)} f(y,s)dy \ .$$

Let us now set

$$u(x,t) = \int_0^t w(x,t;s)ds \ .$$

Differentiating we obtain ([55, Capitolo 8, Corollario 1.17]):

$$u_t(x,t) = w(x,t;t) + \int_0^t w_t(x,t;s)ds = \int_0^t w_t(x,t;s)ds$$

$$u_{tt}(x,t) = w_t(x,t;t) + \int_0^t w_{tt}(x,t;s)ds = f(x,t) + \int_0^t w_{tt}(x,t;s)ds$$

$$u_{tt}(x,t) - c^2 u_{xx}(x,t) = f(x,t) + \int_0^t [w_{tt}(x,t;s) - c^2 w_{xx}(x,t;s)]ds = f(x,t) \ .$$

Moreover $u(x,0) = u_t(x,0) = 0$. Therefore, the solution of $(3.28)_2$ is given by

$$u(x,t) = \frac{1}{2c}\int_0^t \int_{x-cs}^{x+cs} f(y,t-s)dy\, ds \ . \tag{3.35}$$

In conclusion, summing (3.31) and (3.35), we have proved the

Theorem 3.4.2. *Let $g \in C^2(\mathbb{R})$, $h \in C^1(\mathbb{R})$ and $f \in C^2(\mathbb{R}\times[0,\infty))$. Then the unique solution $u \in C^2(\mathbb{R}\times[0,\infty))$ to problem (3.27) is given by*

$$u(x,t) = \frac{1}{2}[g(x+ct)+g(x-ct)] + \frac{1}{2c}\int_{x-ct}^{x+ct} h(y)dy + \frac{1}{2c}\int_0^t\int_{x-cs}^{x+cs} f(y,t-s)dy\, ds \, .$$

Theorem 3.4.2 shows that the solution u evaluated at (x,t) is the superposition of three effects. The first one is the mean value of u at time $t=0$ at the endpoints of $[x-ct, x+ct]$. The second one is (up to a factor t) the average of u_t at time $t=0$ over the interval $[x-ct, x+ct]$. The third one is (up to a factor $t^2/2$) the average of the forcing term f over the triangle delimited by the interval $[x-ct, x+ct]$ and the two characteristics based at the endpoints (characteristic triangle).

Let us now pass to the **Cauchy-Dirichlet problem** for the vibrating string of length ℓ and clamped endpoints, that is:

$$\begin{cases} u_{tt} - c^2 u_{xx} = f(x,t) & (x,t) \in (0,\ell)\times(0,\infty) \\ u(x,0) = g(x) & x \in [0,\ell] \\ u_t(x,0) = h(x) & x \in [0,\ell] \\ u(0,t) = u(\ell,t) = 0 & t \in [0,\infty) \ . \end{cases} \tag{3.36}$$

In the first characteristic triangle (delimited by the three straight lines with equations $t = 0$, $x - ct = 0$ and $x + ct = \ell$) the solution is given by Theorem 3.4.2. In the remaining part of the stripe $(0, \ell) \times (0, \infty)$ one can proceed with the characteristic quadrilaterals method, suitably adapted to the case $f \neq 0$. With such a procedure, we see that in order to have a solution of class C^2, besides the regularity of the data, some *compatibility conditions* have to be satisfied:

$$g(0) = g(\ell) = h(0) = h(\ell) = 0, \qquad g''(0) = -\frac{1}{c^2} f(0,0), \qquad g''(\ell) = -\frac{1}{c^2} f(\ell, 0) \,.$$

Without these conditions, a *propagation of discontinuity* occurs along the characteristics.

Exercise 3.4.3. Consider problem (3.36) with $c = \ell = 1$, $f \equiv 0$, $g \equiv 1$, $h \equiv 0$. Construct the solution using the characteristic quadrilaterals method; if the solution is correct, you will notice discontinuities along the characteristics passing through the points $(0, 0)$ and $(1, 0)$. □

Assuming for simplicity that the forcing term $f(x,t)$ vanishes, we deal with problem (3.36) adopting the separation of variables method. Such procedure allows to obtain useful information about the oscillation frequency of the string as a function of the tension, the length and the mass density. If we think of the string as a part of a musical instrument, this method allows to understand which is the note produced when we make the string vibrate. To compute the oscillation frequency it is not restrictive to assume that the initial velocity $u_t(x, 0)$ is zero in $(0, \ell)$. Thus, assuming that $f \equiv 0$ and $h \equiv 0$, we look for solutions u to the equation in (3.36) of the form $u(x,t) = X(x)T(t)$: we obtain

$$\frac{T''(t)}{c^2 T(t)} = \frac{X''(x)}{X(x)}$$

from which we deduce that there exists $\lambda \in \mathbb{R}$ such that

$$T''(t) + c^2\lambda^2 T(t) = 0, \qquad X''(x) + \lambda^2 X(x) = 0. \tag{3.37}$$

By (3.37) we have

$$X(x) = A\cos(\lambda x) + B\sin(\lambda x), \qquad T(t) = C\cos(c\lambda t) + D\sin(c\lambda t)$$

with $A, B, C, D \in \mathbb{R}$. Imposing on X the boundary conditions $X(0) = X(\ell) = 0$, we obtain $A = 0$, $\lambda = m\pi/\ell$ with $m \in \mathbb{N} \setminus \{0\}$ and $X(x) = \sin\left(\frac{m\pi}{\ell} x\right)$ (we notice that there exist *non trivial* solutions X because the coefficient $c^2\lambda^2$ is strictly positive: this justifies the "curious choice" of writing λ^2 instead of λ).

Let us now look for solutions to (3.36) which are superposition of solutions with separated variables:

$$u(x,t) = \sum_{m=1}^{\infty} \left[\alpha_m \cos\left(\frac{cm\pi}{\ell} t\right) + \beta_m \sin\left(\frac{cm\pi}{\ell} t\right)\right] \sin\left(\frac{m\pi}{\ell} x\right). \tag{3.38}$$

Differentiating (formally) the series we get

$$u_t(x,t) = \sum_{m=1}^{\infty} \left[-\alpha_m \frac{cm\pi}{\ell} \sin\left(\frac{cm\pi}{\ell}t\right) + \beta_m \frac{cm\pi}{\ell} \cos\left(\frac{cm\pi}{\ell}t\right) \right] \sin\left(\frac{m\pi}{\ell}x\right)$$

and imposing the condition $u_t(x,0) = 0$ we obtain

$$\sum_{m=1}^{\infty} \beta_m \frac{cm\pi}{\ell} \sin\left(\frac{m\pi}{\ell}x\right) = 0 \qquad \forall x \in (0,\ell)$$

which implies $\beta_m = 0$ for every $m \geq 1$. Therefore we arrive to

$$u(x,t) = \sum_{m=1}^{\infty} \alpha_m \cos\left(\frac{cm\pi}{\ell}t\right) \sin\left(\frac{m\pi}{\ell}x\right). \tag{3.39}$$

It remains to determine the coefficients α_m. Exploiting the initial condition $u(x,0) = g(x)$, by (2.31) with $T = 2\ell$, we obtain

$$\alpha_m = \frac{2}{\ell} \int_0^\ell g(x) \sin\left(\frac{m\pi}{\ell}x\right) dx. \tag{3.40}$$

We observe that, since $\beta_m = 0$ for every m, we need to develop g into a series which contains only sinusoidal terms; hence we may consider g as an odd function over the interval $(-\ell,\ell)$ and this explains the doubled coefficient in (3.40) with respect to (2.31).

We have then proved the

Proposition 3.4.4. *Assume that $f \equiv 0$ and $h \equiv 0$. Then the unique solution u to* (3.36) *is given by* (3.39) *with α_m defined in* (3.40).

The first term u_1 in the series expansion of the solution u is called **fundamental harmonic** and it is the one which determines the oscillation frequency of the string. Since

$$u_1(x,t) = \alpha_1 \cos\left(\frac{c\pi}{\ell}t\right) \sin\left(\frac{\pi}{\ell}x\right),$$

the period T and the oscillation frequency ν read

$$T = \frac{2\ell}{c}, \qquad \nu = \frac{1}{T} = \frac{c}{2\ell}.$$

Recalling that c^2 is given by τ/ρ_0, we deduce that

$$\nu = \frac{\sqrt{\tau}}{2\ell\sqrt{\rho_0}}.$$

This means that the more stretched the string is, the higher is the oscillation frequency; on the other hand, the longer the string, the lower is the oscillation frequency.

3.4.2 Wave equation in dimension $n = 3$

Wave propagation in elastic media

A solid body which is subject to external loads changes its shape and volume. Let us describe this phenomenon in mathematical terms [45]. In a Cartesian orthogonal coordinate system $Ox_1x_2x_3$, letting the vector $\mathbf{r}$ denote the position of a particle P of a body when it is in the initial undeformed configuration C and letting $\mathbf{r}'$ denote the position occupied by the same point P in a deformed configuration C', we will set $\mathbf{u} = \mathbf{r}' - \mathbf{r}$ and $\mathbf{u}$ will be called displacement vector. Let us now define the **strain tensor E** at a given point of the body:

$$\mathbf{E} = \frac{1}{2}\left(\nabla\mathbf{u} + (\nabla\mathbf{u})^T + (\nabla\mathbf{u})^T\nabla\mathbf{u}\right), \tag{3.41}$$

where with the word tensor we mean a matrix valued function. We recall that $\mathbf{u}$ is a vector function and hence $\nabla\mathbf{u}$ is the *Jacobian matrix* of $\mathbf{u}$. The generic component e_{ij} of the tensor $\mathbf{E}$ is the following:

$$e_{ij} = \frac{1}{2}\left(\frac{\partial u_i}{\partial x_j} + \frac{\partial u_j}{\partial x_i} + \sum_{\alpha=1}^{3}\frac{\partial u_\alpha}{\partial x_i}\frac{\partial u_\alpha}{\partial x_j}\right) \qquad i,j = 1,2,3. \tag{3.42}$$

By (3.42), we deduce that the tensor $\mathbf{E}$ is symmetric and, as such, it can be diagonalized. From a geometrical view point, this matrix transformation defines at each point of the body a triple of mutually orthogonal coordinate axes (*principal axes of deformation*) such that the components of the deformation (in this case, called *principal deformations*) are just unit stretches along these axes: indeed, if $\mathbf{E}$ is diagonal, all the shears related to the extra-diagonal terms of the tensor are zero [6, 13].

The meaning of $\mathbf{E}$ is the following: given two particles P_1 and P_2 very close to each other, we denote by (dx_1, dx_2, dx_3) the vector $P_2 - P_1$ in the configuration C and with (dx'_1, dx'_2, dx'_3) the vector $P_2 - P_1$ in the deformed configuration C'. Thus we have $dx'_i = dx_i + du_i$ and, setting $dl = \sqrt{dx_1^2 + dx_2^2 + dx_3^2}$ and $dl' = \sqrt{{dx'_1}^2 + {dx'_2}^2 + {dx'_3}^2}$, we obtain

$$(dl')^2 = (dl)^2 + 2\sum_{i,j=1}^{3} e_{ij}dx_idx_j.$$

This entails that the tensor $\mathbf{E}$ is related to the variation of distance of two points before and after the deformation.

We introduce another tensor, called **stress tensor**, that we denote by $\mathbf{T}$. Let us explain its physical meaning. When a solid is deformed, because of the intermolecular forces, the body tends to return to its equilibrium configuration. If we consider an arbitrary portion of volume Ω of a solid, the resultant of all the forces which act on this part is due essentially to the intermolecular forces exerted by the portions of material external to Ω. Since these intermolecular forces are short-range, namely they act between molecules which are very close to each other, the major contribution is given by the molecules exterior to Ω which are very close to the boundary $\partial\Omega$. Setting $\mathbf{F} = (F_1, F_2, F_3)$ such force per unit volume, we can think that the total force acting on Ω, given by

$$\int_{\Omega} \mathbf{F} dx,$$

can be expressed as a surface integral. Thus it is reasonable to assume the existence of a tensor $\mathbf{T}$ whose divergence div$\mathbf{T}$ coincides with $\mathbf{F}$, namely

$$F_i = \sum_{j=1}^{3} \frac{\partial \sigma_{ij}}{\partial x_j} \qquad \text{per } i = 1, 2, 3\,,$$

where σ_{ij} denote the components of the tensor $\mathbf{T}$. It can be proved that also for this (symmetric) tensor, defined at every point of the given body, it is possible to find a triple of mutually orthogonal axes (*principal stress axes*) such that the stress components (called in this case *principal*) are only normal, parallel to such axes: in fact, in case $\mathbf{T}$ is diagonal, the tangential stresses related to the extra-diagonal terms of the tensor are always zero ([6, 13]).

In view of the divergence Theorem, we can conclude that the total force due to the deformation is given by

$$\int_{\partial\Omega} \mathbf{T}\nu \; dS\;, \tag{3.43}$$

where ν is the exterior unit normal to $\partial\Omega$ and $\mathbf{T}\nu$ is the usual product between a matrix and a column vector. Now, we have to relate the strain and stress tensors by means of a constitutive law. Under the assumption of small deformations, the tensor $\mathbf{E}$ can be approximated by its "linearization"

$$\mathbf{E}_L = \frac{1}{2}\left(\nabla \mathbf{u} + (\nabla \mathbf{u})^T\right).$$

For thermodynamic reasons, the relation between $\mathbf{T}$ and $\mathbf{E}_L$ is given by

$$\mathbf{T} = \lambda \mathrm{tr}(\mathbf{E}_L)\mathbf{I} + 2\mu \mathbf{E}_L\,, \tag{3.44}$$

where λ and μ are called the Lamé coefficients of the material. We have denoted by $\mathbf{I}$ and $\mathrm{tr}(\cdot)$ the identity matrix and the trace of a matrix respectively. Formula (3.44) is called *Hooke law* .

Let $\mathbf{f}(x, t)$ be an external force per unit volume applied to the body and let ρ be the density of the body. Then, by the balance of the linear momentum, we infer that for every region Ω

$$\frac{d}{dt}\int_{\Omega} \rho \mathbf{u}_t \; dx = \int_{\partial\Omega} \mathbf{T}\nu \; dS + \int_{\Omega} \mathbf{f} dx.$$

Exploiting the divergence Theorem and differentiating under the integral sign, we get

$$\int_\Omega [\rho \mathbf{u}_{tt} - \mathrm{div}\mathbf{T} - \mathbf{f}]\, dx = 0$$

and, by the arbitrariness of Ω, we conclude that

$$\rho \mathbf{u}_{tt} - \mathrm{div}\mathbf{T} = \mathbf{f}. \tag{3.45}$$

Finally, by combining (3.44) and (3.45), we have

$$\rho \mathbf{u}_{tt} - \mu\Delta\mathbf{u} - (\lambda + \mu)\nabla(\mathrm{div}\mathbf{u}) = \mathbf{f}.$$

We remark that the last equation is a vector equation and that each component u_i of the displacement does not satisfy exactly the classical wave equation since, in addition to the Laplacian, other second order terms appear. However, in some particular circumstances, we can reduce to the usual wave equation, as for instance when the displacement is of the form $\mathbf{u} = u\mathbf{k}$, with $\mathbf{k}$ the unit vector corresponding to the third coordinate x_3 and u a scalar function depending only on the first two variables x_1 and x_2. Under these assumptions the function $u = u(x_1, x_2)$ satisfies in $\mathbb{R}^2$ the equation

$$\rho u_{tt} - \mu\Delta u = f_3.$$

Sound wave propagation in a gas

We denote by $\mathbf{v} = \mathbf{v}(x_1, x_2, x_3, t)$, $p = p(x_1, x_2, x_3, t)$, $\rho = \rho(x_1, x_2, x_3, t)$ respectively the velocity, the pressure and the density of a fluid. We specify that by velocity of the fluid $\mathbf{v}(x_1, x_2, x_3, t)$ we mean the speed of the particle which is located in (x_1, x_2, x_3) at time t.

We now introduce two important laws in fluid mechanics [42, 44], which we will meet again in the discussion of Navier-Stokes equations. The first one is the mass conservation law, which reads

$$\frac{d}{dt}\int_\Omega \rho\, dx = -\int_{\partial\Omega} \rho\mathbf{v}\nu\, dS$$

for every portion of space Ω. Exploiting the divergence Theorem and differentiating under the integral sign, we obtain

$$\int_\Omega [\rho_t + \mathrm{div}(\rho\mathbf{v})]\, dx = 0$$

and, by the arbitrariness of Ω, we arrive to the **continuity equation**

$$\frac{\partial\rho}{\partial t} + \mathrm{div}(\rho\mathbf{v}) = 0. \tag{3.46}$$

Let us decompose p and ρ as follows:

$$p = p_0 + p_1 \qquad \text{and} \qquad \rho = \rho_0 + \rho_1 \,, \tag{3.47}$$

where the constants p_0 and ρ_0 denote respectively the pressure and the density of the fluid in the unperturbed state. Since $\nabla\rho_1 \cdot \mathbf{v}$ and $\rho_1 \mathrm{div}\mathbf{v}$ are products of small quantities, they turn out to be negligible with respect to $\frac{\partial\rho_1}{\partial t}$ and $\rho_0\mathrm{div}\mathbf{v}$. For this reason the continuity equation (3.46) becomes

$$\frac{\partial\rho_1}{\partial t} + \rho_0 \mathrm{div}\mathbf{v} = 0. \tag{3.48}$$

We now introduce the balance of the linear momentum in the framework of fluid dynamics. Given a generic portion of fluid Ω, we know that the resultant of the forces due to the pressure acting on Ω is given by

$$-\int_{\partial\Omega} p\nu \, dS.$$

Applying the divergence Theorem (in vector form!) we have that

$$-\int_{\partial\Omega} p\nu \, dS = -\int_{\Omega} \nabla p \, dx \,,$$

therefore, the linear momentum equation per unit volume becomes

$$\rho \frac{d\mathbf{v}}{dt} = -\nabla p \tag{3.49}$$

which is called *Euler equation* and represents the equation of motion of the fluid. It is worth to remark that the symbol $\frac{d\mathbf{v}}{dt}$ represents the acceleration of the particle which is located at (x_1, x_2, x_3) at time t and it must not be confused with $\frac{\partial\mathbf{v}}{\partial t}$, which instead denotes the partial derivative of the function $\mathbf{v} = \mathbf{v}(x_1, x_2, x_3, t)$ with respect to t. If $\mathbf{r}(t) = (x_1(t), x_2(t), x_3(t))$ identifies the motion of a generic particle of the fluid, then

$$\frac{d}{dt}\mathbf{v}(x_1(t), x_2(t), x_3(t), t) = \frac{\partial\mathbf{v}}{\partial t} + \sum_{i=1}^{3} \frac{dx_i}{dt}\frac{\partial\mathbf{v}}{\partial x_i} = \frac{\partial\mathbf{v}}{\partial t} + \sum_{i=1}^{3} v_i \frac{\partial\mathbf{v}}{\partial x_i} = \frac{\partial\mathbf{v}}{\partial t} + (\mathbf{v}\cdot\nabla)\mathbf{v} \tag{3.50}$$

where we have denoted by v_i the components of the velocity and by $\mathbf{v}\cdot\nabla$ the operator

$$\sum_{i=1}^{3} v_i \frac{\partial}{\partial x_i}.$$

In order to clarify the difference between $\frac{d\mathbf{v}}{dt}$ and $\frac{\partial\mathbf{v}}{\partial t}$, we cite the following example, taken from [62]:

Example 3.4.5. Let us consider a fluid which rotates uniformly with constant angular velocity ω. Since $\frac{d\mathbf{v}}{dt}$ is the acceleration of the particle, it holds

$$\frac{d\mathbf{v}}{dt} = -|\omega|^2 \mathbf{r}\,,$$

where $\mathbf{r} = \mathbf{r}(t)$ denotes the position of the particle in function of time. On the other hand, if we fix a generic position $\mathbf{r}$, the speed of a generic particle that at a certain time occupies the position $\mathbf{r}$ is given by $\omega \times \mathbf{r}$, and, since ω and $\mathbf{r}$ do not vary with time, it holds

$$\frac{\partial \mathbf{v}}{\partial t} = 0. \qquad \square$$

We now pass to the problem of sound wave propagation in a gas. By (3.47) and (3.49) we obtain

$$\frac{d\mathbf{v}}{dt} = -\frac{1}{\rho_0}\nabla p_1. \tag{3.51}$$

For thermodynamic reasons, it is possible to assume that small variations of density entail small variations of pressure. Assume we are able to express the pressure in terms of the density by means of the function $p = f(\rho)$; then, by linearization, we obtain

$$p = f(\rho) \approx f(\rho_0) + f'(\rho_0)(\rho - \rho_0) = p_0 + f'(\rho_0)\rho_1\,,$$

that is

$$p_1 = p - p_0 \approx f'(\rho_0)\rho_1.$$

Then, multiplying both sides of (3.48) by $f'(\rho_0)$, we infer

$$\frac{\partial p_1}{\partial t} + \rho_0 f'(\rho_0)\mathrm{div}\mathbf{v} = 0. \tag{3.52}$$

If the fluid motion is irrotational, that is $\mathrm{curl}\,\mathbf{v} = 0$, then the velocity field $\mathbf{v}$ admits a potential φ, namely there exists a scalar function φ such that $\mathbf{v} = \nabla\varphi$. Thus, by (3.50) and (3.51), we have

$$\nabla\left(\frac{\partial\varphi}{\partial t}\right) = \frac{\partial(\nabla\varphi)}{\partial t} = \frac{\partial \mathbf{v}}{\partial t} = \frac{d\mathbf{v}}{dt} - (\mathbf{v}\cdot\nabla)\mathbf{v}$$

$$= -\frac{1}{\rho_0}\nabla p_1 - (\mathbf{v}\cdot\nabla)\mathbf{v} = \nabla\left(-\frac{1}{\rho_0}p_1\right) - (\mathbf{v}\cdot\nabla)\mathbf{v}.$$

Assuming $\mathbf{v}$ small with respect to the sound velocity in the fluid under consideration, we can neglect the nonlinear term $(\mathbf{v}\cdot\nabla)\mathbf{v}$ in the previous identity and we obtain

$$\nabla\left(\frac{\partial\varphi}{\partial t}\right) = \nabla\left(-\frac{1}{\rho_0}p_1\right),$$

whence

$$p_1 = -\rho_0 \frac{\partial \varphi}{\partial t} + c$$

for some suitable constant $c \in \mathbb{R}$. Inserting the last equation into (3.52), we get

$$\frac{\partial^2 \varphi}{\partial t^2} - f'(\rho_0)\Delta\varphi = 0. \tag{3.53}$$

Maxwell equations and electromagnetic wave propagation

We begin by introducing the famous Maxwell equations for the electric field **E** and the magnetic field **B** in a vacuum [48]. Maxwell theory, entirely published in 1873, represents one of the most important scientific results in classical physics: indeed, it states that the electric and magnetic phenomena cannot exist separately, but are different aspects of a single fundamental interaction, related to the existence of the electric charge.

Let ρ be the charge density and let $\mathbf{j}$ be the vector of electric current density. Let us begin with the differential version of the Gauss Theorem for **E**:

$$\text{div}\mathbf{E} = \frac{\rho}{\varepsilon_0}, \tag{3.54}$$

where ε_0 denotes the dielectric constant (or permittivity) of vacuum. This first equation shows the relation between the charge and the existing electric field. In turn, for the magnetic field, there holds the following equation:

$$\text{div}\mathbf{B} = 0 \tag{3.55}$$

which shows that the flux lines of **B** are closed and therefore there can not exist magnetic monopoles, but only dipoles, namely pairs of magnetic poles with the same value and opposite sign. The next equation concerns the curl of the electric field and is the differential version of the Faraday law:

$$\text{curl}\,\mathbf{E} = -\frac{\partial \mathbf{B}}{\partial t} \tag{3.56}$$

which locally relates the variation in time of the magnetic field to the induced electric field. This equation states that also a time-varying magnetic field generates an electric field. Finally, it holds

$$\text{curl}\,\mathbf{B} = \mu_0 \left(\mathbf{j} + \varepsilon_0 \frac{\partial \mathbf{E}}{\partial t} \right), \tag{3.57}$$

where μ_0 denotes the magnetic permeability of vacuum. This law states that both the conduction currents and a time-varying electric field can be sources of a magnetic field.

Let us now deduce from the previous equations the law that governs the propagation of the electromagnetic waves. For the sake of simplicity, we consider their propagation in the vacuum, in absence of electric charge density ρ and current density $\mathbf{j}$. Under these assumptions, differentiating with respect to t the equation (3.56), in view of (3.57) we obtain

$$\frac{\partial^2 \mathbf{B}}{\partial t^2} = -\mathrm{curl}\left(\frac{\partial \mathbf{E}}{\partial t}\right) = -\frac{1}{\varepsilon_0 \mu_0}\mathrm{curl}(\mathrm{curl}\,\mathbf{B}).$$

Taking into account that for every vector field $\mathbf{F}$ it holds $\mathrm{curl}(\mathrm{curl}\,\mathbf{F}) = -\Delta \mathbf{F} + \nabla(\mathrm{div}\mathbf{F})$, using the equation (3.55) we infer

$$\frac{\partial^2 \mathbf{B}}{\partial t^2} - \frac{1}{\varepsilon_0 \mu_0}\Delta \mathbf{B} = 0.$$

Similarly, by combining (3.54), (3.56) and (3.57) we get

$$\frac{\partial^2 \mathbf{E}}{\partial t^2} - \frac{1}{\varepsilon_0 \mu_0}\Delta \mathbf{E} = 0.$$

Note that the constant $1/\sqrt{\varepsilon_0 \mu_0}$ is the propagation speed of the wave and it coincides with the speed of light in a vacuum $(3 \cdot 10^8 m/s)$.

Solution of the wave equation in dimension $n = 3$

In view of the mathematical models described above, we are led to study the solutions of the equation

$$u_{tt} - c^2 \Delta u = f\,, \tag{3.58}$$

with $c > 0$, $u = u(x,t)$ a real function with $x \in \mathbb{R}^3$ and $t \in \mathbb{R}$. Then we impose the initial conditions: first, we assume that the forcing term f is zero, so that we arrive to the following Cauchy problem:

$$\begin{cases} u_{tt} - c^2 \Delta u = 0 & (x,t) \in \mathbb{R}^3 \times (0,\infty) \\ u(x,0) = g(x) & x \in \mathbb{R}^3 \\ u_t(x,0) = h(x) & x \in \mathbb{R}^3. \end{cases} \tag{3.59}$$

Given a surface $\Sigma \subset \mathbb{R}^3$ with finite area $|\Sigma|$ and given an integrable function f over Σ, we will denote by

$$⨍_\Sigma f := \frac{1}{|\Sigma|}\int_\Sigma f$$

the average of f on Σ.

For the solution of this problem, we will use the *method of spherical means*. Given a generic function $w = w(x,t)$, we define its **spherical mean**:

$$W(x;r,t) = ⨍_{\partial B_r(x)} w(\sigma,t)\, dS(\sigma) = \frac{1}{4\pi}\int_{|\sigma|=1} w(x + r\sigma, t)\, dS(\sigma)$$

for every $x \in \mathbb{R}^3,\ t > 0,\ r > 0$. Moreover we set $W(x;0,t) = w(x,t)$ and $W(x;r,t) = W(x;-r,t)$ for $r < 0$, so that $r \mapsto W(x;r,t)$ is even.

Exercise 3.4.6. Let $u \in C^2(\mathbb{R}^3 \times [0,\infty))$ be a solution to (3.59). Let $U = U(x;r,t)$ be the spherical mean of u and let $G = G(r)$ and $H = H(r)$ be the spherical means (independent of time!) of g and h respectively. Prove that for every fixed $x \in \mathbb{R}^3$ there hold $U(x;\cdot,\cdot) \in C^2(\mathbb{R} \times [0,\infty))$ and

$$\begin{cases} \frac{1}{c^2} U_{tt} = U_{rr} + \frac{2}{r} U_r & (r,t) \in \mathbb{R} \times (0,\infty) \\ U(r,0) = G(r), \quad U_t(r,0) = H(r) & r \in \mathbb{R}. \end{cases} \tag{3.60}$$

The equation (3.60) is called **Euler-Poisson equation**. For more details about the derivation of the Laplacian of a radial function, see (3.121). □

We are thus led to solve (3.60). Setting $V = rU$, we obtain $V_r = U + rU_r$, $V_{rr} = U_r + U_r + rU_{rr} = 2U_r + rU_{rr}$, $V_{tt} = rU_{tt}$; by substitution, problem (3.60) reduces to the following one-dimensional wave equation:

$$\begin{cases} V_{tt} - c^2 V_{rr} = 0 & (r,t) \in \mathbb{R} \times (0,\infty) \\ V(r,0) = rG(r) & r \in \mathbb{R} \\ V_t(r,0) = rH(r) & r \in \mathbb{R}. \end{cases}$$

By the D'Alembert formula (3.31) we derive:

$$rU(x;r,t) = V(r,t) = \frac{1}{2}\{(r+ct)\ G(r+ct) + (r-ct)\ G(r-ct)\} + \frac{1}{2c}\int_{r-ct}^{r+ct} sH(s)\ ds$$

and, since the spherical means are extended as even functions, we deduce

$$U(x;r,t) = \frac{(ct+r)\ G(ct+r) - (ct-r)\ G(ct-r)}{2r} + \frac{1}{2cr}\int_{ct-r}^{ct+r} sH(s)\ ds.$$

Hence, passing to the limit as $r \to 0$, we find

$$u(x,t) = t\ H(ct) + \frac{d}{dt}[t\ G(ct)]$$

which gives

$$u(x,t) = \frac{1}{4\pi c^2 t}\int_{|\sigma - x| = ct} h(\sigma)\ dS(\sigma) + \frac{\partial}{\partial t}\frac{1}{4\pi c^2 t}\int_{|\sigma - x| = ct} g(\sigma)\ dS(\sigma)\ , \tag{3.61}$$

called **Kirchhoff formula**. Let us try to rewrite (3.61) in a more explicit form. Setting $B_\rho(y) = \{x \in \mathbb{R}^3;\ |x - y| < \rho\}$, we observe that

$$\frac{\partial}{\partial t} \fint_{\partial B_{ct}(x)} g(\sigma)\ dS(\sigma) = \frac{1}{4\pi}\frac{\partial}{\partial t}\int_{\partial B_1(0)} g(x + ct\sigma)\ dS(\sigma)$$

$$= \frac{c}{4\pi} \int_{\partial B_1(0)} \nabla g(x + ct\sigma) \cdot \sigma \, dS(\sigma) = \frac{1}{t} \fint_{\partial B_{ct}(x)} \nabla g(\sigma) \cdot (\sigma - x) \, dS(\sigma)$$

and hence, differentiating the product,

$$\frac{\partial}{\partial t} \frac{1}{4\pi c^2 t} \int_{|\sigma - x| = ct} g(\sigma) \, dS(\sigma) = \frac{\partial}{\partial t} \left[t \fint_{\partial B_{ct}(x)} g(\sigma) \, dS(\sigma) \right]$$

$$= \fint_{\partial B_{ct}(x)} g(\sigma) \, dS(\sigma) + \fint_{\partial B_{ct}(x)} \nabla g(\sigma) \cdot (\sigma - x) \, dS(\sigma).$$

Thus, by (3.61) we infer

$$u(x,t) = \fint_{\partial B_{ct}(x)} \{ t \, h(\sigma) + g(\sigma) + \nabla g(\sigma) \cdot (\sigma - x) \} \, dS(\sigma). \tag{3.62}$$

We have then proved the

Theorem 3.4.7. *Let $g \in C^3(\mathbb{R}^3)$ and $h \in C^2(\mathbb{R}^3)$. Then the unique solution $u \in C^2(\mathbb{R}^3 \times [0, \infty))$ to* (3.59) *is given by formula* (3.62).

Formula (3.62) has many important consequences:

• (3.62) reveals that u may be *less regular* than the data: if $u(\cdot, 0) = g \in C^k(\mathbb{R}^3)$ and $u_t(\cdot, 0) = h \in C^{k-1}(\mathbb{R}^3)$, then for $t > 0$ we only have $u(\cdot, t) \in C^{k-1}(\mathbb{R}^3)$ and $u_t(\cdot, t) \in C^{k-2}(\mathbb{R}^3)$.

• *Domain of dependence.* Formula (3.62) shows that $u(x,t)$ only depends on the values of g and h on the spherical surface $\partial B_{ct}(x)$. For this reason, such spherical surface is called **domain of dependence**.

• *Range of influence.* The value of the data g and h at a point y belonging to the hyperplane $t = 0$ influence the points (x,t) on the *conic surface* given by the equation $|x - y| = ct$. For this reason, such conic surface is called **range of influence**. An initial datum with support contained into a ball $B_\rho(x_0)$, at time t, affects the "spherical shell" centered at x_0, with interior radius $ct - \rho$ and exterior radius $ct + \rho$. Actually, for $0 < t < \rho/c$, the spherical shell is a ball. The spherical shell expands with speed c (Huygens principle).

Let us now consider the Cauchy problem for the wave equation in the presence of a forcing term. Clearly, in order to determine a solution of

$$\begin{cases} u_{tt} - c^2 \Delta u = f(x,t) & (x,t) \in \mathbb{R}^3 \times (0, \infty) \\ u(x,0) = g(x) & x \in \mathbb{R}^3 \\ u_t(x,0) = h(x) & x \in \mathbb{R}^3, \end{cases} \tag{3.63}$$

it suffices to sum a solution of (3.59) and a solution of

$$\begin{cases} u_{tt} - c^2 \Delta u = f(x,t) & (x,t) \in \mathbb{R}^3 \times (0, \infty) \\ u(x,0) = 0 & x \in \mathbb{R}^3 \\ u_t(x,0) = 0 & x \in \mathbb{R}^3. \end{cases} \tag{3.64}$$

To solve (3.64) we will exploit the Duhamel principle, that we have already used for the equation $(3.28)_2$. Let $w(x,t;s)$ be the solution of

$$\begin{cases} w_{tt}(x,t;s) - c^2\Delta w(x,t;s) = 0 & (x,t) \in \mathbb{R}^3 \times (s,\infty) \\ w(x,s;s) = 0 & x \in \mathbb{R}^3 \\ w_t(x,s;s) = f(x,s) & x \in \mathbb{R}^3 . \end{cases} \tag{3.65}$$

We claim that the function u defined as

$$u(x,t) := \int_0^t w(x,t;s)\, ds \tag{3.66}$$

is a solution to (3.64). Indeed,

$$u_t(x,t) = w(x,t;t) + \int_0^t w_t(x,t;s)\, ds = \int_0^t w_t(x,t;s)\, ds \ ,$$

$$u_{tt}(x,t) = w_t(x,t;t) + \int_0^t w_{tt}(x,t;s)\, ds = f(x,t) + \int_0^t w_{tt}(x,t;s)\, ds \ ,$$

$$u_{tt}(x,t) - c^2\Delta u(x,t) = f(x,t) + \int_0^t [w_{tt}(x,t;s) - c^2\Delta w(x,t;s)]\, ds = f(x,t) \ .$$

Moreover, u satisfies the initial conditions in (3.64). Therefore, by combining Theorem 3.4.7 applied to problem (3.65) and (3.66), it is proved the following

Theorem 3.4.8. *Let $f \in C^2(\mathbb{R}^3 \times [0,\infty))$. Then the unique solution $u \in C^2(\mathbb{R}^3 \times [0,\infty))$ to problem (3.64) is given by*

$$u(x,t) = \frac{1}{4\pi c^2} \int_{B_{ct}(x)} \frac{f(y, t - c^{-1}|x-y|)}{|x-y|}\, dy.$$

Furthermore, summing the solutions obtained in the Theorems 3.4.7-3.4.8, we find a solution to (3.63).

In accordance to what we have done for the wave equation in dimension $n = 1$, we give the definition of fundamental solution.

Definition 3.4.9. *It is defined* **fundamental solution of the wave equation** *in dimension $n = 3$ the distribution $K(z,t) \in \mathcal{D}'(\mathbb{R}^3)$, depending on the two parameters $z \in \mathbb{R}^3$ and $t \geq 0$, defined implicitly by*

$$\langle K(z,t), \varphi \rangle = \int_{\partial B_{ct}(z)} \frac{\varphi(y)}{4\pi c|z-y|} dS(y) \qquad \forall \varphi \in \mathcal{D}(\mathbb{R}^3). \tag{3.67}$$

We observe that the right hand side of (3.67) makes sense for more general test functions, for instance we can compute the duality product $\langle K(z,t), \varphi\rangle$ for every $\varphi \in C^0(\mathbb{R}^3)$.

If $g = 0$, in view of Theorem 3.4.7, the solution u of (3.59) reads

$$u(x,t) = \fint_{\partial B_{ct}(x)} t\, h(y)\, dS(y).$$

Exploiting the equivalence $y \in \partial B_{ct}(x) \Leftrightarrow ct = |y - x|$, we obtain

$$u(x,t) = \int_{\partial B_{ct}(x)} \frac{h(y)}{4\pi c|x-y|} dS(y) = \langle K(x,t), h\rangle.$$

Therefore, as in the case $n = 1$, the solution u can be written in terms of a suitable duality product between the fundamental solution and the initial data. Moreover, the distribution $K(z,t)$ can be seen as a solution of the Cauchy problem

$$\begin{cases} u_{tt} - c^2\Delta u = 0 & \text{in } \mathbb{R}^3 \times (0,\infty) \\ u(x,0) = 0 & \text{in } \mathbb{R}^3 \\ u_t(x,0) = \delta_z & \text{in } \mathbb{R}^3. \end{cases} \tag{3.68}$$

In order to better understand how to interpret (3.68), consider a sequence of initial conditions depending on one continuous parameter ε which approximates the Dirac delta function δ_z. More precisely, for every $\varepsilon > 0$, let u_ε be the solution of the Cauchy problem

$$\begin{cases} u_{tt} - c^2\Delta u = 0 & \text{in } \mathbb{R}^3 \times (0,\infty) \\ u(x,0) = 0 & \text{in } \mathbb{R}^3 \\ u_t(x,0) = \varepsilon^{-3}\pi^{-3/2}\, e^{-\left(\frac{|x-z|}{\varepsilon}\right)^2} & \text{in } \mathbb{R}^3. \end{cases} \tag{3.69}$$

Proceeding as in Example 2.6.9, one verifies that the initial datum for u_t converges to δ_z in $\mathcal{D}'(\mathbb{R}^3)$ as $\varepsilon \to 0^+$. Since the initial condition is radially symmetric with respect to z, we can look for the solution to (3.69) in this class of functions. Let v_ε be a radially symmetric function with respect to the spatial variables, which is a solution to the Cauchy problem

$$\begin{cases} u_{tt} - c^2\Delta u = 0 & \text{in } \mathbb{R}^3 \times (0,\infty) \\ u(x,0) = 0 & \text{in } \mathbb{R}^3 \\ u_t(x,0) = \varepsilon^{-3}\pi^{-3/2}\, e^{-\left(\frac{|x|}{\varepsilon}\right)^2} & \text{in } \mathbb{R}^3. \end{cases}$$

Then $v_\varepsilon = v_\varepsilon(r,t)$ solves

$$\begin{cases} \frac{1}{c^2}u_{tt} = u_{rr} + \frac{2}{r}u_r & (r,t) \in \mathbb{R} \times (0,\infty) \\ u(r,0) = 0, \quad u_t(r,0) = \varepsilon^{-3}\pi^{-3/2}e^{-\left(\frac{r}{\varepsilon}\right)^2} & r \in \mathbb{R}. \end{cases} \tag{3.70}$$

Proceeding as in the solution of problem (3.60), we get

$$v_\varepsilon(x,t) = \frac{1}{2c|x|}\int_{ct-|x|}^{ct+|x|} s\varepsilon^{-3}\pi^{-3/2}e^{-\left(\frac{s}{\varepsilon}\right)^2}\,ds = \frac{1}{4\varepsilon c\pi^{3/2}|x|}\left[e^{-\left(\frac{|x|-ct}{\varepsilon}\right)^2} - e^{-\left(\frac{|x|+ct}{\varepsilon}\right)^2}\right]$$

and hence

$$u_\varepsilon(x,t) = \frac{1}{4\varepsilon c\pi^{3/2}|x-z|}\left[e^{-\left(\frac{|x-z|-ct}{\varepsilon}\right)^2} - e^{-\left(\frac{|x-z|+ct}{\varepsilon}\right)^2}\right].$$

We claim that for every $\varphi \in \mathcal{D}(\mathbb{R}^3)$ it holds

$$\lim_{\varepsilon\to 0^+}\int_{\mathbb{R}^3} u_\varepsilon(y,t)\varphi(y)\,dy = \int_{\partial B_{ct}(z)}\frac{\varphi(\sigma)}{4\pi c|\sigma-z|}\,dS(\sigma) = \langle K(z,t),\varphi\rangle\,, \tag{3.71}$$

namely that $u_\varepsilon \to K(z,t)$ in $\mathcal{D}'(\mathbb{R}^3)$ as $\varepsilon\to 0^+$. The latter assertion means that $K(z,t)$ "solves" (3.68). Let us now prove (3.71). There holds

$$\int_{\mathbb{R}^3} u_\varepsilon(y,t)\varphi(y)\,dy = \int_{\mathbb{R}^3}\frac{1}{4\varepsilon c\pi^{3/2}|y-z|}\left[e^{-\left(\frac{|y-z|-ct}{\varepsilon}\right)^2} - e^{-\left(\frac{|y-z|+ct}{\varepsilon}\right)^2}\right]\varphi(y)\,dy \tag{3.72}$$

$$= \int_{\partial B_1(0)}\left\{\int_0^\infty \frac{r}{4\varepsilon c\pi^{3/2}}\left[e^{-\left(\frac{r-ct}{\varepsilon}\right)^2} - e^{-\left(\frac{r+ct}{\varepsilon}\right)^2}\right]\varphi(z+r\sigma)\,dr\right\}dS(\sigma).$$

We show that for every $t>0$ and for every $\psi \in C^0([0,\infty))\cap L^\infty(0,\infty)$ it holds

$$\lim_{\varepsilon\to 0^+}\int_0^\infty \pi^{-1/2}\varepsilon^{-1}\left[e^{-\left(\frac{r-ct}{\varepsilon}\right)^2} - e^{-\left(\frac{r+ct}{\varepsilon}\right)^2}\right]\psi(r)\,dr = \psi(ct). \tag{3.73}$$

Indeed, extending the function ψ to zero in $(-\infty,0)$, using suitable changes of variables and applying Theorem 2.3.7, we obtain

$$\begin{aligned}
&\lim_{\varepsilon\to 0^+}\int_0^\infty \pi^{-1/2}\varepsilon^{-1}\left[e^{-\left(\frac{r-ct}{\varepsilon}\right)^2} - e^{-\left(\frac{r+ct}{\varepsilon}\right)^2}\right]\psi(r)\,dr\\
&= \lim_{\varepsilon\to 0^+}\left[\int_{\mathbb{R}}\pi^{-1/2}\varepsilon^{-1}e^{-\left(\frac{r-ct}{\varepsilon}\right)^2}\psi(r)\,dr - \int_{\mathbb{R}}\pi^{-1/2}\varepsilon^{-1}e^{-\left(\frac{r+ct}{\varepsilon}\right)^2}\psi(r)\,dr\right]\\
&= \lim_{\varepsilon\to 0^+}\left[\int_{\mathbb{R}}\pi^{-1/2}e^{-r^2}\psi(\varepsilon r+ct)\,dr - \int_{\mathbb{R}}\pi^{-1/2}e^{-r^2}\psi(\varepsilon r-ct)\,dr\right]\\
&= [\psi(ct)-\psi(-ct)]\int_{\mathbb{R}}\pi^{-1/2}e^{-r^2}\,dr = \psi(ct)
\end{aligned}$$

which implies (3.73).

Using (3.73) in (3.72) with $\psi(r) = \frac{r\varphi(z+r\sigma)}{4c\pi}$, a version of the dominated convergence Theorem on $\partial B_1(0)$ and a change of variables, we get

$$\lim_{\varepsilon\to 0^+}\int_{\mathbb{R}^3} u_\varepsilon(y,t)\varphi(y)\,dy = \int_{\partial B_1(0)}\frac{ct\varphi(z+ct\sigma)}{4\pi c}\,dS(\sigma) = \int_{\partial B_{ct}(z)}\frac{1}{ct}\frac{\varphi(\sigma)}{4\pi c}\,dS(\sigma).$$

Finally, recalling that in the last integral $|\sigma - z| = ct$, we infer

$$\lim_{\varepsilon \to 0^+} \int_{\mathbb{R}^3} u_\varepsilon(y,t)\varphi(y)\,dy = \int_{\partial B_{ct}(z)} \frac{\varphi(\sigma)}{4\pi c|\sigma - z|}\,dS(\sigma)$$

which concludes the proof of (3.71).

We conclude this section with a few words about the wave equation set in a bounded domain $\Omega \subset \mathbb{R}^3$, under suitable boundary conditions on $\partial\Omega$. For instance, it is possible to impose the Dirichlet condition over $\partial\Omega$ and thus consider the **Cauchy-Dirichlet problem for the wave equation**

$$\begin{cases} u_{tt} - c^2\Delta u = f(x,t) & (x,t) \in \Omega \times (0,\infty) \\ u(x,0) = g(x) & x \in \Omega \\ u_t(x,0) = h(x) & x \in \Omega \\ u(x,t) = 0 & (x,t) \in \partial\Omega \times [0,\infty). \end{cases} \tag{3.74}$$

Concerning the existence of solutions to (3.74), we refer the reader to Chapter 7. The uniqueness of the solution is obtained formally as follows. By contradiction, suppose that there exist two solutions u_1 and u_2 to (3.74), then their difference $v = u_1 - u_2$ satisfies the homogeneous problem

$$\begin{cases} v_{tt} - c^2\Delta v = 0 & (x,t) \in \Omega \times (0,\infty) \\ v(x,0) = 0 & x \in \Omega \\ v_t(x,0) = 0 & x \in \Omega \\ v(x,t) = 0 & (x,t) \in \partial\Omega \times [0,\infty). \end{cases} \tag{3.75}$$

Let us define the energy functional:

$$E(t) := \frac{1}{2}\int_\Omega [v_t^2(x,t) + c^2|\nabla v(x,t)|^2]\,dx \qquad \forall t \geq 0.$$

Integrating by parts and exploiting (3.75), we get

$$E'(t) = \int_\Omega (v_t\,v_{tt} + c^2\nabla v_t \cdot \nabla v)\,dx = \int_\Omega v_t(v_{tt} - c^2\Delta v)\,dx = 0.$$

Therefore E is constant and, since $E(0) = 0$, it necessarily holds $E \equiv 0$: this proves that $v \equiv 0$ and hence the uniqueness of the solution of (3.75).

In some particular cases, as for instance when Ω is a rectangular parallelepiped, solutions to (3.74) can be searched with the method of *separation of variables*. The procedure is fully analogous to the one previously illustrated for the string clamped in both extremes, hence we omit it: it amounts to look for solutions which are series of solutions with *separated variables*, namely of the kind $u(x_1, x_2, x_3, t) = X_1(x_1)X_2(x_2)X_3(x_3)T(t)$.

3.4.3 Wave equation in dimension $n = 2$

Wave propagation in an elastic membrane

We want to study the transverse oscillations of an elastic membrane subject to its own weight and possible external loads [72]. A membrane with a generic curvature is usually characterized by the medium surface S, the boundary and the thickness, which we assume to be constant and equal to $2h$: this choice makes the upper and lower surfaces that enclose the membrane have distance h from S.
Let us consider a closed curve γ lying in S and let us construct the surface C generated by the straight lines which are orthogonal to S, pass through the points of γ and are delimited by the external surfaces of the membrane. Under suitable assumptions, it is possible to define, at each point of the curve γ, three mutually orthogonal unit vectors μ (outer normal lying in the tangent plane to S), τ (tangent to γ), ν (normal to S), in order that they form a right-handed triple.

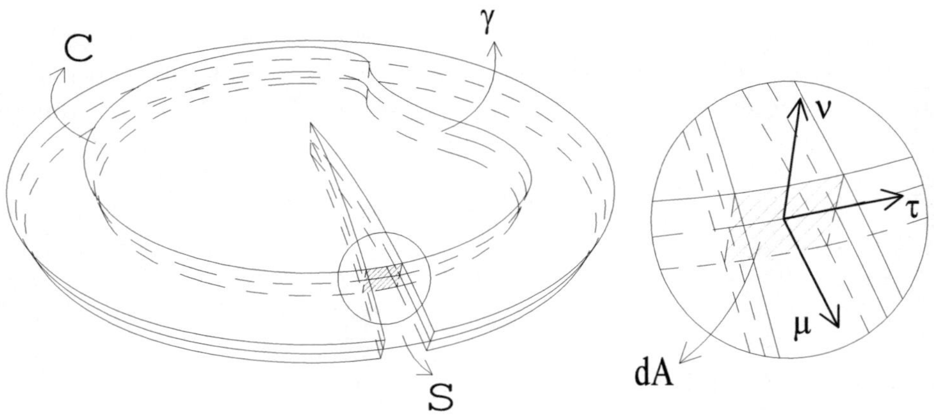

Curved membrane

Let us consider the actions (traction forces) acting on the infinitesimal area dA of C, corresponding to the arc element $d\gamma$ of γ. One can observe that they are equivalent to a force applied at the centroid of dA and a torque. We denote by dM, dT, dN and by dH, dG, dK the components of the force and of the torque respectively, in the directions μ, τ, ν. Since the membranes do not transmit neither forces normal to the medium nor torques, $dN/d\gamma$ and $dH/d\gamma$, $dG/d\gamma$, $dK/d\gamma$ become infinitesimal as $d\gamma$ tends to zero; only $dM/d\gamma$ and $dT/d\gamma$ tend to a finite non zero value M (tension) and T (transversal force) respectively. In this sense we can affirm that the membrane is the two-dimensional version of the string treated in Section 3.4.1: the only forces of interaction between its parts are tangential to the membrane itself. At this point, taking into account the assumption of small deformations (linear approximation), we can identify the deformed configuration of the membrane with the initial undeformed one, obtaining the relationship between the forces M and T and the

single components of the stress tensor defined in the points of the three-dimensional continuum which represents the membrane.
We consider the case of a membrane initially flat, namely such that the medium surface S lies on the plane Oxy and is delimited by a closed and continuously differentiable curve δ. We consider the following assumptions:

- since the membrane is so thin that it can be considered a two-dimensional object, in place of the stresses defined locally in each point of the membrane, we consider the forces M and T defined on the medium surface;
- the motion of the membrane is described by the sole function $z = u(x, y, t)$ that represents the transversal displacement of the points of the medium surface;
- the gradients of the displacements and the velocities in the x and y directions are negligible;
- the membrane, in its initial position, is subject to a constant traction force M_0 per unit length, orthogonal to δ in the Oxy plane;
- in the interior points of the membrane the stress is such that $T = 0$ and $M = M_0$, and this condition remains unchanged during the motion.

With a procedure similar to the one introduced for the vibrating string, we infer that the vertical component of the tension acting on a generic portion of the membrane, which at rest occupies a rectangle $[x_1, x_2] \times [y_1, y_2]$, is given by

$$\int_{y_1}^{y_2} M_0[u_x(x_2, y, t) - u_x(x_1, y, t)]dy + \int_{x_1}^{x_2} M_0[u_y(x, y_2, t) - u_y(x, y_1), t]dx$$

$$= \int_{[x_1,x_2]\times[y_1,y_2]} M_0 \Delta u(x, y, t) dx dy \,. \tag{3.76}$$

Let us denote by ρ_0 the surface mass density of the membrane at rest, and by $F(x, y, t)$ an external load per unit mass parallel to z. By generalizing to the two-dimensional case the procedure introduced to prove (3.24) and (3.25), we obtain that the resultant of the external forces acting on the considered membrane element is given by

$$\int_{[x_1,x_2]\times[y_1,y_2]} \rho_0(-g + F)\; dx\, dy \,, \tag{3.77}$$

while the linear momentum of the same part of the membrane reads

$$\int_{[x_1,x_2]\times[y_1,y_2]} \rho_0 u_t \; dx\, dy \,. \tag{3.78}$$

By (3.76), (3.77), (3.78) and by the balance of the linear momentum, we obtain

$$\int_{[x_1,x_2]\times[y_1,y_2]} [\rho_0 u_{tt} - M_0 \Delta u - \rho_0(-g+F)]\, dx\, dy = 0$$

and hence, by the arbitrariness of the rectangle $[x_1, x_2] \times [y_1, y_2]$, we are led to

$$\rho_0 u_{tt} - M_0 \Delta u = \rho_0(-g+F). \tag{3.79}$$

Setting $c^2 = M_0/\rho_0$ and $f = -g + F$, we arrive to the wave equation (3.58).

Wave propagation on the surface of a liquid

We denote by $\mathbf{v} = \mathbf{v}(x_1, x_2, x_3, t)$, $p = p(x_1, x_2, x_3, t)$ and $\rho = \rho(x_1, x_2, x_3, t)$ the velocity, the pressure and the density of the fluid respectively. By $\mathbf{v}(x_1, x_2, x_3, t)$ we mean the velocity of the particle which is located at position (x_1, x_2, x_3) at time t.

If we assume that, in addition to the pressure, also the gravity acts on the fluid, then (3.49) becomes

$$\rho \frac{d\mathbf{v}}{dt} = -\nabla p + \rho \mathbf{g}\,, \tag{3.80}$$

with $\mathbf{g} = (0, 0, -g)$ and g the acceleration due to the gravity.

If we assume that the vertical components of the velocity v_3 and the vertical component of the acceleration $\frac{dv_3}{dt}$ are negligible with respect to the corresponding horizontal components, the third component of the vector equation (3.80) becomes

$$\frac{1}{\rho}\frac{\partial p}{\partial x_3} = -g\,. \tag{3.81}$$

Now we want to derive the equation which governs the propagation of a wave on the surface of an incompressible fluid, which in the unperturbed configuration has constant depth h_0. We shall assume that this liquid extends without limits and that the bottom is horizontal and characterized by the equation $x_3 = 0$.

If $h = h(x_1, x_2, t)$ denotes the level of a point of the free surface of coordinates (x_1, x_2) at time t, then we define the function $u = h - h_0$ which represents the displacement of the liquid surface with respect to the unperturbed configuration.

If p_0 is the (constant) pressure on the liquid surface, equation (3.81) becomes

$$p(x_1, x_2, x_3, t) = p_0 + g\rho(h(x_1, x_2, t) - x_3) = p_0 + g\rho(h_0 + u(x_1, x_2, t) - x_3)\,. \tag{3.82}$$

Assuming that the amplitude is small compared to the wavelength, we can neglect the nonlinear term $(\mathbf{v}\cdot\nabla)\mathbf{v}$ in (3.50); therefore, by (3.49) and (3.82), we obtain

$$\frac{\partial v_1}{\partial t} + g\frac{\partial u}{\partial x_1} = 0 \quad \text{and} \quad \frac{\partial v_2}{\partial t} + g\frac{\partial u}{\partial x_2} = 0\,. \tag{3.83}$$

Let Ω be a domain contained into the plane $x_3 = 0$, with boundary $\partial\Omega$. Being ρ constant, the mass conservation law becomes

$$\frac{d}{dt}\int_{\Omega} h\,dx_1 dx_2 = -\int_{\partial\Omega} h\,\mathbf{v}\cdot\nu\,ds\,,$$

where $\nu \in \mathbb{R}^2$ is the outer unit normal vector to $\partial\Omega$ and ds is the arc element. Using the divergence Theorem in the plane and exploiting the arbitrariness of Ω we derive the equation

$$\frac{\partial h}{\partial t} + \frac{\partial(hv_1)}{\partial x_1} + \frac{\partial(hv_2)}{\partial x_2} = 0\,. \tag{3.84}$$

Since $h = h_0 + u$ with h_0 constant, then, for $i = 1, 2$, it holds

$$\frac{\partial(hv_i)}{\partial x_i} = h_0\frac{\partial v_i}{\partial x_i} + \frac{\partial u}{\partial x_i}v_i + u\frac{\partial v_i}{\partial x_i}\,. \tag{3.85}$$

We may disregard the terms $\frac{\partial u}{\partial x_i}v_i$ and $u\frac{\partial v_i}{\partial x_i}$ since they are the product of small quantities, then, by (3.84) and (3.85), we obtain

$$\frac{\partial u}{\partial t} + h_0\left(\frac{\partial v_1}{\partial x_1} + \frac{\partial v_2}{\partial x_2}\right) = 0\,.$$

Finally, differentiating the last equation with respect to t and inserting (3.83), we arrive to

$$\frac{\partial^2 u}{\partial t^2} - gh_0\Delta u = 0.$$

Solution of the wave equation in dimension $n = 2$

Consider the Cauchy problem

$$\begin{cases} u_{tt} - c^2\Delta u = 0 & (x,t) \in \mathbb{R}^2 \times (0,\infty) \\ u(x,0) = g(x) & x \in \mathbb{R}^2 \\ u_t(x,0) = h(x) & x \in \mathbb{R}^2\,. \end{cases} \tag{3.86}$$

As in [41], we will consider a solution $u(x_1, x_2, t)$ of the wave equation in dimension $n = 2$ as a solution of the same equation in dimension $n = 3$, with data

$$g = g(x_1, x_2) \qquad h = h(x_1, x_2).$$

The functions g and h can be seen as functions defined over all $\mathbb{R}^3$ depending only on the first two variables. We introduce the following notations: by $x = (x_1, x_2, x_3)$ we denote the generic point of $\mathbb{R}^3$, by $x' = (x_1, x_2)$ the generic point of $\mathbb{R}^2$, by ∇ the gradient in $\mathbb{R}^3$ and, finally, by $\nabla_{x'}$ the gradient in $\mathbb{R}^2$.

Setting $w(x_1, x_2, x_3, t) := u(x_1, x_2, t)$, by Kirchhoff formula (3.62) and since w does not depend on x_3, at a point $x = (x_1, x_2, 0)$ it holds:

$$w(x,t) = \fint_{\partial B_{ct}(x)} \{t\ h(\sigma) + g(\sigma) + \nabla g(\sigma) \cdot (\sigma - x)\}\ dS(\sigma)$$

$$= \frac{1}{4\pi c^2 t^2} \int_{\partial B_{ct}(x)} \{t\ h(\sigma) + g(\sigma) + \nabla_{x'} g(\sigma) \cdot (\sigma' - x')\}\ dS(\sigma).$$

Moreover, in view of the independence from x_3 of the integrands and taking into account that the sphere $\partial B_{ct}(x)$ is symmetric with respect to the plane $x_3 = 0$, we obtain

$$u(x',t) = w(x,t) = \frac{2}{4\pi c^2 t^2} \int_{\{|\sigma - x| = ct, \sigma_3 > 0\}} \{t\ h(\sigma) + g(\sigma) + \nabla_{x'} g(\sigma) \cdot (\sigma' - x')\}\ dS(\sigma)$$

$$= \frac{1}{2\pi c^2 t^2} \int_{|y'-x'|<ct} \{t\ h(y') + g(y') + \nabla_{x'} g(y') \cdot (y' - x')\} \frac{ct}{\sqrt{c^2t^2 - |y'-x'|^2}}\ dy'$$

$$= \frac{1}{2} \fint_{B_{ct}(x')} \frac{ct\{t\ h(y') + g(y') + \nabla_{x'} g(y') \cdot (y' - x')\}}{\sqrt{c^2t^2 - |y'-x'|^2}}\ dy'. \tag{3.87}$$

We point out that $B_{ct}(x')$ represents the *two-dimensional disk* centered at $x' \in \mathbb{R}^2$, with radius ct. Then we can state the following

Theorem 3.4.10. *Let $g \in C^3(\mathbb{R}^2)$ and $h \in C^2(\mathbb{R}^2)$. Then the unique solution $u \in C^2(\mathbb{R}^2 \times [0, \infty))$ to problem* (3.86) *is given by formula* (3.87).

Equation (3.87) entails that Huygens principle *does not hold* when $n = 2$: unlike the case $n = 3$, we do not have a boundary integral, but an integral over a disk. As a consequence, we infer that a disturbance (signal) continues to have effects indefinitely (if there are no dispersions), as it happens for the waves on a liquid surface (signal damped by the gravity as $t \to \infty$).

As already done in dimension $n = 1$ and $n = 3$, starting from formula (3.87) we can define also for $n = 2$ the domain of dependence and the range of influence. Formula (3.87) shows that $u(x,t)$ depends only on the values of g and h in the disk $B_{ct}(x)$. For this reason, such circle is called **domain of dependence**.

On the other hand, the value of the data g and h at a point y lying on the plane $t = 0$ influence the points (x,t) which belong to the set

$$\{(x,t) \in \mathbb{R} \times (0, \infty);\ |x - y| \le ct\}.$$

For this reason, such set is called **range of influence**.

We now give the following

Definition 3.4.11. *Calling H the Heaviside function introduced in (2.40), it is defined* **fundamental solution of the wave equation** *in dimension $n = 2$ with pole at $z \in \mathbb{R}^2$ the function*

$$K(x, z, t) = \frac{1}{2\pi c} \frac{H(c^2t^2 - |x - z|^2)}{\sqrt{c^2t^2 - |x - z|^2}}.$$

If $g = 0$, according to Theorem 3.4.10, the solution to (3.86) reads

$$u(x, t) = \int_{\mathbb{R}^2} K(x, y, t)h(y)dy.$$

Hence, provided that we replace the integral with two brackets, for every fixed $z \in \mathbb{R}^2$ we can interpret the function $K(x, z, t)$ as the solution of the Cauchy problem

$$\begin{cases} u_{tt} - c^2\Delta u = 0 & (x, t) \in \mathbb{R}^2 \times (0, \infty) \\ u(x, 0) = 0 & x \in \mathbb{R}^2 \\ u_t(x, 0) = \delta_z & x \in \mathbb{R}^2 \,. \end{cases} \tag{3.88}$$

Remark 3.4.12. In order to solve the wave equation in dimension $n = 2$ in presence of an external load, one can still use the Duhamel principle and proceed exactly as we have done in dimension $n = 3$, see Theorem 3.4.8. □

We conclude with some remarks about the **Cauchy-Dirichlet problem**

$$\begin{cases} u_{tt} - c^2\Delta u = f & (x, t) \in \Omega \times (0, \infty) \\ u(x, 0) = g(x) & x \in \Omega \\ u_t(x, 0) = h(x) & x \in \Omega \\ u(x, t) = 0 & (x, t) \in \partial\Omega \times [0, \infty) \,, \end{cases} \tag{3.89}$$

where $\Omega \subset \mathbb{R}^2$ is an open bounded set. Problem (3.89) is particularly significant since it models the oscillations of an elastic membrane which is clamped at the boundary.

Concerning the existence of a solution u to (3.89), we refer the reader to Chapter 7. The uniqueness of the solution is obtained as for $n = 3$, arguing by contradiction and applying the energy method to (3.75). Finally, in case Ω is a rectangle, it is possible to solve (3.89) using the method of *separation of variables.* Drawing inspiration from the problem of the string clamped in both extremes, an useful exercise consists in computing the solution with this method and derive some conclusions on the oscillation frequency of the membrane as the tension, the surface density and the dimensions of the rectangle vary (see equation (3.79)).

3.5 Heat equation

3.5.1 Diffusion models

In this section we describe some physical phenomena of diffusion which are modeled by the equation

$$u_t - \Delta u = f. \tag{3.90}$$

We are concerned with the heat diffusion inside a body (with density ρ) due to the temperature gradient and with the molecular diffusion due to a non uniform distribution of the volumic mass ρ.

We start considering the problem of thermal diffusion. For $x \in \mathbb{R}^n$ and $t \in \mathbb{R}$, we denote by $\vartheta := \vartheta(x,t)$ the temperature, by f a heat source (per unit time and volume) and by η the surface density of energy flow; η is a vector quantity such that the amount of energy that flows in unit time through the boundary of a generic portion of Ω is given by

$$\int_{\partial\Omega} \eta \cdot \nu \, dS ,$$

where, as usual, ν denotes the outer unit normal vector to $\partial\Omega$. If $\mathcal{U}$ denotes the internal energy per unit volume, then by the law of conservation of energy, we know that for every bounded domain $\Omega \subset \mathbb{R}^n$ it holds

$$\frac{d}{dt}\int_\Omega \mathcal{U} \, dx = -\int_{\partial\Omega} \eta \cdot \nu \, dS + \int_\Omega f \, dx ,$$

where we adopt the following convention: $f > 0$ in the case of heat supplied to the system. In an isochoric process (volume kept constant), $\mathcal{U}$ is given by the following relation:

$$\partial\mathcal{U} = \rho C_V \partial\vartheta ,$$

where C_V is the specific heat at constant volume. The previous identity holds true with good approximation in solids. Finally, the density of the energy flow η is represented by

$$\eta = -k\nabla\vartheta ,$$

where k denotes the thermal conductivity. This last law shows that the difference of temperature between two points corresponds to a flux of energy from the point with higher temperature to the point with lower temperature.

It follows that

$$\frac{d}{dt}\int_\Omega \rho C_V \vartheta \, dx = \int_{\partial\Omega} k\nabla\vartheta \cdot \nu \, dS + \int_\Omega f \, dx$$

and, using the divergence Theorem and differentiating under the integral sign, we obtain

$$\int_\Omega \left(\frac{\partial(\rho C_V \vartheta)}{\partial t} - \operatorname{div}(k\nabla\vartheta) - f \right) dx = 0.$$

Finally, by the arbitrariness of the domain, we conclude that

$$\frac{\partial(\rho C_V \vartheta)}{\partial t} - \operatorname{div}(k\nabla\vartheta) = f.$$

In case we assume that ρ, C_V and k are constant, we arrive to the simplified equation

$$\rho C_V \frac{\partial \vartheta}{\partial t} - k\Delta\vartheta = f. \tag{3.91}$$

Once the model equation is obtained, it is possible to consider equation (3.91) both in the whole space or in a bounded domain Ω. The Dirichlet problem ($\vartheta = h$ on $\partial\Omega$) models the behavior of the temperature in a body Ω with prescribed value on the boundary, while the homogeneous Neumann problem ($\frac{\partial\vartheta}{\partial\nu} = 0$ on $\partial\Omega$) models the behavior of the temperature in the body Ω if there is no exchange of heat with the external environment.

As for the heat diffusion, similarly we can study the diffusion of a certain substance due to the variation of density of the substance itself: in fact, it is verified that a flux of particles occurs from the regions in which ρ is higher to the regions in which ρ is lower. If we define by $\mathbf{j}$ the surface density vector of particle current, then we may say that

$$\mathbf{j} = -D\nabla\rho\,,$$

where D denotes the diffusion coefficient. If a mass source f is present, by the conservation of mass, we can affirm that for every portion of space Ω it holds

$$\frac{d}{dt}\int_\Omega \rho\, dx = -\int_{\partial\Omega} \mathbf{j}\cdot\nu\, dS + \int_\Omega f\, dx\,,$$

with the conventions adopted so far. As for the heat diffusion, we arrive to the following conclusion:

$$\frac{\partial\rho}{\partial t} - \operatorname{div}(D\rho) = f\,,$$

moreover, if we assume that D is constant, we obtain

$$\frac{\partial\rho}{\partial t} - D\Delta\rho = f\,.$$

3.5.2 Heat equation in dimension $n = 1$

Let us consider the heat equation (3.91) on the real line (infinite bar); once we assume for simplicity that $\rho = C_V = k = 1$, we arrive to the problem

$$\begin{cases} u_t - u_{xx} = f(x,t) & (x,t) \in \mathbb{R}\times(0,\infty) \\ u(x,0) = g(x) & x \in \mathbb{R}\,, \end{cases} \tag{3.92}$$

where we shall assume $f \in C^0(\mathbb{R}\times[0,\infty))\cap L^\infty(\mathbb{R}\times[0,\infty))$ and $g \in C^0(\mathbb{R})\cap L^\infty(\mathbb{R})$. We split problem (3.92) into the two following problems:

$$\begin{cases} u_t - u_{xx} = 0 & (x,t) \in \mathbb{R}\times(0,\infty) \\ u(x,0) = g(x) & x \in \mathbb{R} \end{cases} \qquad \begin{cases} u_t - u_{xx} = f(x,t) & (x,t) \in \mathbb{R}\times(0,\infty) \\ u(x,0) = 0 & x \in \mathbb{R}\,. \end{cases} \tag{3.93}$$

We start showing that

$$u(x,t) = \frac{1}{2\sqrt{\pi t}}\int_{\mathbb{R}} e^{-\frac{(x-y)^2}{4t}} g(y)\,dy \tag{3.94}$$

is a solution to $(3.93)_1$. Differentiating under the integral sign and making some computations, we find

$$u_t(x,t) = u_{xx}(x,t) = -\frac{1}{4t^{3/2}\sqrt{\pi}}\int_{\mathbb{R}} e^{-\frac{(x-y)^2}{4t}} g(y)\,dy$$

$$+\frac{1}{8t^{5/2}\sqrt{\pi}}\int_{\mathbb{R}} (x-y)^2 e^{-\frac{(x-y)^2}{4t}} g(y)\,dy$$

which proves that u in (3.94) solves the equation in $(3.93)_1$. The check of the initial condition appearing in $(3.93)_1$ is a more delicate matter, because of the singularity at $t=0$; however, we can prove that

$$\lim_{(x,t)\to(x_0,0)} \frac{1}{2\sqrt{\pi t}}\int_{\mathbb{R}} e^{-\frac{(x-y)^2}{4t}} g(y)\,dy = g(x_0) \qquad \forall x_0 \in \mathbb{R}\,. \tag{3.95}$$

To this aim, we observe that

$$\frac{1}{2\sqrt{\pi t}}\int_{\mathbb{R}} e^{-\frac{(x-y)^2}{4t}}\,dy = 1. \tag{3.96}$$

Indeed, it suffices to use the change of variable $z = (y-x)/(2\sqrt{t})$ and exploit the fact that $\int_{\mathbb{R}} e^{-z^2}dz = \sqrt{\pi}$. For a fixed $x_0 \in \mathbb{R}$, in view of the continuity of g, we can affirm that for every $\varepsilon > 0$ there exists $\delta > 0$ such that

$$|y - x_0| < \delta \qquad \Longrightarrow \qquad |g(y) - g(x_0)| < \varepsilon.$$

If we fix also $\varepsilon > 0$ and the corresponding δ, by (3.96) we deduce that, for $|x - x_0| < \delta$,

$$\left|\frac{1}{2\sqrt{\pi t}}\int_{\mathbb{R}} e^{-\frac{(x-y)^2}{4t}} g(y)\,dy - g(x_0)\right| \le \frac{1}{2\sqrt{\pi t}}\int_{\mathbb{R}} e^{-\frac{(x-y)^2}{4t}} |g(y) - g(x_0)|\,dy$$

$$\le \frac{\varepsilon}{2\sqrt{\pi t}}\int_{x_0-\delta}^{x_0+\delta} e^{-\frac{(x-y)^2}{4t}}\,dy + \frac{1}{2\sqrt{\pi t}}\int_{\mathbb{R}\setminus(x_0-\delta,x_0+\delta)} e^{-\frac{(x-y)^2}{4t}} |g(y)-g(x_0)|\,dy$$

$$\leq \varepsilon + \frac{2\|g\|_{L^\infty}}{2\sqrt{\pi t}} \int_{-\infty}^{x_0-\delta} e^{-\frac{(x-y)^2}{4t}}\, dy + \frac{2\|g\|_{L^\infty}}{2\sqrt{\pi t}} \int_{x_0+\delta}^{+\infty} e^{-\frac{(x-y)^2}{4t}}\, dy$$

$$\leq \varepsilon + \frac{2\|g\|_{L^\infty}}{\sqrt{\pi}} \int_{-\infty}^{\frac{x_0-x-\delta}{2\sqrt{t}}} e^{-z^2}\, dz + \frac{2\|g\|_{L^\infty}}{\sqrt{\pi}} \int_{\frac{x_0-x+\delta}{2\sqrt{t}}}^{+\infty} e^{-z^2}\, dz\,,$$

where we have used the change of variable $z = (y-x)/(2\sqrt{t})$. Since

$$\lim_{(x,t)\to(x_0,0)} (x_0 - x - \delta)/(2\sqrt{t}) = -\infty, \qquad \lim_{(x,t)\to(x_0,0)} (x_0 - x + \delta)/(2\sqrt{t}) = +\infty,$$

and since $e^{-z^2} \in L^1(\mathbb{R})$, we obtain

$$\lim_{(x,t)\to(x_0,0)} \left| \frac{1}{2\sqrt{\pi t}} \int_{\mathbb{R}} e^{-\frac{(x-y)^2}{4t}} g(y)\, dy - g(x_0) \right|$$

$$\leq \varepsilon + \lim_{(x,t)\to(x_0,0)} \left(\frac{2\|g\|_{L^\infty}}{\sqrt{\pi}} \int_{-\infty}^{\frac{x_0-x-\delta}{2\sqrt{t}}} e^{-z^2}\, dz + \frac{2\|g\|_{L^\infty}}{\sqrt{\pi}} \int_{\frac{x_0-x+\delta}{2\sqrt{t}}}^{+\infty} e^{-z^2}\, dz \right) = \varepsilon.$$

By the arbitrariness of $\varepsilon > 0$, we immediately deduce (3.95).

Remark 3.5.1. For an alternative proof that (3.94) is a solution to $(3.93)_1$, see (5.43). □

Let us now pass to the solution of $(3.93)_2$: to this purpose we exploit Duhamel principle. Considering $s \geq 0$ as a parameter, let $w = w(x,t;s)$ be the solution to the problem

$$\begin{cases} w_t(x,t;s) - w_{xx}(x,t;s) = 0 & x \in \mathbb{R},\ t > s \\ w(x,s;s) = f(x,s) & x \in \mathbb{R}\,. \end{cases} \tag{3.97}$$

By (3.94), it holds

$$w(x,t;s) = \frac{1}{2\sqrt{\pi(t-s)}} \int_{\mathbb{R}} e^{-\frac{(x-y)^2}{4(t-s)}} f(y,s)\, dy\,.$$

Let us now set

$$u(x,t) = \int_0^t w(x,t;s) ds\,.$$

Differentiating we find:

$$u_t(x,t) = w(x,t;t) + \int_0^t w_t(x,t;s) ds = f(x,t) + \int_0^t w_t(x,t;s) ds\,,$$

$$u_{xx}(x,t) = \int_0^t w_{xx}(x,t;s) ds$$

and hence, since w solves (3.97),

$$u_t(x,t) - u_{xx}(x,t) = f(x,t) + \int_0^t [w_t(x,t;s) - w_{xx}(x,t;s)]ds = f(x,t) \ .$$

Therefore, the solution of $(3.93)_2$ is given by

$$u(x,t) = \frac{1}{2\sqrt{\pi}} \int_0^t \frac{1}{\sqrt{s}} \int_{\mathbb{R}} e^{-\frac{(x-y)^2}{4s}} f(y,t-s)\, dy\, ds \ . \tag{3.98}$$

Since f is continuous and bounded, the function u defined in (3.98) belongs to $C^0(\mathbb{R}\times[0,\infty))$ and satisfies $u(\cdot,0)\equiv 0$ in $\mathbb{R}$. Finally, summing (3.94) and (3.98), we obtain the solution of (3.92):

$$u(x,t) = \frac{1}{2\sqrt{\pi t}} \int_{\mathbb{R}} e^{-\frac{(x-y)^2}{4t}} g(y)\, dy + \frac{1}{2\sqrt{\pi}} \int_0^t \frac{1}{\sqrt{s}} \int_{\mathbb{R}} e^{-\frac{(x-y)^2}{4s}} f(y,t-s)\, dy\, ds \ . \tag{3.99}$$

We can summarize what we have said in the following

Theorem 3.5.2. *Let $g \in C^0(\mathbb{R}) \cap L^\infty(\mathbb{R})$ and $f \in C^2(\mathbb{R}\times[0,\infty)) \cap L^\infty(\mathbb{R}\times[0,\infty))$. Then the function u defined in formula (3.99) belongs to $C^2(\mathbb{R}\times(0,\infty)) \cap C^0(\mathbb{R}\times[0,\infty))$ and solves problem (3.92).*

Let us now consider the case of a bar with finite length and whose ends are kept at zero temperature; for the sake of simplicity we choose a vanishing forcing term:

$$\begin{cases} u_t - u_{xx} = 0 & (x,t) \in (0,1)\times(0,\infty) \\ u(0,t) = u(1,t) = 0 & t \in [0,\infty) \\ u(x,0) = g(x) & x \in (0,1) \ . \end{cases} \tag{3.100}$$

We proceed by separating the variables. Looking for solutions of the kind $u(x,t) = X(x)T(t)$, we obtain

$$\frac{T'(t)}{T(t)} = \frac{X''(x)}{X(x)} \ ,$$

whence

$$\exists \lambda \in \mathbb{R} \ , \quad T'(t) + \lambda T(t) = 0 \ , \quad X''(x) + \lambda X(x) = 0 \ .$$

We associate with the equation in X the boundary conditions $X(0) = X(1) = 0$. The eigenvalues and the eigenfunctions of this problem are $\lambda_m = m^2\pi^2$ and $\varphi_m(x) = \sin(m\pi x)$. For $\lambda = \lambda_m$, the equation in T has solutions proportional to $T_m(t) = e^{-m^2\pi^2 t}$. Thus we look for solutions of (3.100) of the form

$$u(x,t) = \sum_{m=1}^{\infty} c_m e^{-m^2\pi^2 t} \sin(m\pi x) \ . \tag{3.101}$$

In order to determine the coefficients c_m we impose the condition $u(x,0) = g(x)$: we derive

$$g(x) = \sum_{m=1}^{\infty} c_m \sin(m\pi x)$$

and, by (2.31) with $T = 2$, we deduce that the coefficients c_m are given by

$$c_m = 2 \int_0^1 g(x) \sin(m\pi x)\, dx \ .$$

Notice that, since we have to develop g into a series which contains only sinusoidal terms, we may consider g as an odd function over the interval $(-1, 1)$ and this explains why the coefficient is doubled with respect to (2.31).

We recall that, in order to ensure that the Fourier series converges in a "reasonable" sense, it is enough to assume that g^2 is integrable over $(0, 1)$; indeed, if so, the Parseval identity applies (see Theorem 2.5.2). Equation (3.101) shows that under the sole assumption of integrability of g^2, the solution u belongs to C^∞ for $t > 0$. This is called *regularization effect* of the heat equation. Moreover, for any choice of the initial temperature $g(x)$ of the bar, since the bar is assumed to have ends with zero temperature, the temperature tends to zero for large values of time ($t \to \infty$).

3.5.3 Heat equation in the general case $n \geq 1$

Consider the Cauchy problem with $n \geq 1$

$$\begin{cases} u_t - \Delta u = f(x,t) & (x,t) \in \mathbb{R}^n \times (0, \infty) \\ u(x,0) = g(x) & x \in \mathbb{R}^n . \end{cases} \tag{3.102}$$

First, let us introduce the fundamental solution of the heat equation:

Definition 3.5.3. *For every* $n \geq 1$ *it is defined the* **fundamental solution of the heat equation** *with pole at* $z \in \mathbb{R}^n$ *the function*

$$K(x,z,t) = \frac{1}{(4\pi t)^{n/2}} e^{-\frac{|x-z|^2}{4t}} . \tag{3.103}$$

Extending to the case $n \geq 1$ the computations performed to obtain formula (3.99), we prove that the unique solution to problem (3.102) is given by

$$u(x,t) = \int_{\mathbb{R}^n} K(x,y,t) g(y)\, dy + \int_0^t \left(\int_{\mathbb{R}^n} K(x,y,s) f(y, t-s)\, dy \right) ds. \tag{3.104}$$

Then we can state the following

Theorem 3.5.4. *Let $g \in C^0(\mathbb{R}^n) \cap L^\infty(\mathbb{R}^n)$ and $f \in C^2(\mathbb{R}^n \times [0,\infty)) \cap L^\infty(\mathbb{R}^n \times [0,\infty))$. Then the function u defined in formula (3.104) belongs to $C^2(\mathbb{R}^n \times (0,\infty)) \cap C^0(\mathbb{R}^n \times [0,\infty))$ and solves problem (3.102).*

If $f \equiv 0$, by (3.104) we derive

$$u(x,t) = \int_{\mathbb{R}^n} K(x,y,t)g(y)\,dy. \tag{3.105}$$

Consequently, provided that we replace the integral with two brackets, $K(x,z,t)$ can be seen as the solution to the problem

$$\begin{cases} u_t - \Delta u = 0 & (x,t) \in \mathbb{R}^n \times (0,\infty) \\ u(x,0) = \delta_z & x \in \mathbb{R}^n \end{cases}$$

and this justifies the name of fundamental solution (for the definition of δ_z see Example 2.6.4). In Chapter 5, by means of the Fourier transform, we will show how it is possible to obtain (3.105) (the solution to problem (3.102) with $f \equiv 0$) directly, see formula (5.43). Arguing by contradiction, this procedure also gives the proof that the unique solution to problem (3.102) is given by (3.104).

The remaining part of this section is devoted to the study of the heat equation on bounded domains. Let us begin with the following

Definition 3.5.5. *Let $\Omega \subset \mathbb{R}^n$ be an open set and let for every $T > 0$ $\Omega_T = \Omega \times (0,T)$. Then the set defined by*

$$\partial_p \Omega_T = (\partial\Omega \times [0,T]) \cup (\Omega \times \{t = 0\})$$

is called **parabolic boundary** *of Ω_T.*

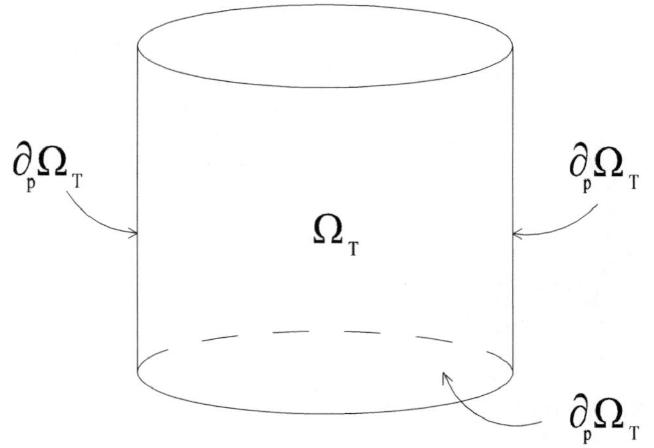

Parabolic boundary

We can state the fundamental

Theorem 3.5.6. (strong maximum principle). *Let $\Omega \subset \mathbb{R}^n$ be an open bounded set and let $u \in C^2(\overline{\Omega}_T)$ be a function satisfying the differential inequality*

$$u_t - \Delta u \leq 0 \qquad \text{in } \Omega_T. \tag{3.106}$$

Then the following conclusions hold true:

(i) $\max\limits_{\overline{\Omega}_T} u = \max\limits_{\partial_p \Omega_T} u,$

(ii) if Ω is connected and there exists $(x_0, t_0) \in \overline{\Omega}_T \setminus \partial_p \Omega_T$ such that $\max\limits_{\overline{\Omega}_T} u =$ $u(x_0, t_0)$, *then u is constant in $\overline{\Omega}_{t_0}$.*

Proof. We limit ourselves to prove assertion (i), which corresponds to the *weak* version of the maximum principle. The proof of (ii) would require the introduction of some preliminary results, therefore we refer the interested reader directly to [18, Section 7.1.4, Theorem 11].

We observe that, being $\partial_p \Omega_T \subset \overline{\Omega}_T$, it holds

$$\max_{\partial_p \Omega_T} u \leq \max_{\overline{\Omega}_T} u. \tag{3.107}$$

We start by considering the particular case in which (3.106) holds with the strict inequality sign:

$$u_t - \Delta u < 0 \qquad \text{in } \Omega_T. \tag{3.108}$$

Let $(x_0, t_0) \in \overline{\Omega}_T$ be a point such that $u(x_0, t_0) = \max\limits_{\overline{\Omega}_T} u$. If we are able to show that $(x_0, t_0) \in \partial_p \Omega_T$, then by (3.107) we immediately deduce that

$$\max_{\overline{\Omega}_T} u = \max_{\partial_p \Omega_T} u.$$

Let us assume by contradiction that $(x_0, t_0) \in \overline{\Omega}_T \setminus \partial_p \Omega_T$. Since u has a local maximum at (x_0, t_0), it necessarily holds

$$u_t(x_0, t_0) \geq 0. \tag{3.109}$$

More precisely, it holds $u_t(x_0, t_0) = 0$ if $t_0 \in (0, T)$, while if $t_0 = T$ we can only affirm that $u_t(x_0, t_0) \geq 0$.

On the other hand, the function $x \mapsto u(x, t_0)$ has a local maximum at x_0 and, since $x_0 \in \Omega$, the Hessian matrix $D_x^2 u(x_0, t_0)$ (computed with respect to the sole spatial variable x) is negative semidefinite, in particular its trace must be non positive:

$$\Delta u(x_0, t_0) = \operatorname{tr}(D_x^2 u(x_0, t_0)) \leq 0. \tag{3.110}$$

By combining (3.109) and (3.110), we get

$$u_t(x_0, t_0) - \Delta u(x_0, t_0) \geq 0\,,$$

in contrast with (3.108).

Let us now pass to the proof of (i) in the case in which (3.106) holds as such. For this purpose, consider the function $v(x,t) = |x|^2$ which depends only on the variable x. Then $v_t \equiv 0$ and $\Delta v = 2n$, thus

$$v_t - \Delta v = -2n < 0. \tag{3.111}$$

Therefore, for every $\varepsilon > 0$, the function $u_\varepsilon = u + \varepsilon v$ satisfies the strict inequality

$$(u_\varepsilon)_t - \Delta u_\varepsilon < 0 \qquad \text{in } \Omega_T$$

and, in view of the first part of the proof, we deduce that for every $\varepsilon > 0$ it holds

$$\max_{\overline{\Omega}_T} u_\varepsilon = \max_{\partial_p \Omega_T} u_\varepsilon\ .$$

Since $v \geq 0$, we infer

$$\max_{\overline{\Omega}_T} u \leq \max_{\overline{\Omega}_T} u_\varepsilon = \max_{\partial_p \Omega_T} u_\varepsilon \leq \max_{\partial_p \Omega_T} u + \varepsilon \max_{\partial_p \Omega_T} v.$$

Letting $\varepsilon \to 0^+$, we readily obtain $\max_{\overline{\Omega}_T} u \leq \max_{\partial_p \Omega_T} u$ which, together with (3.107), concludes the proof of (i). □

An immediate consequence of Theorem 3.5.6 is the following

Theorem 3.5.7. *Let $\Omega \subset \mathbb{R}^n$ be an open bounded set. Then there exists at most a function $u \in C^2(\overline{\Omega}_T)$ which solves the Cauchy-Dirichlet problem*

$$\begin{cases} u_t - \Delta u = f & \text{in } \Omega_T \\ u = g & \text{on } \partial_p \Omega_T\,, \end{cases} \tag{3.112}$$

with $f \in C^0(\overline{\Omega}_T)$ and $g \in C^2(\partial_p \Omega_T)$.

Proof. Let u_1 and u_2 be two solutions of (3.112) and let $w = u_1 - u_2$. Then w solves

$$\begin{cases} w_t - \Delta w = 0 & \text{in } \Omega_T \\ w = 0 & \text{on } \partial_p \Omega_T. \end{cases} \tag{3.113}$$

Theorem 3.5.6 states that w attains its maximum on $\partial_p \Omega_T$ and hence $w \leq 0$ in $\overline{\Omega}_T$. Similarly, by applying Theorem 3.5.6 to the function $-w$, we obtain that $w \geq 0$ in $\overline{\Omega}_T$. Thus $w \equiv 0$ and $u_2 = u_1$. □

The uniqueness for the Cauchy-Dirichlet problem can be also obtained using the energy method, under some assumptions that are slightly different from those considered in Theorem 3.5.7. Assume that Ω has Lipschitz boundary (see Definition 3.1.2) and let $u_1, u_2 \in C^2(\Omega_T) \cap C^1(\overline{\Omega}_T)$ be two solutions of (3.112). We set $w = u_1 - u_2$, so that $w \in C^2(\Omega_T) \cap C^1(\overline{\Omega}_T)$ solves (3.113). Let

$$E(t) = \int_\Omega w^2(x,t)\, dx = \|w(\cdot,t)\|^2_{L^2} \qquad (0 \le t \le T); \tag{3.114}$$

then, by (3.113) and an integration by parts, we derive

$$E'(t) = 2\int_\Omega w\, w_t\, dx = 2\int_\Omega w\,\Delta w\, dx = 2\int_{\partial\Omega} w\frac{\partial w}{\partial \nu}\, dS - 2\int_\Omega |\nabla w|^2\, dx \le 0\,,$$

where the last inequality follows from the fact that the boundary integral vanishes since $w(\cdot,t) \equiv 0$ on $\partial\Omega$ for every $t \in [0,T]$. Therefore, the function E is non increasing and

$$E(t) \le E(0) = 0 \qquad \forall t \in [0,T].$$

This implies that $w \equiv 0$ in $\overline{\Omega}_T$ and hence $u_1 \equiv u_2$.

An interesting property of the heat equation is the *backward uniqueness*. Let u and v satisfy

$$\begin{cases} u_t - \Delta u = v_t - \Delta v & \text{in } \Omega_T \\ u = v & \text{on } \partial\Omega \times (0,T). \end{cases} \tag{3.115}$$

Then there holds the following

Theorem 3.5.8. *Let Ω be an open bounded set with Lipschitz boundary. Let u and v be two functions satisfying (3.115) and belonging to the space $C^2(\overline{\Omega}_T)$. If $u(x,T) = v(x,T)$ for every $x \in \Omega$, then $u \equiv v$ in $\overline{\Omega}_T$.*

Proof. Set $w = u - v$ and let E be defined according to (3.114), then

$$E'(t) = -2\int_\Omega |\nabla w|^2\, dx. \tag{3.116}$$

Moreover, integrating by parts and taking into account that $w_t(\cdot,t) \equiv 0$ on $\partial\Omega$ for every $t \in [0,T]$, it holds

$$E''(t) = -4\int_\Omega \nabla w \nabla(w_t)\, dx = -4\int_{\partial\Omega} \frac{\partial w}{\partial \nu} w_t\, dS + 4\int_\Omega w_t\,\Delta w dx = 4\int_\Omega |\Delta w|^2 dx. \tag{3.117}$$

Integrating by parts and using (2.17), we get

$$\int_\Omega |\nabla w|^2\, dx = -\int_\Omega w\,\Delta w dx \le \left(\int_\Omega w^2 dx\right)^{\frac{1}{2}} \left(\int_\Omega |\Delta w|^2 dx\right)^{\frac{1}{2}}.$$

Hence, by (3.116) and (3.117), we deduce

$$(E'(t))^2 \leq E(t) \cdot E''(t) \qquad \forall t \in [0,T].$$

If we suppose by contradiction that $E(t) > 0$ for $t \in [0,T)$, then $f(t) := \log E(t)$ turns out to be convex in $[0,T)$. Indeed

$$f'(t) = \frac{E'(t)}{E(t)}, \qquad f''(t) = \frac{E''(t)}{E(t)} - \frac{(E'(t))^2}{(E(t))^2} \geq 0.$$

Thus, if $0 < \lambda < 1$ and $0 < t < T$, we have

$$\log E(\lambda t) = f((1-\lambda)0 + \lambda t) \leq (1-\lambda)f(0) + \lambda f(t) = (1-\lambda)\log E(0) + \lambda \log E(t)\ ,$$

namely

$$E(\lambda t) \leq E(0)^{1-\lambda} \cdot E(t)^{\lambda} \qquad \forall \lambda \in (0,1).$$

Passing to the limit as $t \to T^-$ and exploiting the fact that $E(T) = 0$ since $w(\cdot,T) \equiv 0$ in Ω, we obtain

$$E(\lambda T) \leq E(0)^{1-\lambda} \cdot E(T)^{\lambda} = 0 \qquad \forall \lambda \in (0,1)$$

which implies

$$E(t) = 0 \qquad \forall t \in (0,T].$$

This entails that $w \equiv 0$ in $\overline{\Omega}_T$ and hence $u \equiv v$ in $\overline{\Omega}_T$. □

For further results concerning the heat equation in bounded domains we refer to Chapter 7, in which we deal with existence and uniqueness of solutions to the problems

$$\begin{cases} u_t - \Delta u = f(x,t) & (x,t) \in \Omega \times (0,\infty) \\ u(x,0) = g(x) & x \in \Omega \\ u(x,t) = 0 & (x,t) \in \partial\Omega \times [0,\infty) \end{cases} \tag{3.118}$$

and

$$\begin{cases} u_t - \Delta u = f(x,t) & (x,t) \in \Omega \times (0,\infty) \\ u(x,0) = g(x) & x \in \Omega \\ \dfrac{\partial u}{\partial \nu}(x,t) = 0 & (x,t) \in \partial\Omega \times [0,\infty)\,, \end{cases} \tag{3.119}$$

with Ω an open set. Problems (3.118) and (3.119) are called respectively **Cauchy-Dirichlet problem** and **Cauchy-Neumann problem**.

3.6 Laplace and Poisson equations

This section is devoted to the **Poisson equation**

$$-\Delta u = f \tag{3.120}$$

which, in case $f = 0$, is called **Laplace equation**. A function $u \in C^2$ which solves the Laplace equation is called **harmonic**.

If we look at stationary (time-independent) solutions to the wave equation (3.19) or heat equation (3.90) we are led to equation (3.120). From this follows the considerable importance of the Poisson equation in describing physical phenomena of various nature. For instance, if we consider the problem of a string clamped in both extremes and subject to a vertical load f, and we want to determine the equilibrium configuration, we have to study the Dirichlet problem

$$\begin{cases} -u'' = f & \text{in } (0, \ell) \\ u(0) = u(\ell) = 0. \end{cases}$$

Analogously, if we want to study the equilibrium configuration of a membrane clamped at the boundary and subject to a vertical load f, we are led to the problem

$$\begin{cases} -\Delta u = f & \text{in } \Omega \\ u = 0 & \text{on } \partial\Omega. \end{cases}$$

Another application comes from electrostatic. Assuming that the magnetic field $\mathbf{B}$ is constant in time, by (3.56) it holds $\operatorname{curl} \mathbf{E} = 0$ and, from the irrotational behavior of the electric field, it follows the existence of an electric potential V, namely a scalar function such that

$$\mathbf{E} = \nabla V.$$

Inserting the latter equation into (3.54), we reduce to the Poisson equation

$$\Delta V = \frac{\rho}{\varepsilon_0}.$$

3.6.1 Fundamental solution of the Laplace equation

We look for radially symmetric solutions u to the Laplace equation $\Delta u = 0$ in $\mathbb{R}^n \setminus \{0\}$

$$u(x) = v(r) \qquad r = |x| = \left(\sum_{j=1}^{n} x_j^2 \right)^{\frac{1}{2}}.$$

Since

$$\frac{\partial r}{\partial x_i} = \frac{x_i}{r}, \qquad \frac{\partial^2 r}{\partial x_i^2} = \frac{1}{r} - \frac{x_i^2}{r^3},$$

$$u_{x_i} = v'(r)\frac{x_i}{r}\,, \qquad u_{x_i x_i} = v''(r)\frac{x_i^2}{r^2} + v'(r)\left(\frac{1}{r} - \frac{x_i^2}{r^3}\right)\,,$$

we obtain the ordinary differential equation

$$\Delta u = v''(r) + \frac{n-1}{r}v'(r) = 0 \qquad (r > 0). \tag{3.121}$$

Such equation can be rewritten as $(r^{n-1}v'(r))' = 0$ and, integrating, we get

$$v(r) = \begin{cases} b\,\log r + c & n = 2 \\ b\,r^{2-n} + c & n \geq 3. \end{cases} \tag{3.122}$$

After a suitable choice of the constants b and c, we can give the

Definition 3.6.1. *The function*

$$\Phi(x,y) = \begin{cases} -\frac{1}{2\pi}\,\log|x-y| & n = 2 \\ \frac{1}{n(n-2)\omega_n}\,|x-y|^{2-n} & n \geq 3, \quad \omega_n = \frac{\pi^{\frac{n}{2}}}{\Gamma(\frac{n}{2}+1)} = |B_1| \end{cases} \tag{3.123}$$

is called **fundamental solution** *of the Laplace equation. The function* Γ *is defined as* $\Gamma(\alpha) = \int_0^\infty t^{\alpha-1}e^{-t}dt$ *for every* $\alpha > 0$.

Exercise 3.6.2. Prove that for every fixed $y \in \mathbb{R}^n$, it holds $\Phi(\cdot, y) \in L^1_{loc}(\mathbb{R}^n)$ for every $n \geq 2$. □

The choice of the constants b and c in (3.122) has been done so that

$$-\Delta_x \Phi(x,0) = \delta_0 \qquad \text{in } \mathcal{D}'(\mathbb{R}^n)\,,$$

with δ_0 the Dirac mass at $x = 0$, see Example 2.6.4. In other words, there holds

$$-\int_{\mathbb{R}^n} \Phi(x,0)\Delta\varphi\, dx = \varphi(0) \qquad \forall \varphi \in \mathcal{D}(\mathbb{R}^n), \tag{3.124}$$

which, according to Exercise 3.6.2, makes sense. Similarly, it holds

$$-\Delta_x \Phi(x,y) = \delta_y\,,$$

with δ_y the Dirac delta function centered at y. For every "reasonable" f, the map $x \mapsto \Phi(x,y)\,f(y)$ is harmonic in $\mathbb{R}^n \setminus \{y\}$. This suggests that the function

$$u(x) = \int_{\mathbb{R}^n} \Phi(x,y)\; f(y)\; dy \tag{3.125}$$

solves $-\Delta u = f$. In fact, there holds the following result:

Theorem 3.6.3. *Let* $f \in C^2(\mathbb{R}^n)$ *have compact support; then if* u *is defined by* (3.125)*, there hold* $u \in C^2(\mathbb{R}^n)$ *and* $-\Delta u = f$ *in* $\mathbb{R}^n$.

Proof. By a change of variable, we may rewrite

$$u(x) = \int_{\mathbb{R}^n} \Phi(y,0)\ f(x-y)\ dy\ , \tag{3.126}$$

from which we derive

$$\frac{u(x+he_i)-u(x)}{h} = \int_{\mathbb{R}^n} \Phi(y,0)\frac{f(x+he_i-y)-f(x-y)}{h}\,dy \qquad \forall i=1,...,n.$$

Letting $h \to 0$, we infer

$$\frac{f(x+he_i-y)-f(x-y)}{h} \to \frac{\partial f}{\partial x_i}(x-y)$$

uniformly with respect to $y \in \mathbb{R}^n$. As a consequence, in view of Theorem 2.3.7, we can pass to the limit under the integral sign and get

$$u_{x_i}(x) = \int_{\mathbb{R}^n} \Phi(y,0)\ f_{x_i}(x-y)\,dy \qquad \forall i=1,...,n$$

and, similarly,

$$u_{x_ix_j}(x) = \int_{\mathbb{R}^n} \Phi(y,0)\ f_{x_ix_j}(x-y)\ dy \qquad \forall i,j=1,...,n$$

from which we deduce $u \in C^2(\mathbb{R}^n)$. Moreover, by (3.126),

$$-\Delta u(x) = -\int_{\mathbb{R}^n} \Phi(y,0)\Delta_x f(x-y)\,dy = -\int_{\mathbb{R}^n} \Phi(y,0)\Delta_y f(x-y)\,dy = f(x)\ ,$$

where we have also used (3.124). □

Remark 3.6.4. The strategy adopted in this section shows how we should proceed in the case of a general bounded domain Ω. First, we should find a function which plays the role the fundamental solution in Ω (called Green function), then we should combine it with the source f and the boundary datum (Dirichlet or Neumann). However, the difficulty consists precisely in finding the Green function: except for a few elementary cases, it is not possible to write it explicitly. □

3.6.2 Uniqueness for the Dirichlet and Neumann problems

Here we analyze in depth the Poisson equation, combined with suitable boundary conditions.

Theorem 3.6.5. (uniqueness for the Dirichlet problem). *Let $\Omega \subset \mathbb{R}^n$ be an open bounded set with Lipschitz boundary and let $f \in C^0(\Omega)$ and $g \in C^0(\partial\Omega)$. Then there exists at most one solution $u \in C^2(\Omega) \cap C^1(\overline{\Omega})$ to the Dirichlet problem for the Poisson equation*

$$\begin{cases} -\Delta u = f & \text{in } \Omega \\ u = g & \text{on } \partial\Omega\ . \end{cases} \tag{3.127}$$

Proof. By contradiction, let u_1 and u_2 be two solutions of (3.127). Then the difference function $v = u_1 - u_2$ satisfies

$$\begin{cases} \Delta v = 0 & \text{in } \Omega \\ v = 0 & \text{on } \partial\Omega \ . \end{cases} \tag{3.128}$$

Multiplying the equation in (3.128) by v and integrating by parts, we obtain

$$\int_\Omega |\nabla v|^2 = -\int_\Omega v\Delta v + \int_{\partial\Omega} v\frac{\partial v}{\partial \nu} = 0 \ ,$$

where the boundary term cancels out since $v = 0$ on $\partial\Omega$. By the previous equality, we deduce that v is constant over the connected components of Ω; since v vanishes on $\partial\Omega$ we deduce that $v \equiv 0$ in Ω. Thus $u_1 = u_2$. □

Problem (3.127) is associated to a *variational principle.* Indeed, consider the (convex) set

$$X = \{u \in C^2(\Omega) \cap C^1(\overline{\Omega});\ u = g \text{ on } \partial\Omega\}$$

and define over X the functional

$$J(u) = \frac{1}{2}\int_\Omega |\nabla u|^2 - \int_\Omega fu.$$

Even if it is defined on a larger class of functions, we consider J as a functional defined only on X. We prove the following

Theorem 3.6.6. (Dirichlet principle). *A function $u \in X$ is a solution to* (3.127) *if and only if it minimizes J over X.*

Proof. Assume that $u \in X$ solves (3.127) and let $w \in X$. Multiplying (3.127) by $u - w$ and integrating, we get

$$0 = \int_\Omega (-\Delta u - f)(u - w) = \int_\Omega [\nabla u \nabla(u - w) - f(u - w)] \ ,$$

whence

$$\int_\Omega [|\nabla u|^2 - fu] = \int_\Omega [\nabla u \nabla w - fw] \le \frac{1}{2}\int_\Omega |\nabla u|^2 + \frac{1}{2}\int_\Omega |\nabla w|^2 - \int_\Omega fw$$

which proves $J(u) \le J(w)$. Conversely, assume that u minimizes the functional J over X. Let us fix $v \in C_0^2(\overline{\Omega})$ (the function vanishes on $\partial\Omega$) and let us set

$$\Psi(t) = J(u + tv) = J(u) + t\int_\Omega [\nabla u \nabla v - fv] + \frac{t^2}{2}\int_\Omega |\nabla v|^2 \ .$$

We observe that $u + tv \in X$ for every $t \in \mathbb{R}$. By assumption, Ψ attains a minimum at $t = 0$; hence $\Psi'(0) = 0$. Thus, by the arbitrariness of v, we deduce that

$$0 = \int_\Omega [\nabla u \nabla v - fv] = \int_\Omega (-\Delta u - f)v = 0 \qquad \forall v \in C_0^2(\overline{\Omega}) .$$

In view of Proposition 2.4.22, the last equation entails $-\Delta u = f$ in Ω. □

For the Poisson equation with Neumann boundary conditions, we never have at the same time existence and uniqueness of solutions:

Theorem 3.6.7. (uniqueness for the Neumann problem). *Let $\Omega \subset \mathbb{R}^n$ be an open bounded connected set with Lipschitz boundary and let $f \in C^0(\Omega)$ and $g \in C^0(\partial\Omega)$. Then the Neumann problem for the Poisson equation*

$$\begin{cases} -\Delta u = f & \text{in } \Omega \\ \dfrac{\partial u}{\partial \nu} = g & \text{on } \partial\Omega \end{cases} \tag{3.129}$$

satisfies the following properties:

(i) if $\int_\Omega f \neq -\int_{\partial\Omega} g$ then (3.129) does not admit solutions $u \in C^2(\Omega) \cap C^1(\overline{\Omega})$;

(ii) if $\int_\Omega f = -\int_{\partial\Omega} g$ then (3.129) admits at most one solution $u \in C^2(\Omega) \cap C^1(\overline{\Omega})$, up to an additive constant.

Proof. Assume that (3.129) admits a solution $u \in C^2(\Omega) \cap C^1(\overline{\Omega})$; then, by the divergence Theorem, it holds

$$\int_\Omega f = -\int_\Omega \Delta u = -\int_{\partial\Omega} \frac{\partial u}{\partial \nu} = -\int_{\partial\Omega} g .$$

This proves (i).

Let us now assume that $\int_\Omega f = -\int_{\partial\Omega} g$, and let u_1 and u_2 be two solutions of (3.129). Then the difference function $v = u_1 - u_2$ satisfies

$$\begin{cases} \Delta v = 0 & \text{in } \Omega \\ \dfrac{\partial v}{\partial \nu} = 0 & \text{on } \partial\Omega . \end{cases} \tag{3.130}$$

Multiplying the equation (3.130) by v and integrating by parts, we obtain

$$\int_\Omega |\nabla v|^2 = -\int_\Omega v\Delta v + \int_{\partial\Omega} v\frac{\partial v}{\partial \nu} = 0 ,$$

where the boundary term cancels out since $\frac{\partial v}{\partial \nu} = 0$ on $\partial\Omega$. In view of the previous equality we deduce that v is constant over Ω. This proves (ii). □

Remark 3.6.8. The condition

$$\int_\Omega f = -\int_{\partial\Omega} g \tag{3.131}$$

for problem (3.129) is to be read as a *compatibility condition* between the data f and g; if this condition fails, there cannot be solutions. This restriction on the data has an easy physical explanation: if we interpret (3.129) as a stationary heat equation, the compatibility condition states that, in order to have solutions, the outward heat flux through the boundary must be compatible with the internal heat source. □

After having examined the uniqueness of the solution to the Poisson equation, it is natural to face the problem of existence. Basically, we can say that (3.127) always admits a solution, while (3.129) admits a solution provided that the compatibility condition (3.131) holds true. However, it is necessary to specify the *regularity* of the searched solutions and in what sense we mean that problems are solved. This makes the existence problem much more delicate. For this reason, we limit ourselves to present the details just in some particular situations (see the next two sections) and we postpone the complete answer to Chapter 7, in which we introduce the weak formulation of these problems.

3.6.3 Separation of variables

As repeatedly noticed, the proof of the **existence** of a solution to (3.127) or (3.129) is a delicate matter. Here we explain a possible procedure in some simple cases of plane geometries, namely when $\Omega = (0,1)^2$ (the unit square in $\mathbb{R}^2$) or $\Omega = B_1(0)$ (the unit disk in $\mathbb{R}^2$).

Laplace equation in the square

Let us first consider the Dirichlet problem

$$\begin{cases} u_{xx} + u_{yy} = 0 & \text{in } (0,1)^2 \\ u(x,0) = f(x) & \text{for } x \in (0,1) \\ u(x,1) = g(x) & \text{for } x \in (0,1) \\ u(0,y) = \varphi(y) & \text{for } y \in (0,1) \\ u(1,y) = \psi(y) & \text{for } y \in (0,1)\ . \end{cases} \tag{3.132}$$

Since the problem is linear, we can deal with the 4 conditions on the sides separately and finally add the solutions so found. Let us then assume that

$$g \equiv \varphi \equiv \psi \equiv 0\ . \tag{3.133}$$

We look for solutions to (3.132) of kind

$$u(x,y) = X(x)Y(y)\ . \tag{3.134}$$

Then $u_{xx}(x,y) + u_{yy}(x,y) = X''(x)Y(y) + X(x)Y''(y) = 0$. Dividing (formally) by $X(x)Y(y)$ the previous equation, we find

$$\frac{X''(x)}{X(x)} = -\frac{Y''(y)}{Y(y)} .$$

But a function depending only on x can be equal to a function depending only on y if and only if both the functions are constant. Therefore

$$\exists \lambda \in \mathbb{R} , \qquad X''(x) + \lambda X(x) = 0 , \quad Y''(y) - \lambda Y(y) = 0 ; \tag{3.135}$$

with the corresponding boundary conditions

$$X(0) = X(1) = Y(1) = 0 . \tag{3.136}$$

Imposing that X and Y are not identically zero, by (3.135) and (3.136) we deduce that necessarily $\lambda = \lambda_m = m^2\pi^2$ ($m \in \mathbb{N} \setminus \{0\}$), moreover $X_m(x) = A\sin(m\pi x)$, $Y_m(y) = Be^{m\pi y} + Ce^{-m\pi y}$. Prescribing $Y_m(1) = 0$, we infer that $C = -Be^{2m\pi}$ and hence $Y_m(y) = D_m \sinh[m\pi(1-y)]$. Looking for a solution u of the form (3.134) with X and Y satisfying (3.135)-(3.136), we find, up to a multiplicative constant, the infinitely many solutions

$$u_m(x,y) = \sin(m\pi x)\sinh[m\pi(1-y)] \qquad (m \in \mathbb{N} \setminus \{0\}) . \tag{3.137}$$

Since problem (3.135)-(3.136) is linear, any linear combination (even infinite!) of solutions of the kind (3.137) is still a solution; thus we are led to look for solutions to (3.132)-(3.133) of the form

$$u(x,y) = \sum_{m=1}^{\infty} c_m \sin(m\pi x)\sinh[m\pi(1-y)] , \tag{3.138}$$

where the constants c_m will be determined by imposing the last condition $u(x,0) = f(x)$, namely

$$f(x) = u(x,0) = \sum_{m=1}^{\infty} c_m \sinh(m\pi)\sin(m\pi x) , \qquad x \in (0,1) ;$$

therefore, $c_m \sinh(m\pi)$ are the Fourier coefficients of f, see (2.31):

$$c_m \sinh(m\pi) = b_m = 2\int_0^1 f(x)\sin(m\pi x)\, dx .$$

In conclusion, the solution to (3.132)-(3.133) (which we know to be unique in view of Theorem 3.6.5) is given by

$$u(x,y) = \sum_{m=1}^{\infty} \frac{b_m}{\sinh(m\pi)} \sin(m\pi x)\sinh[m\pi(1-y)] . \qquad \square$$

Let us now consider the Neumann problem

$$\begin{cases} u_{xx} + u_{yy} = 0 & \text{in } (0,1)^2 \\ -u_y(x,0) = f(x) & \text{for } x \in (0,1) \\ u_y(x,1) = g(x) & \text{for } x \in (0,1) \\ -u_x(0,y) = \varphi(y) & \text{for } y \in (0,1) \\ u_x(1,y) = \psi(y) & \text{for } y \in (0,1) \ , \end{cases} \tag{3.139}$$

where we observe that the $\pm$ signs before the normal derivatives are necessary to give them the *outward* sense. In view of Theorem 3.6.7, problem (3.139) is not solvable unless the compatibility condition (3.131) holds true: here such condition reads

$$\int_0^1 \big[f(s) + g(s) + \varphi(s) + \psi(s)\big]\, ds = 0 \ . \tag{3.140}$$

Once again, problem (3.139) is linear, thus we may assume (3.133); in this case, condition (3.140) becomes

$$\int_0^1 f(s)\, ds = 0 \ . \tag{3.141}$$

Arguing as already done for problem (3.132), we can separate the variables and look for solutions u of the kind (3.134). Once more, we obtain the ordinary differential equations (3.135), but now the corresponding boundary conditions are

$$X'(0) = X'(1) = Y'(1) = 0 \ . \tag{3.142}$$

As for the Dirichlet problem, imposing that X is a non trivial function, we obtain $\lambda_m = m^2\pi^2$ ($m \in \mathbb{N}$) and $X_m(x) = A\cos(m\pi x)$. For $\lambda = \lambda_m$ we get $Y_m(y) = Be^{m\pi y} + Ce^{-m\pi y}$; prescribing $Y_m'(1) = 0$, we deduce that $C = Be^{2m\pi}$ and hence $Y_m(y) = D_m \cosh[m\pi(1-y)]$. Looking for a solution u of the form (3.134), we find, up to a multiplicative constant, the infinitely many solutions

$$u_m(x,y) = \cos(m\pi x)\cosh[m\pi(1-y)] \qquad (m \in \mathbb{N}) \ .$$

Finally, recalling the assumptions (3.133) and (3.141), we are led to look for solutions to (3.139) of the form

$$u(x,y) = \sum_{m=0}^{\infty} c_m \cos(m\pi x)\cosh[m\pi(1-y)] \ . \tag{3.143}$$

The constants c_m can be derived from the condition $u_y(x,0) = -f(x)$, namely

$$f(x) = -u_y(x,0) = \pi \sum_{m=1}^{\infty} m\, c_m \sinh(m\pi)\cos(m\pi x) \ , \qquad x \in (0,1) \ ;$$

therefore $m\pi c_m \sinh(m\pi)$ are the Fourier coefficients of f:

$$m\pi c_m \sinh(m\pi) = b_m = 2\int_0^1 f(x)\cos(m\pi x)\,dx \qquad (m \in \mathbb{N}) .$$

We observe that the previous equality *would not be satisfied* for $m = 0$ if (3.141) failed; therefore, we have a confirmation that the compatibility condition is necessary to ensure the existence of solutions to the Neumann problem associated to the Poisson equation. In conclusion, up to an additive constant (see Theorem 3.6.7), the solution to (3.139) (under the assumptions (3.133) and (3.141)) is given by

$$u(x,y) = \frac{1}{\pi}\sum_{m=1}^{\infty} \frac{b_m}{m\,\sinh(m\pi)} \cos(m\pi x)\cosh[m\pi(1-y)] . \qquad \square$$

Remark 3.6.9. Clearly, with obvious modifications, the method presented above allows to solve the Dirichlet and Neumann problems for the Laplace equation also in the case of rectangular domains. □

Laplace equation in the disk

Let us consider the Dirichlet problem

$$\begin{cases} u_{xx} + u_{yy} = 0 & \text{if } x^2+y^2 < 1 \\ u(x,y) = g(x,y) & \text{if } x^2+y^2 = 1 . \end{cases} \tag{3.144}$$

Let us try to reformulate problem (3.144) in polar coordinates $x = \rho\cos\vartheta$, $y = \rho\sin\vartheta$. Differentiating with respect to ρ and ϑ the function $u(\rho\cos\vartheta, \rho\sin\vartheta)$, we obtain

$$\frac{\partial u}{\partial \rho} = u_x \cos\vartheta + u_y \sin\vartheta , \qquad \frac{\partial^2 u}{\partial \rho^2} = u_{xx}\cos^2\vartheta + 2u_{xy}\cos\vartheta\sin\vartheta + u_{yy}\sin^2\vartheta ,$$

$$\frac{\partial^2 u}{\partial \vartheta^2} = u_{xx}\,\rho^2\sin^2\vartheta - 2u_{xy}\,\rho^2\cos\vartheta\sin\vartheta + u_{yy}\,\rho^2\cos^2\vartheta - u_x\,\rho\cos\vartheta - u_y\,\rho\sin\vartheta .$$

Therefore,

$$u_{xx} + u_{yy} = \frac{\partial^2 u}{\partial \rho^2} + \frac{1}{\rho}\frac{\partial u}{\partial \rho} + \frac{1}{\rho^2}\frac{\partial^2 u}{\partial \vartheta^2}$$

and problem (3.144) becomes

$$\begin{cases} \dfrac{\partial^2 u}{\partial \rho^2} + \dfrac{1}{\rho}\dfrac{\partial u}{\partial \rho} + \dfrac{1}{\rho^2}\dfrac{\partial^2 u}{\partial \vartheta^2} = 0 & \text{if } \rho \in [0,1) \\ u(1,\vartheta) = g(\vartheta) , \end{cases} \tag{3.145}$$

where $u = u(\rho, \vartheta)$ and g has been written in polar coordinates as well: since it is defined just for $\rho = 1$, g depends only on ϑ. We look for solutions to (3.145) of the form $u(\rho, \vartheta) = R(\rho)T(\vartheta)$. By substitution in the equation and division, we arrive to

$$\frac{T''(\vartheta)}{T(\vartheta)} = -\left(\rho^2 \frac{R''(\rho)}{R(\rho)} + \rho \frac{R'(\rho)}{R(\rho)}\right) .$$

Both terms are constant; we look for (non trivial) solutions to the equations

$$T''(\vartheta) = \lambda T(\vartheta) , \qquad \rho^2 R''(\rho) + \rho R'(\rho) + \lambda R(\rho) = 0 , \qquad (\lambda \in \mathbb{R}) .$$

Clearly, T must be 2π-periodic; therefore we obtain $\lambda_m = -m^2$ ($m \in \mathbb{N}$) and $T(\vartheta) = a\cos(m\vartheta) + b\sin(m\vartheta)$. For $\lambda = \lambda_0 = 0$ the equation in R can be easily integrated and gives the general integral $R(\rho) = a\log\rho + b$. For $\lambda = \lambda_m$ ($m \geq 1$) we look for solutions R among the powers $R(\rho) = \rho^\alpha$: we find $\alpha^2 = m^2$, namely $R(\rho) = a\rho^m + b\rho^{-m}$ is the general integral. Since we look for regular solutions, we must ignore the logarithm and the negative powers. In the end, the only possible regular solutions with separate variables are

$$u_0(\rho, \vartheta) = \frac{a_0}{2} , \qquad u_m(\rho, \vartheta) = \rho^m \big(a_m \cos(m\vartheta) + b_m \sin(m\vartheta)\big) .$$

Finally, by the linearity of the problem, we look for solutions of the kind

$$u(\rho, \vartheta) = \sum_{m=0}^{\infty} u_m(\rho, \vartheta) .$$

Imposing the Dirichlet condition $u(1, \vartheta) = g(\vartheta)$ we deduce that a_m and b_m are the Fourier coefficients of g, see (2.30):

$$a_m = \frac{1}{\pi}\int_0^{2\pi} g(\vartheta)\,\cos(m\vartheta)\,d\vartheta , \qquad b_m = \frac{1}{\pi}\int_0^{2\pi} g(\vartheta)\,\sin(m\vartheta)\,d\vartheta$$

and the solution to (3.145) is given by

$$u(\rho, \vartheta) = \frac{a_0}{2} + \sum_{m=1}^{\infty} \rho^m \big(a_m \cos(m\vartheta) + b_m \sin(m\vartheta)\big) . \qquad \square$$

Let us now consider the Neumann problem

$$\begin{cases} u_{xx} + u_{yy} = 0 & \text{if } x^2 + y^2 < 1 \\ \dfrac{\partial u}{\partial \nu}(x, y) = g(x, y) & \text{if } x^2 + y^2 = 1 . \end{cases} \tag{3.146}$$

In polar coordinates, problem (3.146) reads

$$\begin{cases} \dfrac{\partial^2 u}{\partial \rho^2} + \dfrac{1}{\rho}\dfrac{\partial u}{\partial \rho} + \dfrac{1}{\rho^2}\dfrac{\partial^2 u}{\partial \vartheta^2} = 0 & \text{if } \rho \in [0, 1) \\ \\ \dfrac{\partial u}{\partial \rho}(1, \vartheta) = g(\vartheta) , & \end{cases} \tag{3.147}$$

where we assume the validity of the compatibility condition (3.131):

$$\int_0^{2\pi} g(\vartheta)\, d\vartheta = 0 \ . \tag{3.148}$$

Separating the variables and proceeding as in (3.145), we obtain

$$u(\rho, \vartheta) = \sum_{m=1}^{\infty} \rho^m \big(a_m \cos(m\vartheta) + b_m \sin(m\vartheta) \big) \ ,$$

where, without loss of generality, we have set $u(0, \vartheta) = 0$. Such assumption is licit since the solution is defined up to an additive constant, see Theorem 3.6.7. Then, imposing the Neumann condition we find

$$g(\vartheta) = \frac{\partial u}{\partial \rho}(1, \vartheta) = \sum_{m=1}^{\infty} m \big(a_m \cos(m\vartheta) + b_m \sin(m\vartheta) \big)$$

and hence a_m and b_m are proportional to the Fourier coefficients of g:

$$a_m = \frac{1}{m\,\pi} \int_0^{2\pi} g(\vartheta)\, \cos(m\vartheta)\, d\vartheta \ , \qquad b_m = \frac{1}{m\,\pi} \int_0^{2\pi} g(\vartheta)\, \sin(m\vartheta)\, d\vartheta \ . \qquad \square$$

Remark 3.6.10. In the particular case $f(x, y) = f(x^2 + y^2)$ (namely f radially symmetric), we are also able to solve easily the following Dirichlet problem for the Poisson equation:

$$\begin{cases} u_{xx} + u_{yy} = f(x^2 + y^2) & \text{if } x^2 + y^2 < 1 \\ u(x, y) = k \in \mathbb{R} & \text{if } x^2 + y^2 = 1 \ . \end{cases} \tag{3.149}$$

Indeed, according to Theorem 3.6.5, problem (3.149) admits a unique solution u that must be radially symmetric, namely of the form $u(x, y) = u(x^2+y^2)$ (if not, it would be sufficient to rotate it to obtain a different solution). Consequently, following the same procedure adopted for (3.121), we are led to the problem

$$\begin{cases} u''(r) + \frac{1}{r} u'(r) = f(r) & r \in (0, 1) \\ u(1) = k, \ \ u'(0) = 0 \ . \end{cases}$$

The condition $u(1) = k$ is the Dirichlet condition, while $u'(0) = 0$ is a requirement on the regularity of the solution: if u is radially symmetric, then its gradient necessarily vanishes at the origin. Solving this problem we obtain the solution

$$u(r) = k - \int_r^1 \left(\int_0^s t f(t)\, dt \right) \frac{ds}{s} \ . \qquad \square$$

3.6.4 Properties of harmonic functions

Harmonic functions have remarkable properties. Let us see the main ones.

Theorem 3.6.11. (mean value formula). *Let $\Omega \subset \mathbb{R}^n$ be an open set and let u be harmonic in Ω. Then, for every $x \in \Omega$ and for every $R > 0$ such that $\overline{B_R(x)} \subset \Omega$, there holds*

$$u(x) = \frac{1}{|\partial B_R(x)|} \int_{\partial B_R(x)} u(\sigma) dS(\sigma) = \frac{1}{|B_R(x)|} \int_{B_R(x)} u(y) dy \ . \tag{3.150}$$

Proof. Let $x \in \Omega$ and let $R > 0$ be such that $\overline{B_R(x)} \subset \Omega$. Then, thanks to the divergence Theorem, for every $r \in (0, R]$ it holds

$$0 = \int_{B_r(x)} \Delta u(y) dy = \int_{\partial B_r(x)} \frac{\partial u}{\partial \nu}(\sigma)\, dS(\sigma) = r^{n-1} \int_{|\sigma|=1} \nabla u(x + r\sigma) \cdot \sigma \, dS(\sigma)$$

$$= r^{n-1} \frac{\partial}{\partial r} \int_{|\sigma|=1} u(x + r\sigma)\, dS(\sigma) = r^{n-1} \frac{\partial}{\partial r} \left(\frac{1}{r^{n-1}} \int_{\partial B_r(x)} u(\sigma)\, dS(\sigma) \right)$$

$$= n\omega_n r^{n-1} \frac{\partial}{\partial r} \left(\frac{1}{|\partial B_r(x)|} \int_{\partial B_r(x)} u(\sigma)\, dS(\sigma) \right) .$$

We recall that ω_n denotes the Lebesgue measure of the unit ball of $\mathbb{R}^n$ and hence $n\omega_n$ coincides with the surface measure of the unit sphere in $\mathbb{R}^n$. This proves that the average over the sphere of radius r centered at x is constant with respect to $r \in (0, R]$ and consequently

$$\frac{1}{|\partial B_R(x)|} \int_{\partial B_R(x)} u(\sigma)\, dS(\sigma) = \lim_{r \to 0} \frac{1}{|\partial B_r(x)|} \int_{\partial B_r(x)} u(\sigma)\, dS(\sigma) = u(x) \ ,$$

which proves the first equality in (3.150). On the other hand, for every $r \in (0, R)$ it holds

$$u(x) n\omega_n r^{n-1} = \int_{\partial B_r(x)} u(\sigma)\, dS(\sigma) \ ,$$

hence, integrating over $(0, R)$, we obtain

$$u(x)\omega_n R^n = \int_0^R \left(\int_{\partial B_r(x)} u(\sigma)\, dS(\sigma) \right) dr = \int_{B_R(x)} u(y) dy$$

and this immediately implies the second equality in (3.150). □

Actually, the mean value property characterizes the harmonic functions:

Theorem 3.6.12. (reverse of the mean value property). *Let $\Omega \subset \mathbb{R}^n$ be an open set and let $u \in C^2(\Omega)$. If u satisfies (3.150) for every $B_R(x) \subset \Omega$, then $\Delta u = 0$ in Ω.*

Proof. See [59, Teorema 5.1, Capitolo VI]. □

A sort of mean value property is valid also for superharmonic and subharmonic functions, namely for functions of class C^2 satisfying $\Delta u \leq 0$ and $\Delta u \geq 0$ respectively.

Theorem 3.6.13. *Let $\Omega \subset \mathbb{R}^n$ be an open set and let $u \in C^2(\Omega)$.*
(i) If $\Delta u \leq 0$ (u superharmonic) then, for every $x \in \Omega$ and for every $R > 0$ such that $\overline{B_R(x)} \subset \Omega$, there hold

$$u(x) \geq \frac{1}{|\partial B_R(x)|} \int_{\partial B_R(x)} u(\sigma) dS(\sigma) \qquad \text{and} \qquad u(x) \geq \frac{1}{|B_R(x)|} \int_{B_R(x)} u(y) dy. \tag{3.151}$$

(ii) If $\Delta u \geq 0$ (u subharmonic) then, for every $x \in \Omega$ and for every $R > 0$ such that $\overline{B_R(x)} \subset \Omega$, there hold

$$u(x) \leq \frac{1}{|\partial B_R(x)|} \int_{\partial B_R(x)} u(\sigma) dS(\sigma) \qquad \text{and} \qquad u(x) \leq \frac{1}{|B_R(x)|} \int_{B_R(x)} u(y) dy. \tag{3.152}$$

Proof. It is enough to proceed as already done for the harmonic functions (Theorem 3.6.11) with the appropriate modifications. □

The mean value formulas lead to the following

Theorem 3.6.14. (maximum principle). *Let $\Omega \subset \mathbb{R}^n$ be an open bounded set and let $u \in C^2(\Omega) \cap C^0(\overline{\Omega})$ be subharmonic in Ω. Then*

$$\max_{\overline{\Omega}} u = \max_{\partial\Omega} u .$$

Moreover, if Ω is connected and there exists $x_0 \in \Omega$ such that $u(x_0) = \max_{\overline{\Omega}} u$, then u is constant over Ω.

Proof. Let us assume that there exists $x_0 \in \Omega$ such that $u(x_0) = M = \max_{\overline{\Omega}} u$. By applying (3.152), for every $r > 0$ such that $B_r(x_0) \subset \Omega$ we have

$$M = u(x_0) \leq \frac{1}{|B_r(x_0)|} \int_{B_r(x_0)} u(y) dy \leq M ,$$

where the second inequality follows from $u(y) \leq M$ for every y. But then the previous inequality is an equality. This occurs only if $u \equiv M$ in $B_r(x_0)$. By iterating this procedure at each point of $B_r(x_0)$ and by the arbitrariness of r, we conclude that u is constant over the connected component of Ω which contains x_0. □

As an immediate consequence of Theorem 3.6.14, there hold the following results:

Corollary 3.6.15. *Let $\Omega \subset \mathbb{R}^n$ be an open bounded set and let $u \in C^2(\Omega) \cap C^0(\overline{\Omega})$ be superharmonic over Ω. Then*

$$\min_{\overline{\Omega}} u = \min_{\partial\Omega} u .$$

Moreover, if Ω is connected and there exists $x_0 \in \Omega$ such that $u(x_0) = \min_{\overline{\Omega}} u$, then u is constant in Ω.

Corollary 3.6.16. *Let $\Omega \subset \mathbb{R}^n$ be an open bounded set and let $u \in C^2(\Omega) \cap C^0(\overline{\Omega})$ be harmonic in Ω. Then*

$$\min_{\partial\Omega} u \leq u(x) \leq \max_{\partial\Omega} u \qquad \forall x \in \overline{\Omega}.$$

We can give a physical interpretation of the maximum principles. Let us consider the problem of the membrane clamped at the boundary:

$$\begin{cases} -\Delta u = f & \text{in } \Omega \\ u = 0 & \text{on } \partial\Omega , \end{cases}$$

where u describes the vertical displacement with respect to the equilibrium position and f represents the vertical load which is applied on the membrane. If we assume for instance that the vertical load is directed upwards, then the function f is non negative in Ω. Then it is clear that the function u is superharmonic over Ω and, applying the maximum principle and taking into account that u vanishes on the boundary, we deduce that u must be non negative on the whole Ω. Actually, here we have applied the minimum principle stated in Corollary 3.6.15, but in the literature it is customary to use the terminology “maximum principle”. From the physical view point, this means that the membrane deforms upwards when it is subject to an upward vertical load.

The maximum principle provides also an alternative proof of the uniqueness of the solution to (3.127). Indeed, if by contradiction there were two solutions, then their difference v would satisfy (3.128): since v is harmonic, it attains its maximum and minimum on $\partial\Omega$ and hence $v \equiv 0$.

Remark 3.6.17. A harmonic function u in $\mathbb{R}$ satisfies $u'' = 0$ and hence it is of the form $u(x) = mx + q$, for some $m, q \in \mathbb{R}$. In this case, Corollary 3.6.16 is trivial. Moreover, Theorem 3.6.11 states that

$$u\left(\frac{a+b}{2}\right) = \frac{1}{b-a}\int_a^b u(x)\,dx = \frac{1}{2}[u(b) + u(a)] \qquad \forall a < b .$$

Replacing $u(x) = mx + q$, we have the following geometric interpretations:

– the value of u at the midpoint of the interval $[a, b]$ equals the arithmetic mean of the values of u at the endpoints of the interval;

– if $u \geq 0$ in $[a, b]$ and if we fix the bases in the segment $[a, b]$, then the area of the rectangle with height computed at the midpoint equals the area of the rectangle trapezoid defined by $0 \leq y \leq u(x)$. □

Note that, unlike the one-dimensional case, harmonic functions in dimension $n \geq 2$ do not reduce to the sole affine functions (of degree 1 in the independent variables):

Exercise 3.6.18. Prove that the following functions are harmonic in $\mathbb{R}^2$ (actually, in any dimension $n \geq 2$):

$$\text{(i) } u(x,y) = xy \qquad \text{(ii) } u(x,y) = x^3 - 3xy^2 \qquad \text{(iii) } u(x,y) = e^x \cos y\ .$$

Then, exploiting Corollary 4.1.5 (in the next chapter) and the subsequent exercises, provide other examples of harmonic functions in $\mathbb{R}^2$: the harmonic functions in the plane are the real part or the imaginary part of a holomorphic function in $\mathbb{C}$. □

Chapter 4

Complex Analysis

The notion of derivative of a real valued function of a real variable finds a natural extension in the framework of complex valued functions of a complex variable. Whereas, it seems less obvious to extend the concept of integral to a function of a complex variable. As it will be clear from the reading of this chapter, some results of Complex Analysis cannot be compared with those of Real Analysis: one of the results that at first glance may seem almost paradoxical is the fact that the differentiable functions in the complex sense on some open set $A \subseteq \mathbb{C}$ are even differentiable infinitely many times on A, in complete contrast with the real case. Complex Analysis turns out to be a powerful tool for the computation of integrals extended to half-lines of $\mathbb{R}$ or to the whole real line. As an immediate consequence of this fact, Complex Analysis proves essential also in the computation of integral transforms, that will be presented in the next chapter.

4.1 Holomorphic functions

Let $A \subset \mathbb{C}$ be a non empty open set, let $z \in A$ and let $f : A \to \mathbb{C}$ be a function; we will say that f is **entire** if it is defined in all $\mathbb{C}$. Depending on the situations, it may be convenient to identify $\mathbb{C}$ with $\mathbb{R}^2$; in other words, f may be seen as a function from $\mathbb{R}^2$ to $\mathbb{R}^2$. This identification is frequently convenient, but sometimes it is misleading since, as we will see, in a function of a complex variable $z = x + iy$, the variables x and y are not completely independent, as it happens instead for the functions in $\mathbb{R}^2$.

We will say that f is **differentiable** at z if there exists $\alpha \in \mathbb{C}$ such that

$$f(z+h) - f(z) = \alpha h + o(|h|) \qquad \text{as } |h| \to 0 \ ;$$

obviously, in this case also the limit of the incremental ratio exists (pay attention! it is not a real increment, the set $\mathbb{C}$ is not ordered!) and we will set $f'(z) = \alpha$. We observe also that if f is differentiable at z, for every $v \in \mathbb{C}$ such that $|v| = 1$, there exists the partial derivative of f at the point z in the direction v and it holds

$$D_v f(z) = \lim_{t \to 0} \frac{f(z + tv) - f(z)}{t} = v f'(z) \qquad (t \in \mathbb{R}) \; ; \tag{4.1}$$

this is nothing else but the gradient rule for functions $\mathbb{R}^2 \to \mathbb{R}$.

Definition 4.1.1. *Let $A \subset \mathbb{C}$ be a non empty open set, let $z \in A$ and let $f : A \to \mathbb{C}$ be a function; we will say that f is* **locally holomorphic** *at z if there exists a neighborhood of z on which f is differentiable. We shall say that f is* **holomorphic** *on A if f is locally holomorphic at every point of A.*

We will denote by $H(A)$ the set of holomorphic functions on A: then, to say that f is holomorphic on A, we will simply write $f \in H(A)$.

Exercise 4.1.2. Let $A \subset \mathbb{C}$ be a non empty open set; prove that $(H(A), +, \cdot)$ is a complex vector space. □

Let us go back to (4.1): if we take $v = 1$ or $v = i$ we obtain

$$\frac{\partial f}{\partial x}(z) = f'(z) \qquad \frac{\partial f}{\partial y}(z) = i f'(z) \; ;$$

then if we set $f(z) = f(x + iy) = u(z) + iv(z) = u(x, y) + iv(x, y)$, so that $u(z) = \mathrm{Re}(f(z))$ and $v(z) = \mathrm{Im}(f(z))$, we derive

$$\frac{\partial u}{\partial x}(z) = \frac{\partial v}{\partial y}(z) \qquad \frac{\partial u}{\partial y}(z) = -\frac{\partial v}{\partial x}(z) \; . \tag{4.2}$$

From now on, given a function f, we will always denote by u and v the functions $u(z) = \mathrm{Re}(f(z))$, $v(z) = \mathrm{Im}(f(z))$. The conditions (4.2) are called **Cauchy-Riemann conditions** and they play an important role in the theory of holomorphic functions; the first property of (4.2) is presented in the

Theorem 4.1.3. *Let $A \subset \mathbb{C}$ be a non empty open set, let $z \in A$ and let $f : A \to \mathbb{C}$ be a function; if u and v are differentiable at z and if (4.2) holds true, then f is differentiable at z.*

Proof. See [59, Teorema 2.1 p.16]. □

Remark 4.1.4. To ensure the differentiability of f at z, it is neither sufficient that u and v are differentiable at z, nor that (4.2) holds true even in the case f continuous. As examples, in the first case consider the function $f(z) = \bar{z}$ while in the second case take $z = 0$ and consider the function $g(z)$ which is zero on the axes (real and imaginary) and equals $|z|$ elsewhere. □

As an immediate consequence of Theorem 4.1.3 we obtain the

Corollary 4.1.5. *Let $A \subset \mathbb{C}$ be a non empty open set, then $f \in H(A)$ if and only if u and v are differentiable on A and satisfy conditions (4.2) at every point of A .*

Exercise 4.1.6. (i) By using (4.2), prove that if $A \subset \mathbb{C}$ is a non empty connected open set and if $f \in H(A)$, then the following equivalence holds true:

$$f'(z) = 0 \quad \forall z \in A \qquad \Longleftrightarrow \qquad f \text{ is constant on } A.$$

(ii) Show that (4.2) written in polar coordinates read $u_\vartheta = -\rho v_\rho$ and $v_\vartheta = \rho u_\rho$; note that such formulas are valid for $\rho \neq 0$, since polar coordinates have a singularity at the origin.
(iii) Study the validity of (4.2) for the function $z \mapsto |z|^a$ $(a \in \mathbb{R})$.
(iv) Prove that if $u = u(x, y)$ and $v = v(x, y)$ are C^2 on an open set $A \subset \mathbb{C}$, then $\Delta u = \Delta v = 0$ (u and v are harmonic). □

Remark 4.1.7. Every function g of $C^2(A)$ which satisfies the equation $\Delta g = 0$ is called **harmonic**; the previous exercise shows that both the real and the imaginary parts of a holomorphic function are harmonic: indeed, as we will see in the sequel, every holomorphic function is analytic (and hence of class C^2). □

The the following question arises naturally: *given a function u which is harmonic on an open set A, does it always exist a harmonic function v such that $f = u + iv$ is holomorphic on A?* If so, such function v is said to be an **harmonic conjugate** of u. If the open set is simply connected, the answer is positive:

Theorem 4.1.8. *If A is simply connected and u is harmonic on A, then there exists a harmonic conjugate v of u on A. Such function is unique up to an additive constant.*

Proof. See [59, Teorema 6.2 p.25]. □

Exercise 4.1.9. Determine a harmonic conjugate of the function $u(x, y) = x^4 - 6x^2y^2 + y^4$ and the corresponding holomorphic function f.
Solution: $v(x, y) = 4x^3y - 4xy^3$ and $f(z) = z^4$. □

Other exercises about this section can be found in [9, Capitolo 1] and in [47, Capitolo I].

4.2 Continuation of functions from $\mathbb{R}$ to $\mathbb{C}$

We wish to extend to all $\mathbb{C}$ some functions defined on $\mathbb{R}$ or over subsets of $\mathbb{R}$, more precisely we wonder if:
given an interval $I \subset \mathbb{R}$ and given $f : I \to \mathbb{C}$ differentiable, do they exist $A \subset \mathbb{C}$ and

$F : A \to \mathbb{C}$ holomorphic such that $I \subset A$ and the restriction of F to I coincides with f?

As first answer we claim the

Theorem 4.2.1. *Let $f : I \to \mathbb{R}$ be differentiable; if there exist an open set $A \subset \mathbb{C}$ such that $I \subset A$ and a function $F \in H(A)$ such that the restriction of F to I coincides with f, then F is unique.*

Proof. See [59, Teorema 1.1 p.103]; see also the proof of Theorem 4.9.1 below. □

Remark 4.2.2. The previous theorem is a uniqueness result but does not entail existence, which indeed may fail, as we will see in the next example. Let $I = (-1, 1)$ and let $f(x) = |x|^n x$ for some $n \in \mathbb{N}$, n odd; obviously $f \in C^n(I)$. By applying Theorem 4.2.1 to the subsets $(-1, 0)$ and $(0, 1)$, we deduce that the only possible extension of such a function would verify

$$F(z) = \begin{cases} z^{n+1} & \text{if } \mathrm{Re}(z) > 0 \\ -z^{n+1} & \text{if } \mathrm{Re}(z) < 0 \end{cases}$$

which in turn cannot be continuously extended to the points of the axis $x = 0$; thus none of the extensions of F can be continuous in an open set $A \subset \mathbb{C}$ containing I. □

However, Theorem 4.2.1 is more useful than what it seems at first glance: in fact, as a consequence we can easily deduce the following

Corollary 4.2.3. *The unique holomorphic function whose restriction to the real line coincides with the polynomial*

$$\sum_{k=0}^{n} a_k x^k \qquad (a_k \in \mathbb{C})$$

is the polynomial $\sum_{k=0}^{n} a_k z^k$.

This result allows then to pass to the power series: in particular, the unique holomorphic function whose restriction to the real line coincides with the function $f(x) = e^x$ is the entire function

$$f(z) = \sum_{n=0}^{\infty} \frac{z^n}{n!} \, ,$$

therefore we will set

$$e^z = \sum_{n=0}^{\infty} \frac{z^n}{n!} \, .$$

Similarly, we will have

$$\sin z = \sum_{n=0}^{\infty}(-1)^n \frac{z^{2n+1}}{(2n+1)!} \qquad \cos z = \sum_{n=0}^{\infty}(-1)^n \frac{z^{2n}}{(2n)!}$$
$$\sinh z = \sum_{n=0}^{\infty} \frac{z^{2n+1}}{(2n+1)!} \qquad \cosh z = \sum_{n=0}^{\infty} \frac{z^{2n}}{(2n)!}$$

and so on also for the other elementary power series. Then it is possible to define all the functions which are obtainable from the previous ones by means of algebraic operations, composition or differentiation.

Exercise 4.2.4. Starting from the geometric series

$$\sum_{n=0}^{\infty} z^n = \frac{1}{1-z},$$

deduce the power series of $z \mapsto \arctan z$. We suggest to write the series of $(1+z)^{-1}$, $(1+z^2)^{-1}$ and integrate. □

It is important to remark that the power series behave well with respect to convergence and differentiation:

Theorem 4.2.5. *A power series converges normally in every compact set lying in the interior of the circle of convergence. Moreover, the sum of a power series is holomorphic in the interior of the circle of convergence and its first derivative is the sum of the power series obtained differentiating term by term the original series.*

Proof. See [59, Teorema 2.3 p.29]. □

We recall that also the derivative series and the primitive series have the same radius of convergence of the initial series.

We now list the main properties of the elementary functions that we have just extended to all $\mathbb{C}$:

- $\forall z_1, z_2 \in \mathbb{C} \quad e^{z_1+z_2} = e^{z_1} \cdot e^{z_2}$.

  Indeed, it is enough to multiply (Cauchy product) the two series in the right hand side.
- The function $z \mapsto e^z$ is periodic with period $2\pi i$.

  Indeed, $e^{z+2\pi i} = e^z \cdot e^{2\pi i} = e^z(\cos(2\pi) + i\sin(2\pi)) = e^z$.
- $\forall w \neq 0$ the equation $e^z = w$ admits infinitely many solutions.

  Indeed, the solutions are $z = \log|w| + i(\arg w + 2k\pi)$, $k \in \mathbb{Z}$.
- The function $z \mapsto e^z$ belongs to $C^\infty(\mathbb{C})$ and $\frac{d}{dz}e^z = e^z$.
- The function $z \mapsto \sin z$ is odd, the function $z \mapsto \cos z$ is even.
- $\forall z \in \mathbb{C} \quad \sin z = -i\sinh(iz)$, $\cos z = \cosh(iz)$, $\cos^2 z + \sin^2 z = 1$.
- The addition formulas hold for $\sin(z_1 + z_2)$ and $\cos(z_1 + z_2)$.

Some exercises about this section can be found in [9, Capitolo 2] and in [47, Capitoli II e III].

4.3 Polydrome functions

Here we are concerned with the problem of solving the equation $e^z = w$, where z is the unknown and $w \in \mathbb{C}$ is assigned. As already said in the previous section, we have to assume $w \neq 0$, so that we obtain the infinitely many solutions $z_k = \log|w| + i(\arg w + 2k\pi)$, $k \in \mathbb{Z}$ where $\arg w \in [0, 2\pi)$ denotes the argument of a complex number. One may think that it is somehow possible to "prefer" one of these solutions, for instance z_0; actually, this is the way of defining the square root in $\mathbb{R}$: by solving the equation $x^2 = y$ we find two solutions and, by convention, we call square root of y the unique positive solution of such equation. As we will see, this is *not possible* in $\mathbb{C}$ and for this reason the logarithm function (as well as the square root function) is said **polydrome function**. Concerning the root, the underlying reason is that, unlike what occurs in $\mathbb{R}$, "it is possible to pass from one solution to the other without passing from 0"; this is what happens also with the logarithm.

Suppose then to define $\log w = z_0$ with the notations above. Let us consider for instance $w = 1$ and let us begin to move in counterclockwise sense around the unit circle. Thus, according to the previous definition, we have $\log(e^{i\varphi}) = i\varphi$ for every $\varphi \in [0, 2\pi)$: as $\varphi \to 2\pi$, the point $e^{i\varphi}$ tends to the starting point $w = 1$ and hence "we return to w taking a value different from the starting one". This means that the definition given *is not good.*

Thus we are obliged to take into account all the solutions z_k and then $\log w$ represents a subset of $\mathbb{C}$ *and not* a single value. In some sense we are led not to read a function $f : \mathbb{C} \to \mathbb{C}$ with values in $\mathbb{C}$, but to interpret it with values in the power set of $\mathbb{C}$, $f : \mathbb{C} \to 2^{\mathbb{C}}$; however, since we want f to satisfy some differentiability properties, we will not be interested in dealing with such a generality and we will limit ourselves to consider functions which admit *holomorphic selections.*

Going back to the definition of logarithm, the problem arises since we "make a loop around 0": let us then try to prevent the variable z from making such a path. We remove from the plane $\mathbb{C}$ the positive real half-line (or any other half-line) and we call $\mathbb{C}^*$ the open set that we obtain: it is clear that for every closed path that we follow now in $\mathbb{C}^*$, we return to the initial point with the same starting value.

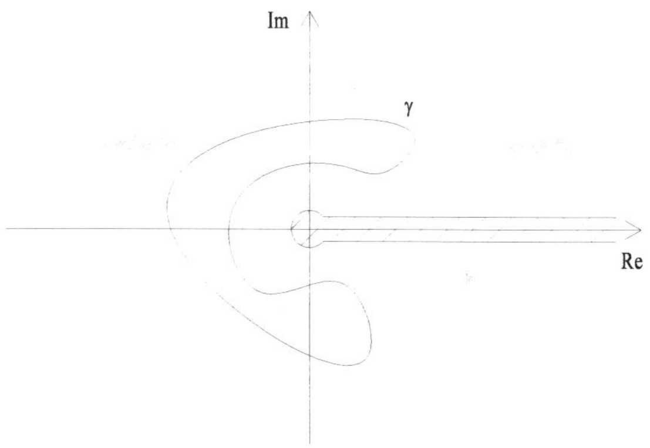

Closed path which does not cross the positive real half-line

Thus we may decide to prefer the solution z_0 and to call it $\log w$; thus obtaining a function f_0 which is holomorphic on $\mathbb{C}^*$. In order not to lose the other solutions, we have to imagine infinitely many complex planes $\mathbb{C}^*$ on each of which lies one of the infinitely many **determinations** f_k of $\log w$; then we may think that such planes are superposed according to the order of the integers k and that the contiguous ones are glued along the boundary of the cut, as shown in the picture below.

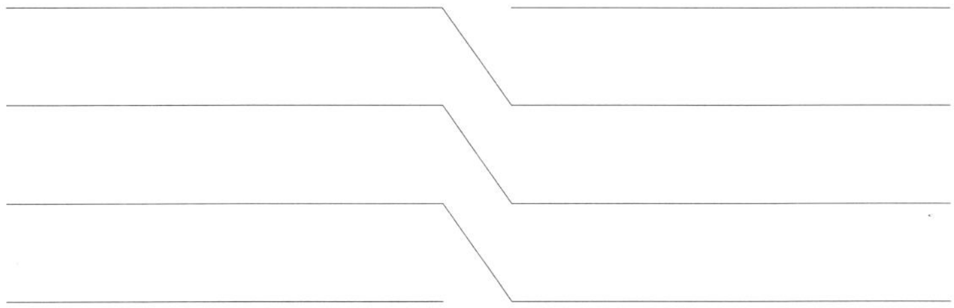

Riemann surface for the function $z \mapsto \log z$

The surface that we obtain with this procedure is called **Riemann surface** of the logarithm. In opposition to the polydrome functions, a function assuming a single value is called **monodrome**.

Besides the logarithm, there are other simple examples of polydrome functions, that we go through more quickly, since the same arguments already seen apply. Given $a \in \mathbb{C}$, let us consider the function $z \mapsto z^a$: we set by definition

$$z^a = e^{a \log z} \qquad \forall z \neq 0 \, .$$

• The case $a \in \mathbb{Z}$.
In this case the definition coincides with the classical one, indeed

$$e^{k \log z} = (e^{\log z})^k = z^k \qquad \forall k \in \mathbb{Z}$$

and hence the function $z \mapsto z^a$ defines a monodrome (holomorphic and, if $a \in \mathbb{N}$, also entire) function. In all the other cases, this does not hold.

• The case $a \in \mathbb{Q} \setminus \mathbb{Z}$.

For simplicity, let $a = \frac{m}{n}$ with m and n relatively prime integers and $n > 1$. It is easy to verify that $z^{m/n}$ assumes, at every point $z \neq 0$, exactly n distinct values and hence the function $z \mapsto z^{m/n}$ is polydrome with n determinations which turn one into the other after each cycle around the origin: after each cycle the starting value is multiplied by $\exp(2m\pi i/n)$ and hence, after n cycles we will be returned to the starting value. In terms of Riemann surface, the representation is now the following:

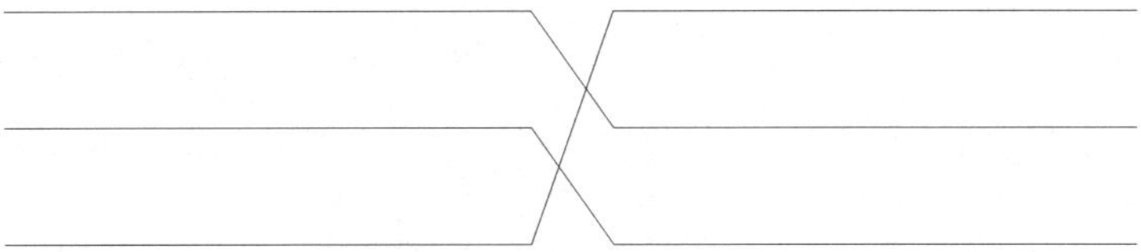

Riemann surface for the function $z \mapsto z^{m/n}$

• The case $a \notin \mathbb{Q}$.

This case displays the same features of the logarithm: in particular, the function $z \mapsto z^a$ admits infinitely many determinations and making a counterclockwise cycle around the origin we return to the starting point, with value equal to the initial one multiplied by $e^{2a\pi i}$.

Exercise 4.3.1. (i) Prove that for every $z_1, z_2 \in \mathbb{C} \setminus \{0\}$ it holds $\log(z_1 z_2) = \log(z_1) + \log(z_2)$ and show that $\log(z_1) + \log(z_1) \neq 2\log(z_1)$.

(ii) Write in algebraic form:
$z_1 = e^i$, $z_2 = \cos(\pi + 3i)$, $z_3 = \sin(i\pi)$, $z_4 = e^{2+i\frac{\pi}{2}}$, $z_5 = \cos(\frac{3}{2}\pi - i)$, $z_6 = \sinh(1+i)$, $z_7 = \cosh(1-i)$.

(iii) Find all the zeros of the functions:
(a) $z \mapsto \sin z$ (b) $z \mapsto \cos z$ (c) $z \mapsto \tan z$ (d) $z \mapsto \sinh z$ (e) $z \mapsto \cosh z$ (f) $z \mapsto e^z$.

Solution: (a) $k\pi$, $k \in \mathbb{Z}$ (b) $\frac{\pi}{2} + k\pi$, $k \in \mathbb{Z}$ (c) $k\pi$, $k \in \mathbb{Z}$ (d) $ik\pi$, $k \in \mathbb{Z}$ (e) $i\frac{\pi}{2} + k\pi i$, $k \in \mathbb{Z}$ (f) impossible.

(iv) Solve the equations for the unknown z:
(a) $e^z = i$ (b) $\sin z = 3$ (c) $\cos^2 z = -1$ (d) $e^z = |z|$ (e) $e^z = e^{\bar{z}} + 2i$ (f) $\sin z = \cos z$ (g) $(1+i)^z = \pi$.

Solution: (a) $(\frac{\pi}{2} + 2k\pi)i$, $k \in \mathbb{Z}$ (b) $-i\log(3 \pm 2\sqrt{2}) + \frac{\pi}{2} + 2k\pi$, $k \in \mathbb{Z}$ (c) $-i\log(\sqrt{2} \pm 1) + \frac{\pi}{2} + k\pi$, $k \in \mathbb{Z}$ (d) $z = x + iy$ with $y = 2k\pi$ and x solution of $e^x = \sqrt{x^2 + 4k^2\pi^2}$, $k \in \mathbb{Z}$ (e) $z = x + iy$ with $x, y \in \mathbb{R}$ satisfying the relation $e^x \sin y = 1$ (f) $z = \frac{\pi}{4} + k\pi$, $k \in \mathbb{Z}$ (g) $\dfrac{\log \pi + 2k\pi i}{\frac{1}{2}\log 2 + i\frac{\pi}{4} + 2h\pi i}$, $h, k \in \mathbb{Z}$. □

Other exercises and complements for this section can be found in [9, Chapetr 7].

4.4 Curves in $\mathbb{C}$

As in the case of $\mathbb{R}^2$, also in $\mathbb{C}$ we talk about **curves**: given a compact interval $I \subset \mathbb{R}$, we will call curve any continuous function $\alpha : I \to \mathbb{C}$. If $\alpha : I \to \mathbb{C}$ and $\beta : J \to \mathbb{C}$ are two curves and if there exists $\varphi : J \to I$ continuous, strictly increasing and surjective such that $\beta = \alpha \circ \varphi$, we shall say that α and β are **equivalent**. If $\alpha \in C^1(I, \mathbb{C})$ we will say that α is **regular**: here we do not require $\alpha'(t) \neq 0$ for every $t \in I$ since the tangent vector will not be used. Finally, given a curve $\alpha : I \to \mathbb{C}$, we will call **support** of α the set $\alpha(I)$; often, with an abuse of terminology, we shall use the term curve meaning the support of the curve itself. We will say that a curve $\alpha : [a, b] \to \mathbb{C}$ is **simple** if the function α is injective and, furthermore, we will say that α is **closed** if $\alpha(a) = \alpha(b)$.

Let $\alpha : [a, b] \to \mathbb{C} \setminus \{z_0\}$ be a curve whose support does not contain z_0; then, setting $\vartheta(t) = \arg(\alpha(t) - z_0) + 2g\pi$ with g equal to the number of turns made by the curve around the point z_0 (thus $\vartheta : [a, b] \to \mathbb{R}$), we can write

$$\alpha(t) = z_0 + |\alpha(t) - z_0| e^{i\vartheta(t)} \qquad \forall t \in [a, b] \; ; \tag{4.3}$$

clearly, ϑ is continuous and, more generally, it will have the same regularity of α.

Definition 4.4.1. *The real number*

$$\mathrm{ind}(\alpha, z_0) = \frac{1}{2\pi}[\vartheta(b) - \vartheta(a)]$$

is called **winding number** *of α around z_0.*

In the particular case in which $\vartheta(t) = t$, it holds $\alpha(t) = z_0 + \rho e^{it}$; if in addition $t \in [\vartheta_1, \vartheta_2]$, it holds

$$\mathrm{ind}(\alpha, z_0) = \frac{\vartheta_2 - \vartheta_1}{2\pi} \quad : \tag{4.4}$$

basically, we can then say that the winding number denotes the portions of turn that the moving point $\alpha(t)$ makes around z_0 counterclockwise. In the particular case in which α is a **Jordan curve**, namely a simple closed curve, the winding number equals 1.

Now, given a continuous function $f : \mathbb{C} \to \mathbb{C}$, we can define its *line integral* along a regular curve $\alpha : I \to \mathbb{C}$ by the formula

$$\int_\alpha f(z)dz = \int_I f(\alpha(t))\alpha'(t)dt \; . \tag{4.5}$$

Clearly, if α is a closed and regular curve (not necessarily simple!) there holds

$$\operatorname{ind}(\alpha, z_0) = \frac{1}{2\pi i}\int_\alpha \frac{dz}{z - z_0} \ ,$$

which is an integer.

Exercise 4.4.2. Recover (4.4) by using the line integral. □

If instead α is not regular (but at least rectifiable) we can define the line integral as the limit of suitable approximations. Let $\alpha : [a, b] \to \mathbb{C}$ be a rectifiable line and let f be continuous in an open set containing the support of α; we take finitely many points z_i ($i = 0, ..., n$) on the support of α such that $z_0 = \alpha(a)$, $z_n = \alpha(b)$ and $z_k = \alpha(t_k)$ with $t_k \in [a, b]$, $t_0 = a$, $t_n = b$ and $t_{k-1} < t_k < t_{k+1}$ for every $k = 1, ..., n-1$. Then we consider the sum

$$\sum_{i=1}^{n} f(\widetilde{z}_i)(z_i - z_{i-1})$$

where $\widetilde{z}_i$ is any point in the segment $[z_{i-1}, z_i]$: we observe that $f(\widetilde{z}_i)$ is certainly defined if $|z_i - z_{i-1}|$ is small enough; setting $\delta = \max_i |z_i - z_{i-1}|$, it can be shown that the limit of the previous sum as $\delta \to 0$ exists and such limit is called integral of f along α. Alternatively, we could consider a sequence of regular curves $\alpha_n : [a, b] \to \mathbb{C}$ which converge uniformly to α and set

$$\int_\alpha f(z)dz = \lim_{n\to\infty} \int_{\alpha_n} f(z)dz \ ;$$

indeed it can be proved that such limit does not depend on the sequence $\{\alpha_n\}$ considered.

Following the same proofs performed for the line integrals in $\mathbb{R}^2$, the following properties are immediate:

- $\displaystyle\int_\alpha (f(z) + g(z))dz = \int_\alpha f(z)dz + \int_\alpha g(z)dz \quad \forall f, g \in C^0(\alpha)$
- $\displaystyle\int_\alpha kf(z)dz = k\int_\alpha f(z)dz \quad \forall k \in \mathbb{C},\ f \in C^0(\alpha)$
- $\displaystyle\int_\alpha f(z)dz = -\int_{-\alpha} f(z)dz \quad \forall f \in C^0(\alpha)$
- $\displaystyle\left|\int_\alpha f(z)dz\right| \le \int_\alpha |f(z)|ds \quad \forall f \in C^0(\alpha)$. (Be careful! $ds = |\alpha'(t)|dt \ge 0$.)

• Let α be rectifiable with length ℓ_α, then for every $f \in C^0(\alpha)$

$$\left| \int_\alpha f(z)dz \right| \leq \ell_\alpha \max_t |f(\alpha(t))| \; . \tag{4.6}$$

In the next section we will see other properties.

4.5 The vanishing integral Theorem of Cauchy

The theorem that we present in this section is definitely the most important in the theory of holomorphic functions: it will allow to prove that *the sole hypothesis of differentiability in complex sense entails analyticity.*

Definition 4.5.1. *We say that an open set $A \subset \mathbb{C}$ is a* **simple** *region if it is the interior region of a Jordan curve.*

Then it holds the

Theorem 4.5.2. (The vanishing integral Theorem of Cauchy). *Let $A \subset \mathbb{C}$ be a simple region and let $f \in H(A)$; then, if γ is any closed path in the interior of A, there holds*

$$\int_\gamma f(z)dz = 0 \; .$$

Proof. The proof of this theorem is rather technique and long: it can be found in extenso in [59, Sections IV.4 and IV.5].

Here we limit ourselves to describe the main steps:
(I) We prove Goursat Lemma, namely Theorem 4.5.2 in the case in which γ is the boundary of a triangle: note that in this case γ *is not regular.* We argue by contradiction, supposing that

$$\left| \int_\gamma f(z)dz \right| > 0 \; ;$$

in order to obtain the absurdity, we split the triangle into a sequence of triangles, each of which is similar to the previous one with scale factor $\frac{1}{2}$.
(II) Then we prove Theorem 4.5.2 in the case in which γ is a closed polygonal curve: this is not too difficult to handle since Goursat Lemma applies.
(III) Finally, we prove Theorem 4.5.2 for an arbitrary closed curve γ: we construct a sequence of closed polygonal curves (not necessarily simple!) inscribed in γ and approximating it; by passing to the limit and by exploiting the definition of line integral, we obtain the desired result.

For an alternative proof which exploits Runge Theorem about the approximation of functions by means of rational functions, see [60, Teorema 13.11]. □

In some textbooks Theorem 4.5.2 is proved assuming that

$$u, v \in C^1(A) \ : \tag{4.7}$$

in such a way, the theory becomes a bit trivial (one loses perhaps the most significant aspect) but fits very well into the theory of exact and closed differential forms. Indeed, we have the

Theorem 4.5.3. *Let $A \subset \mathbb{C}$ be a non empty open set and let* (4.7) *hold true; then $f \in H(A)$ if and only if the differential forms $udx - vdy$ and $vdx + udy$ are closed in A.*

Proof. It is an immediate consequence of Corollary 4.1.5. □

Now we want to study the relationship between the holomorphy (namely the differentiability) and the existence of primitives: as we will see, the situation is completely different from the real case.

Definition 4.5.4. *Let $A \subset \mathbb{C}$ be a non empty open set and let $f \in H(A)$; we shall say that $F \in H(A)$ is a* **primitive** *of f if $F'(z) = f(z)$ for every $z \in A$.*

Note that the primitive, if it exists, is unique up to an additive constant. Moreover, the previous definition may seem restrictive, since the differentiability of f is required; however, the sole continuity *is not enough* to ensure the existence of a primitive:

Theorem 4.5.5. *Let $A \subset \mathbb{C}$ be a non empty open set, let $f \in C^1(A)$ and let* (4.7) *hold true; then f has a primitive in A if and only if the differential forms $udx - vdy$ and $vdx + udy$ are exact in A.*

Proof. Let $F \in H(A)$ be a primitive of f and let $\alpha : [a, b] \to A$ be an arbitrary regular curve; then, in view of (4.5) it holds

$$\int_\alpha f(z)dz = \int_a^b F'(\alpha(t))\alpha'(t)dt = F(\alpha(b)) - F(\alpha(a)) \ ,$$

that is, the integral *does not depend* on the support of α but only on its endpoints. This proves that the differential forms $udx - vdy$ and $vdx + udy$ are exact: in fact, setting $\alpha(t) = \beta(t) + i\gamma(t)$, there holds

$$\int_\alpha f(z)dz = \int_a^b [u(\beta(t), \gamma(t)) + iv(\beta(t), \gamma(t))](\beta'(t) + i\gamma'(t))dt$$

$$= \int_\alpha (udx - vdy) + i \int_\alpha (vdx + udy) \ . \tag{4.8}$$

Let now $udx - vdy$ and $vdx + udy$ be exact: take $z_0 \in A$ and, for every $z \in A$, take an arbitrary regular curve $\alpha_z : [0,1] \to A$ such that $\alpha_z(0) = z_0$ and $\alpha_z(1) = z$; then set

$$F(z) = \int_{\alpha_z} f(w)dw \ .$$

In view of (4.8), this definition makes sense since it does not depend on the choice of α_z; but then, for $|h|$ small enough,

$$F(z+h) - F(z) = \int_{\alpha_{z+h}} f(w)dw - \int_{\alpha_z} f(w)dw = \int_{[z,z+h]} f(w)dw \, ,$$

where $[z, z+h]$ is the segment joining z and $z+h$ (which lies in the interior of A for $|h|$ is small). Therefore, with the parametrization $t \mapsto z + th$, we obtain

$$F(z+h) - F(z) = \int_0^1 f(z+th)h \ dt \ ;$$

dividing both terms of the last identity by h and letting $h \to 0$, we get $F'(z) = f(z)$. □

We recall that if $f \in C^0(\mathbb{R}, \mathbb{R})$ then f has a primitive (of class C^1). In $\mathbb{C}$ this is not true any more: the existence of a primitive turns out to be a stricter condition, since, in view of the previous theorems, it implies the differentiability. Of course, it holds the following

Corollary 4.5.6. *Let $A \subset \mathbb{C}$ be a non empty simply connected open set and let $f \in C^0(A)$ satisfy (4.7); then the three following statements are equivalent:*
(i) f admits a primitive in A
(ii) $f \in H(A)$
(iii) the differential forms $udx - vdy$ and $vdx + udy$ are exact.

The simply connection of A is an essential assumption: indeed, if we consider $A = \{|z| < 1; \ z \neq 0\}$ and $f(z) = \frac{1}{z}$, it turns out that $f \in H(A)$ (with $f'(z) = -\frac{1}{z^2}$) but f does not admit any primitive in A (in view of the polydromy of the complex logarithm).

Theorem 4.5.2 can be extended in many ways; the most important ones are the extension to the boundary of the open set and the extension to the case of multiple boundaries.

Theorem 4.5.7. *Let γ be a simple closed curve and let $A \subset \mathbb{C}$ be its interior region. Let $f \in H(A) \cap C^0(\overline{A})$; then*

$$\int_\gamma f(z)dz = 0 \ .$$

Proof. See [59, Teorema 6.1 p.58]. □

Let now $\gamma_0, \gamma_1, ..., \gamma_n$ be some simple closed curves such that $\gamma_i \cap \gamma_j = \emptyset$ if $i \neq j$ and that γ_i belongs to the interior region of γ_0 for every $i = 1, ..., n$. Such set of curves is called *multiple boundary*. The set A of points interior to γ_0 and exterior to the other γ_i is called region with multiple boundary $\gamma = \gamma_0 - \gamma_1 - ... - \gamma_n$.

Theorem 4.5.8. *Let $\gamma_0, \gamma_1, ..., \gamma_n$ be simple closed curves forming a multiple boundary and let A be the region whose multiple boundary is $\gamma = \gamma_0 - \gamma_1 - ... - \gamma_n$. Let $f \in H(A) \cap C^0(\overline{A})$; then*

$$\int_{\gamma_0} f(z)dz = \sum_{i=1}^{n} \int_{\gamma_i} f(z)dz \ .$$

Proof. See [59, Teorema 7.1 p.60]. □

4.6 Consequences of the Cauchy Theorem

In this section we present the main consequences of Theorem 4.5.2. Let us begin with the

Theorem 4.6.1. (Cauchy integral formula). *Let γ be a simple closed path and let $A \subset \mathbb{C}$ be its interior region. Let $f \in H(A) \cap C^0(\overline{A})$, then for every $z_0 \in A$ there holds*

$$f(z_0) = \frac{1}{2\pi i} \int_{\gamma} \frac{f(z)}{z - z_0} dz \ .$$

Proof. Let $z_0 \in A$ and for every $r < d(z_0, \gamma)$ define $\gamma_r = \{z \in A;\ |z - z_0| = r\}$; setting $z = z_0 + re^{i\vartheta}$, it holds

$$\int_{\gamma_r} \frac{f(z)}{z - z_0} dz = i \int_0^{2\pi} f(z_0 + re^{i\vartheta}) d\vartheta \ .$$

In view of the continuity of f at z_0, we infer that $f(z_0 + re^{i\vartheta}) = f(z_0) + o(1)$ as $r \to 0$; therefore

$$\int_{\gamma_r} \frac{f(z)}{z - z_0} dz = i \int_0^{2\pi} f(z_0) d\vartheta + o(1) = 2\pi i f(z_0) + o(1) \ .$$

The assertion follows from Theorem 4.5.8 with $\gamma_0 = \gamma$ and $\gamma_1 = \gamma_r$ and letting $r \to 0$. □

This theorem states that a holomorphic (and continuous up to the boundary) function is uniquely determined by its restriction to the boundary; we can also state this important result in the following form:

Corollary 4.6.2. *Let γ be a simple closed path and let $A \subset \mathbb{C}$ be its interior region. Let $f, g \in H(A) \cap C^0(\overline{A})$ be such that $f(z) = g(z)$ for every $z \in \gamma$; then $f(z) = g(z)$ for every $z \in \overline{A}$.*

Another consequence of Theorem 4.6.1 is the

Theorem 4.6.3. (mean value Theorem). *Let $D \subset \mathbb{C}$ be the circle with center z_0 and radius $r > 0$; let $f \in H(D) \cap C^0(\overline{D})$, then*

$$f(z_0) = \frac{1}{2\pi} \int_0^{2\pi} f(z_0 + re^{i\vartheta}) d\vartheta \ .$$

Proof. It is enough to apply the Cauchy integral formula with the change of variable $z = z_0 + re^{i\vartheta}$. □

The most important consequence of Theorem 4.5.2 is the

Theorem 4.6.4. (Weierstrass Theorem). *Let $A \subset \mathbb{C}$ be a non empty open set and let $f \in H(A)$. Then, in a neighborhood of each point z_0, f can be represented by the power series*

$$\sum_{n=0}^{\infty} a_n (z - z_0)^n \,,$$

where

$$\forall n \in \mathbb{N} \qquad a_n = \frac{1}{2\pi i} \int_\gamma \frac{f(w)}{(w - z_0)^{n+1}} dw$$

and γ is an arbitrary circumference centered at z_0 and contained in A. Moreover, the radius of convergence of the series is at least $d(z_0, \partial A)$.

Proof. In view of Theorem 4.5.8, it is clear that the integral that defines a_n does not depend on the choice of γ, provided that $\gamma \subset A$. We observe that for every $z, z_0, w \in \mathbb{C}$ such that $|\frac{z-z_0}{w-z_0}| < 1$ we obtain the identity

$$\frac{1}{w - z} = \frac{1}{w - z_0} \frac{1}{1 - \frac{z-z_0}{w-z_0}} = \sum_{n=0}^{\infty} \frac{(z - z_0)^n}{(w - z_0)^{n+1}} \tag{4.9}$$

and the series converges.

Let $r_0 = d(z_0, \partial A)$ and let us take $z \in A$ such that $|z - z_0| < r_0$; then there exists a circumference γ centered at z_0 and with radius $r < r_0$ which contains z. By Theorem 4.6.1 and by (4.9) we have

$$f(z) = \frac{1}{2\pi i} \int_\gamma \frac{f(w)}{w - z} dw = \frac{1}{2\pi i} \int_\gamma f(w) \sum_{n=0}^{\infty} \frac{(z - z_0)^n}{(w - z_0)^{n+1}} dw$$

$$= \frac{1}{2\pi i}\sum_{n=0}^{\infty}(z-z_0)^n \int_\gamma \frac{f(w)}{(w-z_0)^{n+1}}dw = \sum_{n=0}^{\infty} a_n(z-z_0)^n \ ,$$

which proves the assertion. □

This theorem states that *from the sole assumption that f is differentiable on an open set, we deduce that f is analytic on such open set.* As we well know, such property is far from being true for functions of a real variable: for instance, let us consider the function defined by

$$f(x) = \begin{cases} e^{-1/x^2} & \text{if } x \neq 0 \\ 0 & \text{if } x = 0 \ ; \end{cases}$$

then $f \in C^\infty(\mathbb{R})$ but f is not analytic at $x = 0$ since it holds $f^{(n)}(0) = 0$ for every $n \in \mathbb{N}$ and hence the power series of f centered at 0 converges to $f(x)$ just for $x = 0$.

As a consequence of the Weierstrass Theorem we obtain a generalization of the Cauchy integral formula (coinciding in the case $n = 0$):

Corollary 4.6.5. (derivative formula). *Let A be a region with simple or multiple boundary and let $f \in H(A) \cap C^0(\overline{A})$; then*

$$\forall n \in \mathbb{N} \quad \forall z \in A \qquad f^{(n)}(z) = \frac{n!}{2\pi i}\int_{\partial A} \frac{f(w)}{(w-z)^{n+1}}dw \ .$$

We conclude the section with the

Theorem 4.6.6. (maximum modulus Theorem). *Let $A \subset \mathbb{C}$ be an open connected set and let $f \in H(A)$; if the function $z \mapsto |f(z)|$ admits its maximum at a point of A, then f is constant on A.*

Proof. See [59, Teorema 4.1 p.70]. □

In particular, if $f \in H(A) \cap C^0(\overline{A})$, then $|f|$ admits its maximum on ∂A. For the minimum modulus the same theorem does not hold, as it can be easily shown by considering $A = \{z \in \mathbb{C};\ |z| < 1\}$ and $f(z) = z$. Nevertheless, we have the following

Corollary 4.6.7. *Let $A \subset \mathbb{C}$ be an open connected set and let $f \in H(A) \cap C^0(\overline{A})$; if $f(z) \neq 0$ for every $z \in A$, then the function $z \mapsto |f(z)|$ admits its minimum on ∂A.*

Proof. It is enough to apply the maximum modulus Theorem to the function (holomorphic on A) $g(z) = \frac{1}{f(z)}$. □

Exercise 4.6.8. Prove that

$$\int_0^{+\infty} \sin x^2 \, dx = \int_0^{+\infty} \cos x^2 \, dx = \sqrt{\frac{\pi}{8}} \ .$$

These integrals are called Fresnel integrals and appear in optics, in diffraction phenomena. Hint: integrate the function $f(z) = e^{iz^2}$ along the path $\gamma_R = \gamma_R^1 \cup \gamma_R^2 \cup \gamma_R^3$ where $\gamma_R^1 = [0, R]$, γ_R^2 is defined by $z = Re^{i\vartheta}$ with $\vartheta \in [0, \frac{\pi}{4}]$, γ_R^3 is the segment $[R + iR, 0]$. Then take into account Theorem 4.5.2 and let $R \to \infty$. $\square$

Other exercises about the Sections 4.4, 4.5, 4.6 can be found in [9, Capitolo 3] and in [47, Capitolo IV].

4.7 Laurent series

Let $A \subset \mathbb{C}$ be a non empty open set and let $f \in H(A)$; take a point $z_0 \in \partial A$ and a point $\tilde{z} \in A$ and consider a regular curve γ with endpoints z_0 and $\tilde{z}$ such that, setting $\gamma' = \gamma \setminus \{z_0\}$, there holds $\gamma' \subset A$. In view of Theorem 4.6.4, f is represented by a Taylor series in a neighborhood of each point $z \in \gamma'$. Then the following question arises naturally: is there a point $z \in \gamma'$ such that the interior of the maximal disk of convergence of the corresponding Taylor series contains z_0? If the answer is negative, z_0 is called **critical point** or **singular point** of f: moreover, it is proved that such answer *does not depend* on the path γ used to approach z_0. In other words, a critical point z_0 is a point *out of reach* by the convergence disks of the points of any curve γ which has an endpoint at z_0. As an example, consider $A = \{z \in \mathbb{C};\ |z| < 1\}$, $f(z) = \frac{1}{1-z}$ and $z_0 = 1$; for simplicity take $\gamma = [0, 1]$ and fix $z^* \in [0, 1)$: for every $z \in A$ such that $|z - z^*| < |1 - z^*|$ it holds

$$\frac{1}{1-z} = \frac{1}{(1-z^*)[1 - \frac{z-z^*}{1-z^*}]} = \sum_{n=0}^{\infty} \frac{(z-z^*)^n}{(1-z^*)^{n+1}}$$

and it is clear that such representation is actually possible only if we have $|z - z^*| < |1 - z^*|$, which excludes the point z_0.

Therefore, the representation by Taylor series seen in Theorem 4.6.4 is not allowed any more if f is holomorphic only in a neighborhood with a hole at z_0: in this case, f can be represented by a bilateral power series, called **Laurent series**. Indeed, it is proved the

Theorem 4.7.1. *Let $A \subset \mathbb{C}$ be an annulus consisting of the points which are exterior to a circumference γ_1 and interior to a circumference γ_2, both centered at z_0; let $f \in H(A)$. Let γ be an arbitrary circumference lying in the interior of A and containing z_0 in its interior; then*

$$\forall z \in A \qquad f(z) = \sum_{n=-\infty}^{+\infty} a_n (z - z_0)^n ,$$

where

$$a_n = \frac{1}{2\pi i} \int_\gamma \frac{f(w)}{(w - z_0)^{n+1}} dw \qquad \forall n \in \mathbb{Z}.$$

Moreover, the series converges absolutely in A and uniformly in every compact set lying in the interior of A.

Proof. The proof exploits Cauchy integral formula (Theorem 4.6.1) and Weierstrass Theorem (Theorem 4.6.4); see [59] p.124. □

Of course, in view of Theorem 4.5.2, the previous result can be easily extended to more general domains.

Definition 4.7.2. *Let z_0 be a critical point of a holomorphic function f on an open set $A \subset \mathbb{C}$; we will say that z_0 is an* **isolated critical point** *of f if there exists a neighborhood of z_0 (interior point of A) which does not contain other critical points of f. Moreover, we will say that $z_0 = \infty$ is an isolated singular point of f if $f(\frac{1}{z})$ has an isolated singularity at $z = 0$.*

If z_0 is a singular point for f, then there exists $r > 0$ such that f is holomorphic on the annulus (holed neighborhood) defined by $0 < |z - z_0| < r$: in such annulus, the Laurent series of f necessarily contains some terms $(z - z_0)$ with negative power since, otherwise, f would be holomorphic on $|z - z_0| < r$. Then two cases may occur:
(i) the part which contains the negative powers of $(z - z_0)$ is formed by a finite number of terms;
(ii) the part which contains the negative powers of $(z - z_0)$ is a series.

In the first case we will say that z_0 is a **pole of order** m (being m the largest positive integer such that $a_{-m} \neq 0$), in the second case we will say that z_0 is an **essential singularity**.

If f is holomorphic for $|z|$ large enough (for instance for $|z| > R > 0$), then $z_0 = \infty$ is a regular point or an isolated singularity of f: more precisely, we shall say that ∞ is a pole of order m or an essential singularity of $f = f(z)$ if $z^* = 0$ is for the function $z \mapsto f(\frac{1}{z})$. Moreover, f is representable by Laurent power series in the "annulus centered at infinity" $|z| > R$; let then $\sum_{\mathbb{Z}} a_n z^n$ be such series: if f is regular at infinity, clearly it holds $a_n = 0$ for every $n > 0$. In conclusion, if we want to study the singularity at a point, for a point z_0 in the finite plane we have to look at the negative part of the series, while at $z_0 = \infty$ we have to look at the part with positive powers; therefore, if a function does not have singular point neither in the finite plane nor at infinity, it is necessarily constant.

There is also another isolated critical point for a holomorphic function but, as it is easily removable (indeed it is called **removable singularity**) we do not dwell on it; we just see an example. Let $f(z) = \frac{\sin z}{z}$: then f is not defined at $z_0 = 0$; however, there exists the limit $\lim_{z\to 0} f(z) = 1$ and, setting $f(0) = 1$ (namely by removing the singularity), nothing prevents us to write the Taylor series expansion of f centered at 0:

$$f(z) = \sum_{n=0}^{\infty} (-1)^n \frac{z^{2n}}{(2n+1)!} .$$

Thus, a removable singularity of a holomorphic function f is a point in which f is not defined but admits a finite limit.

Exercise 4.7.3. Given the function f, determine the type of singularity at the specified point z_0:
(a) $f(z) = \frac{1}{(z-1)^3}$, $z_0 = 1$ (b) $f(z) = \frac{1-\cos z}{z^2}$, $z_0 = 0$ (c) $f(z) = e^{1/z}$, $z_0 = 0$
(d) $f(z) = \text{cotan}\frac{\pi}{z}$, $z_0 = \frac{1}{k}$ $(k \in \mathbb{Z})$ (e) $f(z) = \text{cotan}\frac{\pi}{z}$, $z_0 = 0$ (f) $f(z) = \frac{z^3}{z+1}$, $z_0 = \infty$. *Solution:* (a) pole of order 3 (b) removable singularity (c) essential singularity (d) pole of order 1 (e) non-isolated singularity (f) pole of order 2. □

We observe that, as the open set A varies, the Laurent series associated to a function f may change: in other words, each time we will have to specify the annulus we are referring to. E.g., let us consider the function $f(z) = \frac{1}{z(1-z)}$; it turns out that

$$\text{in the annulus } 0 < |z| < 1 \qquad f(z) = \sum_{n=-1}^{\infty} z^n$$

$$\text{in the annulus } 0 < |z-1| < 1 \qquad f(z) = \sum_{n=-1}^{\infty} (-1)^n (z-1)^n$$

$$\text{in the annulus } 1 < |z| \qquad f(z) = -\sum_{n=-\infty}^{-2} z^n \ :$$

the last series converges also at $z = \infty$, which therefore is a regular point for f, and $f(\infty) = 0$.

Let us now characterize the poles and the essential singularities; we limit ourselves to state the following properties (the student try to prove them!).

Theorem 4.7.4. *Let z_0 be an isolated singular point of a holomorphic function f; the following properties are equivalent:*
(i) f has a pole of order m at z_0;
(ii) the function $g(z) = (z - z_0)^m f(z)$ has a removable singularity at z_0 and in addition $\lim_{z \to z_0} g(z) \neq 0$;
(iii) $|f(z)| \asymp |z - z_0|^{-m}$ as $z \to z_0$;
(iv) the function $z \mapsto \frac{1}{f(z)}$ has a zero of order m at z_0.

Clearly, an isolated singularity (not removable) z_0 of a function f is an essential singularity if and only if there exist no positive integer m such that one of the (i)-(iv) in the previous theorem holds true.

Theorem 4.7.5. *Let z_0 be an essential singular point of a function f; then z_0 is a singular point also for the function $z \mapsto \frac{1}{f(z)}$: more precisely, it can be an essential singularity or a cluster point of poles (and thus a non-isolated singularity).*

For instance, the functions $f(z) = e^{1/z}$ and $g(z) = \sin\frac{1}{z}$ both have an essential singularity at $z = 0$ but $\frac{1}{f(z)}$ has another essential singularity at 0 while $\frac{1}{g(z)}$ has a cluster point of poles at 0.

Theorem 4.7.6. *Let z_0 be an essential singular point of a function f; then every complex number is a limit value of $f(z)$ as $z \to z_0$.*

In particular, from these properties we deduce the

Corollary 4.7.7. *Let z_0 be an isolated singular point of a holomorphic function f. Then:*
(i) z_0 is a removable singularity if and only if $\lim_{z\to z_0} f(z)$ exists and is finite;
(ii) z_0 is a pole if and only if $\lim_{z\to z_0} f(z) = \infty$;
(iii) z_0 is an essential singularity if and only the limit $\lim_{z\to z_0} f(z)$ does not exist.

Some exercises about the topics of this section can be found in [9, Capitolo 4], [47, Capitolo V] and [70, Capitolo II].

4.8 The residue Theorem

Consider a function f holomorphic on a holed neighborhood of z_0 and let

$$f(z) = \sum_{n=-\infty}^{\infty} a_n (z - z_0)^n \tag{4.10}$$

be its Laurent series expansion in such neighborhood; then it holds the

Theorem 4.8.1. *Let γ be a circumference lying in the interior of the convergence field of* (4.10) *with $z_0 \in \overset{\circ}{\gamma}$; then*

$$\frac{1}{2\pi i}\int_{\gamma} f(z)dz = a_{-1} \ .$$

Proof. It is a particular case of Theorem 4.7.1. □

This theorem justifies the

Definition 4.8.2. *The coefficient a_{-1} is called* **residue** *of f at z_0 and we denote it by* $\mathrm{Res}(f, z_0) = a_{-1}$.

Similarly, at infinity we have the

Theorem 4.8.3. *Let $R > 0$ and let f be holomorphic in $|z| > R$; let*

$$f(z) = \sum_{n=-\infty}^{\infty} a_n z^n$$

be the Laurent expansion of f in a neighborhood of infinity. Let γ be a circumference of radius $\rho > R$ centered at the origin (such that all the possible singularities of f lying in the finite complex plane are interior to it), then

$$\frac{1}{2\pi i}\int_{-\gamma} f(z)dz = -a_{-1}\ .$$

Proof. With the change of variables $z = \rho e^{i\vartheta}$ we get

$$\frac{1}{2\pi i}\int_{-\gamma} f(z)dz = -\frac{\rho}{2\pi}\int_0^{2\pi} f(\rho e^{i\vartheta})e^{i\vartheta}d\vartheta$$

$$= -\frac{\rho}{2\pi}\sum_{n=-\infty}^{\infty} a_n\rho^n\int_0^{2\pi} e^{i(n+1)\vartheta}d\vartheta = -a_{-1}\,,$$

where the last equality follows from the periodicity of the exponential function. □

In accordance to the previous theorem, we will set (mind the sign $-$)

$$\mathrm{Res}(f,\infty) = -a_{-1}\ .$$

Thus, in order to compute the residue of a function at a point, we have to determine the associated Laurent series and find out the coefficient a_{-1}; in some cases it is possible to use particular tricks in order to spare some calculations: let us see some of them (exercise: prove them!).
• If $z_0 \neq \infty$ is a removable singularity or a regular point of f then $\mathrm{Res}(f,z_0) = 0$ (pay attention, this can be false if $z_0 = \infty$, take for instance $f(z) = \frac{1}{z}$).
• If $z_0 \neq \infty$ is a pole of order 1 for f then necessarily $\mathrm{Res}(f,z_0) \neq 0$.
• If ∞ is a zero of order 1 for f then $\mathrm{Res}(f,\infty) = -\lim_{z\to\infty} zf(z) \neq 0$.
• If ∞ is a zero of order greater that 1 for f then $\mathrm{Res}(f,\infty) = 0$.
• If $z_0 \neq \infty$ is a pole of order m for f then, setting $g(z) = (z-z_0)^m f(z)$ and extending g by continuity at z_0, it holds $\mathrm{Res}(f,z_0) = \dfrac{g^{(m-1)}(z_0)}{(m-1)!}$.
• In the computation of the residue at polar singularities it is also useful *de L'Hôpital rule*: let Ω be an open set, let $f, g \in H(\Omega)$ and let $z_0 \in \Omega$ be such that $f(z_0) = g(z_0) = 0$, then

$$\lim_{z\to z_0}\frac{f(z)}{g(z)} = \lim_{z\to z_0}\frac{f'(z)}{g'(z)}.$$

• In particular, if $f(z) = \frac{g(z)}{h(z)}$ with g regular and different from zero at $z_0 \neq \infty$ and h regular with a zero of order 1 at z_0 (and hence f has a pole of order 1), we have $\mathrm{Res}(f,z_0) = \dfrac{g(z_0)}{h'(z_0)}$.
• For the point $z = \infty$, it is useful the formula $\mathrm{Res}(f,\infty) = -\mathrm{Res}(\frac{1}{z^2}f(\frac{1}{z}),0)$.

Exercise 4.8.4. Determine the residues at $z = 0$ of the following functions:
(a) $f(z) = \frac{1}{\sin z}$ (b) $f(z) = \frac{1}{z^2}$ (c) $f(z) = ze^{1/z}$.
Solution: (a) 1 (b) 0 (c) $\frac{1}{2}$. □

Let us now present an important result which will allow us to solve several problems, also in Real Analysis:

Theorem 4.8.5. (residue Theorem). *Let A be an open set with simple or multiple boundary ∂A and let $z_1, ..., z_n \in A$ be singular points of a function $f \in H(A \setminus \cup_j \{z_j\}) \cap C^0(\overline{A} \setminus \cup_j \{z_j\})$; then*

$$\int_{\partial A} f(z) dz = 2\pi i \sum_{j=1}^{n} \operatorname{Res}(f, z_j) \ .$$

Proof. Let γ_i $(i = 1, ..., n)$ be circumferences lying in the interior of A, centered at z_i and such that $\gamma_i \cap \gamma_j = \emptyset$ if $i \neq j$; in view of Theorem 4.5.8 there holds

$$\int_{\partial A} f(z) dz = \sum_{j=1}^{n} \int_{\gamma_j} f(z) dz$$

and the desired result follows by Theorem 4.8.1. □

Similarly, if f is holomorphic on a region exterior to a closed curve γ and it is continuous on $\overline{A}$, except at finitely many points $z_1, ..., z_n$, it is shown that

$$\int_{-\gamma} f(z) dz = 2\pi i \Big(\sum_{j=1}^{n} \operatorname{Res}(f, z_j) + \operatorname{Res}(f, \infty) \Big) \ ;$$

thus, as an immediate consequence, we have the

Corollary 4.8.6. *Let f be a holomorphic function on all $\mathbb{C}$, except at most at finitely many singular points $z_1, ..., z_n$. Then*

$$\sum_{j=1}^{n} \operatorname{Res}(f, z_j) + \operatorname{Res}(f, \infty) = 0 \ .$$

Some exercises about this section may be found in [9, Capitolo 4].

4.9 The logarithmic index Theorem

We begin this section with a generalization of Corollary 4.6.2:

Theorem 4.9.1. (Identity principle for holomorphic functions). *Let $A \subset \mathbb{C}$ be an open connected set and let $f, g \in H(A)$; the following facts are equivalent:*

(i) $f \equiv g$ on A;

(ii) $\exists z_0 \in A$ such that $f \equiv g$ on a neighborhood of z_0;

(iii) $\exists z_0 \in A$ such that $f^{(n)}(z_0) = g^{(n)}(z_0)$ $\forall n \in \mathbb{N}$;

(iv) the set $A_0 = \{z \in A;\ f(z) = g(z)\}$ has a cluster point which lies in the interior of A.

Proof. The implications (i) $\Rightarrow$ (ii) and (ii) $\Rightarrow$ (iv) are trivial.

The equivalence (ii) $\Leftrightarrow$ (iii) follows from the developability into Taylor series proved in Theorem 4.6.4.

Let us now show that (iv) $\Rightarrow$ (iii). Let $z_0 \in A$ be a cluster point of A_0 and assume by contradiction that there exists $m \in \mathbb{N}$ such that $f^{(m)}(z_0) \neq g^{(m)}(z_0)$; then, if $\bar{n}$ is the smallest of such m,

$$f(z) - g(z) = a_{\bar{n}}(z - z_0)^{\bar{n}} + o[(z - z_0)^{\bar{n}}]$$

and hence $f(z) - g(z) \neq 0$ in all the points belonging to a suitable holed neighborhood of z_0, in contrast with (iv).

Finally, let us prove that (iv) $\Rightarrow$ (i). Let A_1 be the subset of A formed by the cluster points of A_0: in view of the equivalence (ii) $\Leftrightarrow$ (iii) and thanks to what we have just proved, A_1 turns out to be open (it is formed by interior points). Let now A_2 be the set of points which are not cluster points of A_0: A_2 is an open set as well. Thus we have

$$A_1 \cap A_2 = \emptyset \qquad A_1 \cup A_2 = A \qquad A_1 \neq \emptyset\ ;$$

since A is connected, this implies that $A_2 = \emptyset$ and completes the proof. □

Thanks to this theorem, we can prove the

Theorem 4.9.2. *Let $A \subset \mathbb{C}$ be a simple region and let $f \in H(A) \cap C^0(\overline{A})$ be a function which does not have any zero on ∂A; then*

$$N_0 = \frac{1}{2\pi i} \int_{\partial A} \frac{f'(z)}{f(z)} dz$$

equals the number of zeros of f in A, counted according to their multiplicity.

Proof. In view of Theorem 4.9.1 and Bolzano-Weierstrass Theorem, the zeros of f in A are finitely many: let then $z_1, ..., z_n$ be the zeros of f in A and $k_1, ..., k_n$ their respective multiplicities; thus we may write $f(z) = (z - z_j)^{k_j} f_j(z)$, with $f_j(z_j) \neq 0$ and f_j locally holomorphic at z_j. Let us then set

$$g(z) = \frac{f'(z)}{f(z)} = \frac{k_j}{z - z_j} + \frac{f_j'(z)}{f_j(z)}\,,$$

where we have pointed out the behavior at z_j: the function g is holomorphic on A and it is continuous on $\overline{A}$, except at the points $z_1, ..., z_n$, at which it has a pole of order 1 with residues $k_1, ..., k_n$ respectively. Then, the result follows from Theorem 4.8.5. □

Now we can prove a result whose name comes from the fact that $\frac{f'(z)}{f(z)}$ is the logarithmic derivative of f:

Theorem 4.9.3. (logarithmic index Theorem). *Let $A \subset \mathbb{C}$ be a simple region; let f be a holomorphic function on A except at finitely many poles $p_1, ..., p_m$ of order $h_1, ..., h_m$ respectively: moreover, let $f \in C^0(\overline{A} \setminus \{p_1, ..., p_m\})$ and let f have neither zeros nor singular points on the boundary ∂A. Then*

$$N_0 - N_\infty = \frac{1}{2\pi i} \int_{\partial A} \frac{f'(z)}{f(z)} dz \,,$$

where $N_\infty = \sum_j h_j$ and N_0 equals the number of zeros of f in A, counted according to their multiplicity.

Proof. It is analogous to the previous one: it is enough to observe that g has a pole of order 1 and residue $-h_j$ at p_j. □

Finally, the previous theorem allows to prove the important

Theorem 4.9.4. *Assume that the hypotheses of Theorem 4.9.3 hold true and let*

$$m = \min_{z \in \partial A} |f(z)| \qquad M = \max_{z \in \partial A} |f(z)| \,.$$

Then the equation $f(z) = w$ has in A

$$N_0 \textit{ solutions if } |w| < m \qquad N_\infty \textit{ solutions if } |w| > M \,.$$

Proof. Let $|w| < m$: then also the function $h(z) = f(z) - w$ satisfies the assumptions of Theorem 4.9.3, moreover $N_\infty(h) = N_\infty(f)$ (with the obvious meaning). Therefore we deduce that

$$N_0(h) = \frac{1}{2\pi i} \int_{\partial A} \frac{h'(z)}{h(z)} dz + N_\infty(h) = \frac{1}{2\pi i} \int_{\partial A} \frac{f'(z)}{f(z) - w} dz + N_\infty(f) \,;$$

now we note that the function

$$\varphi(w) = \frac{1}{2\pi i} \int_{\partial A} \frac{f'(z)}{f(z) - w} dz$$

is continuous on the open disk $\{w \in \mathbb{C};\ |w| < m\}$ and, since $N_0(h)$ and $N_\infty(f)$ are integer numbers, it follows that φ is constant, that is $\varphi(w) = \varphi(0) = N_0(f) - N_\infty(f)$: this allows to conclude that $N_0(h) = N_0(f)$, which proves the first half of the claim.

In a similar way, let $|w| > M$: setting $h(z) = f(z) - w$, we get again $N_\infty(h) = N_\infty(f)$ and, thanks to Theorem 4.9.3, we infer that

$$N_0(h) = \frac{1}{2\pi i}\int_{\partial A} \frac{f'(z)}{f(z) - w} dz + N_\infty(f) \ ;$$

the function $\varphi = \varphi(w)$ defined before is continuous on the open annulus $\{w \in \mathbb{C};\ |w| > M\}$ and, since it assumes integer values and $\lim_{w\to\infty}\varphi(w) = 0$, we deduce that $\varphi(w) \equiv 0$; thus $N_0(h) = N_\infty(f)$, namely the second half of the assertion is achieved. □

Similarly, we obtain the

Theorem 4.9.5. (Rouché Theorem). *Let $A \subset \mathbb{C}$ be a simple region and let $f, g \in H(A) \cap C^0(\overline{A})$; if*

$$|g(z)| < |f(z)| \qquad \forall z \in \partial A \ ,$$

then $N_0(f + g) = N_0(f)$, where N_0 is defined according to the previous theorems.

Proof. Exercise for the reader. □

As an application of this theorem, we purpose to determine the number of solutions of the equation $e^z + 4z^3 = 0$ in the unit disk. We set $f(z) = 4z^3$ and $g(z) = e^z$; then, for every $|z| = 1$ it holds $|g(z)| \le e < 4 = |f(z)|$: since $f(z)$ vanishes only at $z = 0$ (zero of order 3), the equation under consideration has three solutions in the unit disk.

Some exercises about this section can be found in [9, Paragrafo 5.3] and in [47, Capitolo 6].

4.10 Jordan Lemmas

In this section we see some lemmas that will allow the computation of line integrals by means of suitable passages to the limit: their main applications are in the computation of Fourier transforms and inverse Laplace transforms.

In the following it will be useful to consider portions of circumferences that will be denoted by

$$C_R(\vartheta_1, \vartheta_2) = \{z \in \mathbb{C};\ z = Re^{i\vartheta},\ \vartheta \in [\vartheta_1, \vartheta_2]\} \qquad \text{where } R > 0,\ \vartheta_1, \vartheta_2 \in [0, 2\pi] \ :$$

we immediately note that, if f is continuous on $C_R(\vartheta_1, \vartheta_2)$, in view of (4.6) we infer that

$$\left| \int_{C_R(\vartheta_1,\vartheta_2)} f(z)dz \right| \le (\vartheta_2 - \vartheta_1) R \max_{|z|=R} |f(z)| \ . \qquad (4.11)$$

Moreover, by (4.11) we deduce the

Corollary 4.10.1. *Let $K > 0$ and let f be continuous in the annulus $|z| > K$; if there exist $R_0 > K$, $M > 0$ and $\beta > 1$ such that $|f(z)| \leq \frac{M}{|z|^\beta}$ for every $|z| \geq R_0$, then*

$$\lim_{R\to\infty} \int_{C_R(\vartheta_1,\vartheta_2)} f(z)dz = 0 \qquad \forall \vartheta_1, \vartheta_2 .$$

Let $K > 0$ and let f be continuous in the annulus $0 < |z| < K$; if there exist $\varepsilon_0 < K$, $M > 0$ and $\beta < 1$ such that $|f(z)| \leq \frac{M}{|z|^\beta}$ for every $0 < |z| < \varepsilon_0$, then

$$\lim_{\varepsilon\to 0} \int_{C_\varepsilon(\vartheta_1,\vartheta_2)} f(z)dz = 0 \qquad \forall \vartheta_1, \vartheta_2 .$$

Let us now see a general lemma.

Lemma 4.10.2. *Let $\alpha(t) = \rho(t)e^{it}$ with $t \in [0, \pi]$ be a curve such that $\rho(t) > 0$ for every t; let $\omega > 0$ and let f be a continuous function over the support of α. Then*

$$\Big| \int_\alpha e^{i\omega z} f(z)dz \Big| \leq \frac{\pi}{\omega} \frac{\sup_t |\alpha'(t)|}{\inf_t |\alpha(t)|} \sup_t |f(\alpha(t))| .$$

Proof. Let us set for convenience $M = \sup_t |\alpha'(t)|$, $m = \inf_t |\alpha(t)|$ and $K = \sup_t |f(\alpha(t))|$; taking into account that $\sin t \geq \frac{2}{\pi} t$ for every $t \in [0, \frac{\pi}{2}]$, with some easy steps we obtain

$$\Big| \int_\alpha e^{i\omega z} f(z)dz \Big| = \Big| \int_0^\pi e^{i\omega\alpha(t)} f(\alpha(t))\alpha'(t)dt \Big| \leq MK \int_0^\pi e^{-\omega\rho(t)\sin t} dt \leq$$

$$\leq 2MK \int_0^{\pi/2} e^{-m\omega \sin t} dt \leq 2MK \int_0^{\pi/2} e^{-2m\omega t/\pi} dt \leq \frac{\pi}{\omega} \frac{M}{m} K . \qquad \square$$

In particular, from the previous lemma we deduce the following

Corollary 4.10.3. *Let $\vartheta_1, \vartheta_2 \in [0, 2\pi]$, let $a, K > 0$ and let f be continuous in the annulus $|z| > K$; if we assume that*

$$\lim_{R\to\infty} \sup_{z\in C_R(\vartheta_1,\vartheta_2)} |f(z)| = 0 ,$$

then the following results hold true:
(i) if $\vartheta_1 = 0$ and $\vartheta_2 = \pi$, then

$$\lim_{R\to\infty} \int_{C_R(\vartheta_1,\vartheta_2)} e^{iaz} f(z)dz = 0 ;$$

(ii) if $\vartheta_1 = \pi$ and $\vartheta_2 = 2\pi$, then

$$\lim_{R\to\infty} \int_{C_R(\vartheta_1,\vartheta_2)} e^{-iaz} f(z)dz = 0 ;$$

(iii) if $\vartheta_1 = \frac{\pi}{2}$ *and* $\vartheta_2 = \frac{3\pi}{2}$*, then*

$$\lim_{R\to\infty} \int_{C_R(\vartheta_1,\vartheta_2)} e^{az} f(z)dz = 0 \ ;$$

(iv) if $\vartheta_1 = 0$ *and* $\vartheta_2 = \frac{\pi}{2}$ *or* $\vartheta_1 = \frac{3\pi}{2}$ *and* $\vartheta_2 = 2\pi$*, then*

$$\lim_{R\to\infty} \int_{C_R(\vartheta_1,\vartheta_2)} e^{-az} f(z)dz = 0 \ .$$

Exercise 4.10.4. Let $a, K > 0$ and let f be continuous and bounded on the annulus $|z| > K$; prove that

$$\lim_{R\to\infty} \int_{C_R(0,\pi/2)} e^{iaz^2} f(z)dz = 0 \ . \qquad \square$$

Finally, let us see the behavior of the line integral in a neighborhood of a simple pole; the following result generalizes Theorem 4.8.1, indeed it states that, when passing to the limit, "one pays" just the portion of turn done:

Lemma 4.10.5. *Let* $\vartheta_1, \vartheta_2 \in [0, 2\pi]$*, let* $z_0 \in \mathbb{C}$ *and let* f *be a function having a pole of order 1 at* z_0 *(and hence holomorphic on a holed neighborhood of* z_0*); then*

$$\lim_{r\to 0} \int_{C_r} f(z)dz = i(\vartheta_2 - \vartheta_1)\mathrm{Res}(f, z_0) \ ,$$

where $C_r = \{z \in \mathbb{C};\ z - z_0 = re^{i\vartheta},\ \vartheta \in [\vartheta_1, \vartheta_2]\}$.

Proof. The Laurent expansion in the annulus $0 < |z - z_0| < \bar{r}$ (with $\bar{r}$ small enough) has the form

$$f(z) = \frac{a_{-1}}{z - z_0} + g(z) \ ,$$

with $a_{-1} = \mathrm{Res}(f, z_0)$ and g holomorphic on the disk $|z - z_0| < \bar{r}$; then for $r < \bar{r}$ we have

$$\int_{C_r} f(z)dz = a_{-1} \int_{C_r} \frac{dz}{z - z_0} + \int_{C_r} g(z)dz \ .$$

Now we let $r \to 0$: the second integral tends to 0 because g is bounded in $|z - z_0| < \bar{r}$. Setting $z - z_0 = re^{i\vartheta}$, the first integral turns out to be equal to

$$ia_{-1} \int_{\vartheta_1}^{\vartheta_2} d\vartheta \ ,$$

which is independent of r: the result is achieved. $\square$

4.11 Computation of integrals using the residue theory

In this section we present some examples in which the residue theory can be used for the computation of integrals of a real variable.

Proposition 4.11.1. *Let $R(\cos\vartheta, \sin\vartheta)$ be a bounded rational function with $\vartheta \in [0, 2\pi]$. Then, setting $\bar{R}(z) = \frac{1}{z} R(\frac{z+1/z}{2}, \frac{z-1/z}{2i})$, there holds*

$$\int_0^{2\pi} R(\cos\vartheta, \sin\vartheta) d\vartheta = 2\pi \sum_{|w|<1} \mathrm{Res}(\bar{R}, w) \ .$$

Proof. By the assumptions made on R, the integral considered exists. Setting $z = e^{i\vartheta}$, and observing that $\bar{z} = 1/z$ if $|z| = 1$, we obtain

$$\int_0^{2\pi} R(\cos\vartheta, \sin\vartheta) d\vartheta = \int_{|z|=1} R\Big(\frac{z+\bar{z}}{2}, \frac{z-\bar{z}}{2i}\Big)\frac{dz}{iz} = -i \int_{|z|=1} \bar{R}(z) dz \ ;$$

since the function $z \mapsto \bar{R}(z)$ is continuous for $|z| = 1$, in view of Theorem 4.8.5 it holds

$$-i \int_{|z|=1} \bar{R}(z) dz = -i\Big(2\pi i \sum_{|w|<1} \mathrm{Res}(\bar{R}, w)\Big) = 2\pi \sum_{|w|<1} \mathrm{Res}(\bar{R}, w) \ . \qquad \square$$

Proposition 4.11.2. *Let $P(x)$ and $Q(x)$ be two polynomials (of real variable x) such that $\deg(Q) \geq 2 + \deg(P)$ and that $Q(x) \neq 0$ for every $x \in \mathbb{R}$. Then, by defining the rational function of complex variable $f(z) = \frac{P(z)}{Q(z)}$, there holds*

$$\int_{-\infty}^{+\infty} \frac{P(x)}{Q(x)} dx = 2\pi i \sum_{\mathrm{Im}(w)>0} \mathrm{Res}(f, w) \ .$$

Proof. In view of the assumptions, the improper integral considered exists. For every $R > 0$, we apply Theorem 4.8.5 to f on the curve $\gamma_R = [-R, R] \cup C_R(0, \pi)$ (oriented counterclockwise): if R is large enough (say $R > \bar{R}$), all the poles of f with positive imaginary part are contained in the interior of γ_R; then, by Theorem 4.8.5, we infer

$$\int_{\gamma_R} f(z) dz = 2\pi i \sum_{\mathrm{Im}(w)>0} \mathrm{Res}(f, w) \ .$$

Now we note that

$$\int_{\gamma_R} f(z) dz = \int_{-R}^{R} f(x) dx + \int_{C_R(0,\pi)} f(z) dz \ ;$$

in view of the hypotheses, there exists a constant $M > 0$ such that

$$|f(z)| \leq \frac{M}{R^2} \qquad \text{if } |z| = R > \bar{R} + 1 \ :$$

but then, by (4.11) we obtain

$$\Big| \int_{C_R(0,\pi)} f(z)dz \Big| \leq \pi R \frac{M}{R^2} \to 0 \qquad \text{if } R \to \infty \ .$$

Finally, exploiting the relation

$$\lim_{R \to \infty} \int_{-R}^{R} f(x)dx = \int_{-\infty}^{\infty} f(x)dx \, ,$$

we conclude the proof. □

Proposition 4.11.3. *Let $a > 0$ and let $P(x)$ and $Q(x)$ be two polynomials (of real variable x) such that* $\deg(Q) \geq 1 + \deg(P)$ *and that $Q(x) \neq 0$ for every $x \in \mathbb{R}$. Then, by defining the function of complex variable $f(z) = \frac{P(z)}{Q(z)} e^{iaz}$, there holds*

$$\int_{-\infty}^{+\infty} \frac{P(x)}{Q(x)} e^{iax} dx = 2\pi i \sum_{\mathrm{Im}(w)>0} \mathrm{Res}(f, w) \ .$$

Proof. The integrability in the improper sense of the function $f(x) = \frac{P(x)}{Q(x)} e^{iax}$ cannot be deduced directly from a simple upper bound, unlike the previous case. Hence, we will have to prove the existence of the integral by a direct computation: in particular, it will no longer be possible to pass to the limit as $R \to \infty$ over the segments $[-R, R]$, but we will have to consider intervals which are not symmetric with respect to 0. Let us then define the rectangle delimited by the oriented segments

$$\gamma_1 = [x_1, x_2] \qquad \gamma_2 = [x_2, x_2 + iy] \qquad \gamma_3 = [x_2 + iy, x_1 + iy] \qquad \gamma_4 = [x_1 + iy, x_1] \ ,$$

where $x_1 < 0 < x_2$ and $y > 0$. Now we choose $-x_1, x_2, y$ large enough such that all the poles of f (the zeros of Q) with positive imaginary part are contained within the rectangle: setting $\gamma = \cup_i \gamma_i$, in view of Theorem 4.8.5 it holds

$$\int_{\gamma} f(z)dz = 2\pi i \sum_{\mathrm{Im}(w)>0} \mathrm{Res}(f, w)$$

and we are led to estimate the four integrals $\int_{\gamma_i} f$.

The first integral converges to $I = \int_{\mathbb{R}} \frac{P(x)}{Q(x)} e^{iax} dx$:

$$\int_{\gamma_1} f(z)dz = \int_{x_1}^{x_2} \frac{P(x)}{Q(x)} e^{iax} dx \to \int_{-\infty}^{+\infty} \frac{P(x)}{Q(x)} e^{iax} dx \qquad \text{if } -x_1, x_2 \to \infty \ .$$

For the second and fourth ones, we perform the changes of variables $z = x_2 + it$ and $z = x_1 + it$ respectively, and get

$$\int_{\gamma_2} f(z)dz = i\int_0^y \frac{P(x_2+it)}{Q(x_2+it)}e^{ia(x_2+it)}dt$$

$$\int_{\gamma_4} f(z)dz = i\int_y^0 \frac{P(x_1+it)}{Q(x_1+it)}e^{ia(x_1+it)}dt \ ;$$

hence (the case γ_4 is analogous):

$$\Big|\int_{\gamma_2} f(z)dz\Big| \le \frac{M}{x_2}\int_0^y e^{-at}dt \to 0 \qquad \text{if } x_2, y \to \infty \ .$$

Finally,

$$\int_{\gamma_3} f(z)dz = \int_{x_2}^{x_1} \frac{P(x+iy)}{Q(x+iy)}e^{ia(x+iy)}dx \ ,$$

and hence

$$\Big|\int_{\gamma_3} f(z)dz\Big| \le M\int_{x_1}^{x_2} \frac{e^{-ay}}{|x+iy|}dx \le M\frac{e^{-ay}}{y}(x_2 - x_1) \ ;$$

we now let $-x_1, x_2 \to \infty$ "freely" and choose $y = x_2 - x_1$ so that also $y \to \infty$: then the last integral tends to 0. The assertion then follows by the previous four limits.□

Remark 4.11.4. (i) Of course, we are allowed to make y depend on x_1 and x_2, as we have done in the previous proof. Conversely, it is not permitted to make x_1 depend on x_2.

(ii) If P and Q satisfy the assumptions of the previous proposition, we can compute also the integral $\int_{\mathbb{R}} \frac{P(x)}{Q(x)}e^{iax}dx$ with $a < 0$; indeed, it is enough to consider the change of variable $t = -x$.

(iii) Under the same assumptions of the previous proposition, it is also possible to compute the integrals of the type

$$\int_{-\infty}^{+\infty} \frac{P(x)}{Q(x)}\cos(ax)dx \qquad \text{and} \qquad \int_{-\infty}^{+\infty} \frac{P(x)}{Q(x)}\sin(ax)dx \ :$$

indeed, it is enough to consider, respectively, the real part and the imaginary part of $\int_{\mathbb{R}} \frac{P(x)}{Q(x)}e^{iax}dx$. □

We conclude with a significant example showing how the polidromy of the logarithm function can be used (by $\log z$ we will denote the principal determination of the logarithm):

Proposition 4.11.5. *Let $a \in \mathbb{R}$ and let $P(x)$ and $Q(x)$ be two polynomials (of real variable x) such that $\deg(Q) \geq 2+\deg(P)$ and that $Q(x) \neq 0$ for every $x \geq a$. Then, by defining the function of complex variable $f(z) = \frac{P(z+a)}{Q(z+a)} \log z$, there holds*

$$\int_a^{+\infty} \frac{P(x)}{Q(x)} dx = - \sum_{w \in \mathbb{C} \setminus [0,\infty)} \operatorname{Res}(f, w) \ .$$

Proof. We observe that the integral under examination exists and that, setting $t = x - a$, it holds

$$I = \int_a^{+\infty} \frac{P(x)}{Q(x)} dx = \int_0^{+\infty} \frac{P(t+a)}{Q(t+a)} dt \ .$$

Let us consider the curve $\gamma = \gamma(R, r, \vartheta)$, which is the union of the four following curves ($r < R$, $\vartheta < \frac{\pi}{2}$)

$$\gamma_1 = C_R(\vartheta, 2\pi - \vartheta) \quad \text{counterclockwise} \qquad \gamma_2 = C_r(\vartheta, 2\pi - \vartheta) \quad \text{clockwise}$$

$$\gamma_3 = [re^{i\vartheta}, Re^{i\vartheta}] \qquad \gamma_4 = [Re^{-i\vartheta}, re^{-i\vartheta}] \ ;$$

if ϑ and r are small enough and R is sufficiently large, all the poles of the function f defined in the statement are contained in the interior of γ (note that, except at finitely many poles, f is holomorphic on the open set $\mathbb{C} \setminus [0, \infty)$). Then, in view of Theorem 4.8.5, we have

$$\int_\gamma f(z) dz = 2\pi i \sum_{w \in \mathbb{C} \setminus [0,\infty)} \operatorname{Res}(f, w) \ .$$

Setting $z = Re^{i\varphi}$ we obtain

$$\int_{\gamma_1} f(z) dz = \int_\vartheta^{2\pi - \vartheta} \frac{P(a + Re^{i\varphi})}{Q(a + Re^{i\varphi})} \log(Re^{i\varphi}) i Re^{i\varphi} d\varphi \ ;$$

by the assumptions made on P and Q, if R is large enough, it turns out that

$$\left| \frac{P(a + Re^{i\varphi})}{Q(a + Re^{i\varphi})} \right| \leq \frac{M}{R^2}$$

and hence

$$\left| \int_{\gamma_1} f(z) dz \right| \leq \frac{2\pi M}{R^2} (\log R + 2\pi) R \to 0 \qquad \text{as } R \to \infty \ .$$

Similarly, setting $z = re^{i\varphi}$, we get

$$\left| \int_{\gamma_2} f(z) dz \right| \leq Cr(|\log r| + 2\pi) \to 0 \qquad \text{as } r \to 0 ,$$

where we have used the boundedness of $\frac{P(z)}{Q(z)}$ as $|z| < r$ is sufficiently small.

Finally, exploiting the polydromy of the logarithm, we obtain

$$\int_{\gamma_3} f(z)dz \to \int_0^\infty \frac{P(t+a)}{Q(t+a)} \log t \, dt \qquad \text{as } R \to \infty \,, \quad r, \vartheta \to 0$$

$$\int_{\gamma_4} f(z)dz \to -\int_0^\infty \frac{P(t+a)}{Q(t+a)} (\log t + 2\pi i) \, dt \qquad \text{as } R \to \infty \,, \quad r, \vartheta \to 0 \,.$$

Eventually, we have proved that (by lim we mean the limit as $R \to \infty$ and $r, \vartheta \to 0$)

$$I = \int_0^{+\infty} \frac{P(t+a)}{Q(t+a)} dt = -\frac{1}{2\pi i} \Big(\lim \int_{\gamma_3} f(z)dz + \lim \int_{\gamma_4} f(z)dz \Big) =$$

$$= -\frac{1}{2\pi i} \lim \int_\gamma f(z)dz = - \sum_{w \in \mathbb{C} \setminus [0,\infty)} \operatorname{Res}(f, w) \,,$$

that is, the assertion. □

Exercise 4.11.6. Compute the integrals

(a) $\displaystyle\int_0^{2\pi} \frac{d\vartheta}{a + b \sin \vartheta} \quad (a > b > 0)$ (b) $\displaystyle\int_{-\infty}^{+\infty} \frac{dx}{x^6 + 64}$

(c) $\displaystyle\int_{-\infty}^{+\infty} \frac{\cos(mx)}{x^2+1} dx \quad (m \in \mathbb{R})$ (d) $\displaystyle\int_1^{+\infty} \frac{x^2 - 2}{x^2(x^3+1)} dx.$

Solution: (a) $\frac{2\pi}{\sqrt{a^2 - b^2}}$ (b) $\frac{\pi}{48}$ (c) πe^{-m} (d) $-2 + \frac{1}{3} \log 2 + \frac{\sqrt{3}}{3} \pi$. □

We remark that the cases that we have examined in this section are far away from being satisfactory, since we were just able to compute some very particular integrals; nevertheless, they represent examples rather relevant, since they outline a general strategy: one has to choose suitably a function of complex variable, find a sequence of closed paths on which apply the residue Theorem and verify that with suitable passages to the limit one gets the integral under consideration. Actually, we have just given a preview of the enormous potentiality of the residue theory for the computation of integrals of real variable.

Some exercises about this section can be found in [31], see also [9, Capitoli 5 and 7], [47, Capitolo VII], [70].

Chapter 5

Integral transforms

In this chapter we introduce the Fourier and Laplace transforms. For several years now, these integral transforms proved to be a fundamental tool in many fields of Engineering, Physics, Mathematics and, more generally, scientific research. We will deal with the most theoretical aspects and we will describe some applications to ordinary and partial differential equations. The Fourier transform also plays a crucial role in the theory of Sobolev spaces, see Chapter 6.

5.1 Fourier transform

In this section we will deal with complex functional spaces and from now on we will omit to specify it.

Let us consider a function $u \in L^1(\mathbb{R}^n)$; since for every $x, \xi \in \mathbb{R}^n$ it holds $|e^{-i\xi\cdot x}u(x)| = |u(x)|$, it makes sense the following

Definition 5.1.1. *Let $u \in L^1(\mathbb{R}^n)$; we call* **Fourier transform** *of u the function $\widehat{u} : \mathbb{R}^n \to \mathbb{C}$ defined by*

$$\widehat{u}(\xi) = \mathcal{F}(u(x), \xi) = \int_{\mathbb{R}^n} e^{-i\xi\cdot x}u(x)dx \ .$$

Remark 5.1.2. In some texts the definition of $\widehat{u}$ is slightly different: in the integral there may be $e^{i\xi\cdot x}u(x)$ or $e^{-2\pi i\xi\cdot x}u(x)$, see [3], [70]. However, the theory remains unchanged even if some constants are modified. □

We define the following space:

$$C_0^0(\mathbb{R}^n) = \left\{ u \in C^0(\mathbb{R}^n);\ \lim_{|x|\to\infty} u(x) = 0 \right\} \ ;$$

clearly $C_0^0(\mathbb{R}^n) \subset L^\infty(\mathbb{R}^n)$, but *it is not true* that $C^0(\mathbb{R}^n) \subset L^\infty(\mathbb{R}^n)$, because the domain is not compact.

Exercise 5.1.3. Prove that $C^0_0(\mathbb{R}^n)$ is a closed subspace of $L^\infty(\mathbb{R}^n)$. □

We begin by showing the following

Lemma 5.1.4. *Let $f \in L^1(\mathbb{R}^n)$ and for every $y \in \mathbb{R}^n$ set $f_y(x) = f(x+y)$; then* $\lim_{|y|\to 0} \|f_y - f\|_{L^1} = 0$.

Proof. Since $\mathcal{D}(\mathbb{R}^n)$ is dense in $L^1(\mathbb{R}^n)$ (see Proposition 2.4.18), for every $\varepsilon > 0$ there exists $g \in \mathcal{D}(\mathbb{R}^n)$ such that $\|f - g\|_{L^1} < \varepsilon$. Let $K_\varepsilon = \operatorname{supp} g$. In view of the uniform continuity of g, there exists $\delta > 0$ such that

$$x_1, x_2 \in \mathbb{R}^n,\ |x_1 - x_2| < \delta \Longrightarrow |g(x_1) - g(x_2)| < \frac{\varepsilon}{2|K_\varepsilon|} \ :$$

therefore, if $|y| < \delta$ we obtain

$$\int_{\mathbb{R}^n} |g(x+y) - g(x)|dx = \int_{(K_\varepsilon - y)\setminus K_\varepsilon} |g(x+y) - g(x)|dx + \int_{K_\varepsilon} |g(x+y) - g(x)|dx$$
$$\leq |K_\varepsilon|\frac{\varepsilon}{2|K_\varepsilon|} + |K_\varepsilon|\frac{\varepsilon}{2|K_\varepsilon|}$$

being $|(K_\varepsilon - y) \setminus K_\varepsilon| \leq |K_\varepsilon - y| = |K_\varepsilon|$; hence $\|g_y - g\|_{L^1} < \varepsilon$. But then

$$\|f_y - f\|_{L^1} \leq \|f_y - g_y\|_{L^1} + \|g_y - g\|_{L^1} + \|g - f\|_{L^1} < 3\varepsilon$$

where we have used the translation invariance of the $L^1(\mathbb{R}^n)$ norm. In conclusion, we have proved that for every $\varepsilon > 0$ there exists $\delta > 0$ such that if $|y| < \delta$ then $\|f_y - f\|_{L^1} < 3\varepsilon$, namely the desired result. □

We can now prove the

Theorem 5.1.5. (Riemann-Lebesgue Theorem). *Let $u \in L^1(\mathbb{R}^n)$; then $\widehat{u} \in C^0_0(\mathbb{R}^n)$ and $\|\widehat{u}\|_{L^\infty} \leq \|u\|_{L^1}$.*

Proof. Consider $\xi \in \mathbb{R}^n$ and a sequence $\{\xi_m\} \subset \mathbb{R}^n$ such that $\xi_m \to \xi$: then $e^{-i\xi_m\cdot x}u(x) \to e^{-i\xi\cdot x}u(x)$ for a.e. $x \in \mathbb{R}^n$; moreover, $|e^{-i\xi_m\cdot x}u(x)| = |u(x)|$ and $|u| \in L^1(\mathbb{R}^n)$. Thus we may apply Theorem 2.3.7 and we infer $\widehat{u}(\xi_m) \to \widehat{u}(\xi)$, which proves the continuity of $\widehat{u}$. We also remark that

$$\forall \xi \in \mathbb{R}^n \qquad |\widehat{u}(\xi)| \leq \int_{\mathbb{R}^n} |u(x)|dx = \|u\|_{L^1} \ ,$$

which proves the desired inequality between the norms.

Since $e^{i\pi} = -1$, making a change of variables we get

$$\widehat{u}(\xi) = -\int_{\mathbb{R}^n} \exp\left[-i\xi \cdot \left(x - \frac{\pi}{|\xi|^2}\xi\right)\right] u(x)dx = -\int_{\mathbb{R}^n} e^{-i\xi\cdot x} u\left(x + \frac{\pi}{|\xi|^2}\xi\right) dx \ ;$$

thus we can write

$$2\widehat{u}(\xi) = \int_{\mathbb{R}^n} \left(u(x) - u\left(x + \frac{\pi}{|\xi|^2}\xi \right) \right) e^{-i\xi\cdot x} dx$$

and hence $2|\widehat{u}(\xi)| \leq \|u - u_{\pi|\xi|^{-2}\xi}\|_{L^1}$: letting $|\xi| \to \infty$, by Lemma 5.1.4 we deduce that $\widehat{u}(\xi) \to 0$, that is $\widehat{u} \in C_0^0(\mathbb{R}^n)$. □

From Theorems 1.3.2 and 5.1.5 it immediately follows that *the linear function* $\mathcal{F} : L^1(\mathbb{R}^n) \to C_0^0(\mathbb{R}^n)$ *is continuous*, provided that the space $C_0^0(\mathbb{R}^n)$ is endowed with the norm induced by $L^\infty(\mathbb{R}^n)$. In other words, there holds

$$u_m \to u \quad \text{in } L^1(\mathbb{R}^n) \implies \widehat{u}_m \to \widehat{u} \quad \text{in } L^\infty(\mathbb{R}^n) .$$

For the computation of the Fourier transforms, the following properties turn out to be useful:

Theorem 5.1.6. *Let* $u \in L^1(\mathbb{R}^n)$, $y \in \mathbb{R}^n$ *and* A *be a real invertible* $n \times n$ *matrix. Then:*
(i) $\mathcal{F}(u(x-y), \xi) = e^{-i\xi\cdot y}\widehat{u}(\xi)$;
(ii) $\mathcal{F}(e^{iy\cdot x}u(x), \xi) = \widehat{u}(\xi - y)$;
(iii) $\mathcal{F}(u(A^{-1}x), \xi) = |\det A| \cdot \widehat{u}(A^T\xi)$, *where* A^T *denotes the transpose matrix of* A;
(iv) $\mathcal{F}(\overline{u(x)}, \xi) = \overline{\widehat{u}(-\xi)}$;
(v) if u *is radially symmetric, namely* $u(x_1) = u(x_2)$ *for every* $x_1, x_2 \in \mathbb{R}^n$ *such that* $|x_1| = |x_2|$, *then also* $\widehat{u}$ *is radially symmetric;*
(vi) if u *is even, also* $\widehat{u}$ *is even;*
(vii) if u *is odd, also* $\widehat{u}$ *is odd;*
(viii) if u *is real valued and even, also* $\widehat{u}$ *is real valued (and even);*
(ix) if u *is real valued and odd, also* $i\widehat{u}$ *is real valued (and odd).*

Proof. (i) can be obtained with the change of variables $z = x - y$. (ii) is an easy computation. (iii) can be obtained with the change of variables $z = A^{-1}x$ and observing that $\xi \cdot Az = A^T\xi \cdot z$. (iv) is a consequence of

$$\mathcal{F}(\overline{u(x)}, \xi) = \int_{\mathbb{R}^n} \overline{e^{i\xi\cdot x}u(x)}dx = \overline{\int_{\mathbb{R}^n} e^{i\xi\cdot x}u(x)dx} .$$

To prove assertion (v), choose arbitrarily $\xi_1, \xi_2 \in \mathbb{R}^n$ such that $|\xi_1| = |\xi_2|$ and consider an orthogonal matrix A such that $\xi_2 = A\xi_1$. Exploiting the property $|\det A| = 1$ and the fact that the symmetry of u implies $u(Az) = u(z)$ for every $z \in \mathbb{R}^n$, making the change of variables $z = A^{-1}x$, we obtain

$$\int_{\mathbb{R}^n} e^{-i\xi_2\cdot x}u(x)dx = \int_{\mathbb{R}^n} e^{-i\xi_1\cdot A^T x}u(x)dx = \int_{\mathbb{R}^n} e^{-i\xi_1\cdot A^{-1}x}u(x)dx$$

$$= \int_{\mathbb{R}^n} e^{-i\xi_1\cdot z}u(Az)|\det A|dz = \int_{\mathbb{R}^n} e^{-i\xi_1\cdot z}u(z)dz.$$

(vi) and (vii) follow from (iii) with $A = -I$. By combining (vi) and (vii) with (iv) we readily deduce (viii) and (ix). □

Let us now illustrate the behavior of the Fourier transform with respect to differentiation:

Theorem 5.1.7. *Let $u \in L^1(\mathbb{R}^n)$, let $y \in \mathbb{R}^n$ be a unit vector and let $v(x) = (x \cdot y)u(x)$; if $v \in L^1(\mathbb{R}^n)$ then the directional derivative $\partial_y \widehat{u}$ exists and is continuous, and it reads $\partial_y \widehat{u}(\xi) = -i\widehat{v}(\xi)$. Moreover, if also $|x|u(x) \in L^1(\mathbb{R}^n)$, then $\widehat{u} \in C^1(\mathbb{R}^n)$.*

Proof. Let us evaluate the difference quotient

$$\frac{\widehat{u}(\xi + ty) - \widehat{u}(\xi)}{t} = \int_{\mathbb{R}^n} e^{-i\xi \cdot x} \frac{e^{-ity \cdot x} - 1}{t} u(x) dx \ ;$$

letting $t \to 0$, for a.e. $x \in \mathbb{R}^n$ the integrand tends to $-i(y \cdot x)e^{-i\xi \cdot x}u(x)$. Moreover, for every $\lambda \in \mathbb{R}$ it holds $|e^{i\lambda} - 1| \le |\lambda|$ and hence the modulus of the integrand can be bounded from above by

$$\frac{1}{|t|}|ty \cdot x||u(x)| = |(y \cdot x)u(x)|$$

which by assumption belongs to $L^1(\mathbb{R}^n)$: the existence of $\partial_y \widehat{u}$ and its expression follow then from Theorem 2.3.7. Such derivative is continuous in view of Theorem 5.1.5.

If $|x|u(x) \in L^1(\mathbb{R}^n)$, it also turns out that $(x \cdot y)u(x) \in L^1(\mathbb{R}^n)$ for every unit vector $y \in \mathbb{R}^n$ and hence all the partial derivatives of $\widehat{u}$ exist and are continuous, in view of what we have just seen. □

The previous result can be somehow reversed:

Theorem 5.1.8. *Let $u \in C^1(\mathbb{R}^n) \cap L^1(\mathbb{R}^n)$ and let $y \in \mathbb{R}^n$ be a unit vector; if $\partial_y u \in L^1(\mathbb{R}^n)$ and $u(x) = o(|x|^{1-n})$ as $|x| \to \infty$, then*

$$\mathcal{F}(\partial_y u(x), \xi) = i(\xi \cdot y)\widehat{u}(\xi) \ .$$

Here $\partial_y u$ denotes the derivative of u in the direction y.

Proof. Let $R > 0$ and let B_R be the ball of radius R centered at the origin. We recall that, thanks to the divergence Theorem, we can affirm that if $\Omega \subset \mathbb{R}^n$ is a bounded domain with boundary of class C^1 (see Definition 3.1.2) then, for every $u, v \in C^1(\overline{\Omega})$ and for every unit vector $y \in \mathbb{R}^n$, there holds

$$\int_\Omega (\partial_y u)v = \int_{\partial\Omega} uv(y \cdot \nu) - \int_\Omega u(\partial_y v)$$

where ν denotes the outer unit normal to the boundary of Ω. Therefore,

$$\mathcal{F}(\partial_y u(x), \xi) = \lim_{R\to+\infty} \int_{B_R} e^{-i\xi\cdot x} \partial_y u(x) dx$$

$$= \lim_{R\to+\infty} \left[\int_{\partial B_R} e^{-i\xi\cdot x} u(x)(y\cdot\nu) dS_x - \int_{B_R} -i(\xi\cdot y) e^{-i\xi\cdot x} u(x) dx \right]$$

$$= i(\xi\cdot y)\widehat{u}(\xi) + \lim_{R\to+\infty} \int_{\partial B_R} e^{-i\xi\cdot x} u(x)(y\cdot\nu) dS_x. \tag{5.1}$$

Using the decay estimate $u(x) = o(|x|^{1-n})$, we obtain

$$\left| \int_{\partial B_R} e^{-i\xi\cdot x} u(x)(y\cdot\nu) dS_x \right| \le n|B_1|R^{n-1} \cdot \sup_{x\in\partial B_R} |u(x)| \to 0$$

as $R \to +\infty$. The latter estimate, together with (5.1), concludes the proof. □

It is not difficult to foresee that, in some cases, the computation of the Fourier transform may require the usage of Complex Analysis; here we limit ourselves to prove the easy but fundamental

Proposition 5.1.9. *If* $u(x) = e^{-|x|^2}$, *then* $\widehat{u}(\xi) = \pi^{n/2} e^{-|\xi|^2/4}$.

Proof. Thanks to Theorem 2.3.6, we infer

$$\widehat{u}(\xi) = \int_{\mathbb{R}^n} e^{-i\xi\cdot x} e^{-|x|^2} dx = \prod_{j=1}^{n} \int_{-\infty}^{+\infty} e^{-i\xi_j x_j} e^{-x_j^2} dx_j \tag{5.2}$$

and hence the statement holds true for every n if it holds true for $n = 1$. Thus we are led to compute

$$\int_{-\infty}^{+\infty} e^{-i\xi x} e^{-x^2} dx = e^{-\xi^2/4} \int_{-\infty}^{+\infty} e^{-\left(x + i\frac{\xi}{2}\right)^2} dx\ .$$

Let us consider the rectangle $\gamma = [-R, R] \cup [R, R + i\frac{\xi}{2}] \cup [R + i\frac{\xi}{2}, -R + i\frac{\xi}{2}] \cup [-R + i\frac{\xi}{2}, -R]$; we recall that we use such notation to represent segments lying in the complex plane and whose endpoints are written into square brackets. Applying Theorem 4.5.2, we obtain

$$\int_\gamma e^{-z^2} dz = 0\ .$$

Moreover, by (4.6), the integrals along the segments $[R, R + i\frac{\xi}{2}]$ and $[-R + i\frac{\xi}{2}, -R]$ converge to 0 as $R \to +\infty$; therefore, with the change of variables $y = x + i\frac{\xi}{2}$, we get

$$\int_{-\infty}^{+\infty} e^{-\left(x+i\frac{\xi}{2}\right)^2} dx = \lim_{R\to+\infty} \int_{-R}^{R} e^{-\left(x+i\frac{\xi}{2}\right)^2} dx = \lim_{R\to+\infty} \int_{-R+i\frac{\xi}{2}}^{R+i\frac{\xi}{2}} e^{-y^2} dy$$

$$= \lim_{R\to+\infty} \int_{-R}^{R} e^{-y^2} dy = \sqrt{\pi} \ ,$$

and the desired result follows from (5.2). □

Exercise 5.1.10. Using the theorems presented in this section and Proposition 5.1.9, determine the Fourier transform of the following functions of real variable:

$$\text{(i) } u(x) = e^{-(x-1)^2} \quad \text{(ii) } u(x) = e^{-4x^2} \quad \text{(iii) } u(x) = xe^{-x^2} \ .$$ □

5.2 Convolutions and inverse Fourier transform

Provided that we consider the right functional space, the Fourier transform is invertible; in order to treat this issue we need to introduce a new tool:

Theorem 5.2.1. *Let $u, v \in L^1(\mathbb{R}^n)$; then for a.e. $x \in \mathbb{R}^n$ there exists the integral*

$$(u * v)(x) = \int_{\mathbb{R}^n} u(x-t)v(t)dt \ . \tag{5.3}$$

Moreover, $u * v \in L^1(\mathbb{R}^n)$ and $\|u * v\|_{L^1} \le \|u\|_{L^1}\|v\|_{L^1}$.

Proof. Let $H(x,t) = u(x-t)v(t)$; in view of the measurability of u and v over $\mathbb{R}^n$, the function H turns out to be measurable over $\mathbb{R}^{2n}$. By Theorem 2.3.6, we have

$$\int_{\mathbb{R}^n}\int_{\mathbb{R}^n} |H(x,t)|dxdt = \int_{\mathbb{R}^n} |v(t)| \int_{\mathbb{R}^n} |u(x-t)|dxdt = \|u\|_{L^1}\|v\|_{L^1} \ ;$$

therefore, by Theorem 2.3.6, $H \in L^1(\mathbb{R}^{2n})$ and the integral in (5.3) exists. By applying again Theorem 2.3.6, we deduce that $(u * v)(x)$ exists for a.e. x and it is integrable; finally, the norm inequality follows from

$$\|u * v\|_{L^1} \le \int_{\mathbb{R}^n}\int_{\mathbb{R}^n} |H(x,t)|dxdt \ .$$ □

Therefore, given two functions $u, v \in L^1(\mathbb{R}^n)$, (5.3) defines a third function $u * v \in L^1(\mathbb{R}^n)$, called **convolution** of u and v.

Exercise 5.2.2. Prove that the convolution operation $*$ is commutative and associative. □

In the next result we show that, after the application of the Fourier transform, the convolution becomes the usual product of functions.

Theorem 5.2.3. *Let $u, v \in L^1(\mathbb{R})$. Then there holds*

$$\mathcal{F}((u * v)(x), \xi) = \mathcal{F}(u(x), \xi) \cdot \mathcal{F}(v(x), \xi) \, .$$

Proof. We leave to the reader the proof of this result as an exercise. We suggest to exploit Theorem 2.3.6. □

Before stating the inversion Theorem, we introduce two lemmas:

Lemma 5.2.4. *Let $v \in L^1(\mathbb{R}^n, \mathbb{R})$ satisfy $v(x) \geq 0$ a.e. and $\|v\|_{L^1} = 1$; for every $\varepsilon > 0$ set $v_\varepsilon(x) = \varepsilon^{-n} v(x/\varepsilon)$. Then, for every $u \in L^1(\mathbb{R}^n)$ there holds*

$$\lim_{\varepsilon \to 0} \|(v_\varepsilon * u) - u\|_{L^1} = 0 \ .$$

Proof. Letting $y = x - \varepsilon z$, for a.e. x we obtain

$$(v_\varepsilon * u)(x) = \int_{\mathbb{R}^n} u(y) v_\varepsilon(x-y)dy$$

$$= \varepsilon^{-n} \int_{\mathbb{R}^n} u(y) v\left(\frac{x-y}{\varepsilon}\right) dy = \int_{\mathbb{R}^n} u(x - \varepsilon z) v(z) dz \ ;$$

thus we are led to estimate the difference (pointwise a.e.)

$$\left| u(x) - \int_{\mathbb{R}^n} u(x-\varepsilon z) v(z) dz \right|$$

$$= \left| \int_{\mathbb{R}^n} [u(x-\varepsilon z) - u(x)] v(z) dz \right| \leq \int_{\mathbb{R}^n} |u(x-\varepsilon z) - u(x)| v(z) dz \ .$$

Let us now pass to the $L^1(\mathbb{R}^n)$ norm: in view of the pointwise inequality that we have just proved, we infer that

$$\|(v_\varepsilon * u) - u\|_{L^1} \leq \int_{\mathbb{R}^n} \int_{\mathbb{R}^n} |u(x-\varepsilon z) - u(x)| v(z) dz dx$$

$$= \int_{\mathbb{R}^n} v(z) \int_{\mathbb{R}^n} |u(x-\varepsilon z) - u(x)| dx dz \,, \tag{5.4}$$

where the equality follows from Theorem 2.3.6. If we set

$$\psi_\varepsilon(z) = v(z) \int_{\mathbb{R}^n} |u(x-\varepsilon z) - u(x)| dx = v(z) \|u_{-\varepsilon z} - u\|_{L^1} \ ,$$

In view of Lemma 5.1.4 it turns out that, as $\varepsilon \to 0$, $\psi_\varepsilon(z) \to 0$ for a.e. z; moreover, $|\psi_\varepsilon(z)| \leq 2\|u\|_{L^1} v(z)$ for a.e. z: thus, applying Theorem 2.3.7 to (5.4), we obtain the desired result. □

Lemma 5.2.5. *Let $v(x) = \pi^{-n/2} e^{-|x|^2}$ and for every $\varepsilon > 0$ let $v_\varepsilon(x) = \varepsilon^{-n} v(x/\varepsilon)$. Let $u \in L^1(\mathbb{R}^n)$ satisfy $\widehat{u} \in L^1(\mathbb{R}^n)$, then*

$$\lim_{\varepsilon \to 0} (v_\varepsilon * u)(x) = \frac{1}{(2\pi)^n} \int_{\mathbb{R}^n} e^{i\xi \cdot x} \widehat{u}(\xi) d\xi \qquad \text{for a.e. } x \in \mathbb{R}^n \ .$$

Proof. By Theorem 5.1.6 (iii) and Proposition 5.1.9, we infer

$$\widehat{v}_\varepsilon(\xi) = \varepsilon^{-n}\varepsilon^n\widehat{v}(\varepsilon\xi) = e^{-\varepsilon^2|\xi|^2/4} \ :$$

therefore $\widehat{v}_\varepsilon(\xi) \to 1$ in pointwise sense as $\varepsilon \to 0$; moreover, for every $x \in \mathbb{R}^n$ and for every $\varepsilon > 0$ we have

$$|e^{i\xi\cdot x}\widehat{v}_\varepsilon(\xi)\widehat{u}(\xi)| \le |\widehat{u}(\xi)| \qquad \text{for a.e. } \xi \in \mathbb{R}^n \ .$$

We leave to the reader the check that

$$v_\varepsilon(x) = \frac{1}{(2\pi)^n}\int_{\mathbb{R}^n} e^{i\xi\cdot x}\widehat{v}_\varepsilon(\xi)d\xi \qquad \forall x \in \mathbb{R}^n \ . \tag{5.5}$$

Let us fix $x \in \mathbb{R}^n$: applying Theorem 2.3.7 we obtain

$$\int_{\mathbb{R}^n} e^{i\xi\cdot x}\widehat{u}(\xi)d\xi = \lim_{\varepsilon\to 0}\int_{\mathbb{R}^n} e^{i\xi\cdot x}\widehat{v}_\varepsilon(\xi)\widehat{u}(\xi)d\xi = \lim_{\varepsilon\to 0}\int_{\mathbb{R}^n} e^{i\xi\cdot x}\widehat{v}_\varepsilon(\xi)\int_{\mathbb{R}^n} e^{-i\xi\cdot y}u(y)dy\, d\xi$$

which, in view of Theorem 2.3.6 and (5.5), reads

$$\lim_{\varepsilon\to 0}\int_{\mathbb{R}^n} u(y)\int_{\mathbb{R}^n} e^{i\xi\cdot(x-y)}\widehat{v}_\varepsilon(\xi)d\xi\, dy = (2\pi)^n \lim_{\varepsilon\to 0}\int_{\mathbb{R}^n} u(y)v_\varepsilon(x-y)dy$$

and the proof is achieved. □

We are now in position to prove the inversion Theorem:

Theorem 5.2.6. *Let $u \in L^1(\mathbb{R}^n)$ satisfy $\widehat{u} \in L^1(\mathbb{R}^n)$, then $u \in C^0_0(\mathbb{R}^n)$ and*

$$u(x) = \frac{1}{(2\pi)^n}\int_{\mathbb{R}^n} e^{i\xi\cdot x}\widehat{u}(\xi)d\xi \qquad \textit{for every } x \in \mathbb{R}^n \ .$$

Proof. In view of Proposition 5.1.9, the function $v(x) = \pi^{-n/2}e^{-|x|^2}$ introduced in Lemma 5.2.5 satisfies the assumptions of Lemma 5.2.4; thus

$$\lim_{\varepsilon\to 0}\|(v_\varepsilon * u) - u\|_{L^1} = 0$$

and, up to subsequences, $(v_\varepsilon * u)(x) \to u(x)$ for a.e. x (see Theorem 2.4.11). Therefore, in view of the uniqueness of the limit and by Lemma 5.2.5, we infer that

$$u(x) = \frac{1}{(2\pi)^n}\int_{\mathbb{R}^n} e^{i\xi\cdot x}\widehat{u}(\xi)d\xi \qquad \text{for a.e. } x \in \mathbb{R}^n \ .$$

Finally, arguing as in Theorem 5.1.5, we deduce that $u \in C^0_0(\mathbb{R}^n)$ and hence that the previous equality actually holds true for every $x \in \mathbb{R}^n$. □

Let us introduce the space

$$V = \{u \in L^1(\mathbb{R}^n);\ \widehat{u} \in L^1(\mathbb{R}^n)\}. \tag{5.6}$$

Theorem 5.2.6 states that the function $\mathcal{F} : V \to V$ is invertible and that

$$\mathcal{F}^{-1}(u(\xi), x) = \frac{1}{(2\pi)^n} \int_{\mathbb{R}^n} e^{i\xi \cdot x} u(\xi) d\xi \qquad \forall u \in V.$$

This last formula clearly shows that

$$\mathcal{F}(u(x), \xi) = (2\pi)^n \mathcal{F}^{-1}(u(x), -\xi). \tag{5.7}$$

By Theorem 5.2.3 and by (5.7), we deduce the validity of the following equality

$$\mathcal{F}^{-1}((\widehat{u} * \widehat{v})(\xi), x) = (2\pi)^n \mathcal{F}^{-1}(\widehat{u}(\xi), x) \cdot \mathcal{F}^{-1}(\widehat{v}(\xi), x)\,.$$

5.3 Fourier transform in $L^2(\mathbb{R}^n)$ and in $\mathcal{S}'(\mathbb{R}^n)$

Another space in which $\mathcal{F}$ establishes a one-to-one correspondence is $L^2(\mathbb{R}^n)$:

Theorem 5.3.1. (Plancherel Theorem). *Let $u \in L^1(\mathbb{R}^n) \cap L^2(\mathbb{R}^n)$. Then $\widehat{u} \in L^2(\mathbb{R}^n)$ and $\|\widehat{u}\|_{L^2} = (2\pi)^{n/2} \|u\|_{L^2}$.*

Proof. If we prove the theorem under the assumption $u \in \mathcal{D}(\mathbb{R}^n)$, it will be true also for $u \in L^1(\mathbb{R}^n) \cap L^2(\mathbb{R}^n)$, due to the usual density argument (see Theorem 2.4.18). Let us then consider $u \in \mathcal{D}(\mathbb{R}^n)$. In view of Theorems 5.1.5 and 5.1.8, we infer that $P(\xi)\widehat{u}(\xi) \in L^\infty(\mathbb{R}^n)$ for every polynomial P, namely for every $P(\xi) = \sum_\beta c_\beta \xi^\beta$ with $\beta = (\beta_1, ..., \beta_n) \in \mathbb{N}^n$ a multi-index, $\xi^\beta := \xi_1^{\beta_1} \cdot ... \cdot \xi_n^{\beta_n}$ and $c_\beta \in \mathbb{C}$. Indeed, setting $P(D) := \sum_\beta i^{-|\beta|} c_\beta D^\beta$ the differential operator associated to the polynomial P, there hold $P(D)u \in L^1(\mathbb{R}^n)$ and

$$\mathcal{F}(P(D)u) = \sum_\beta i^{-|\beta|} c_\beta \mathcal{F}(D^\beta u) = \sum_\beta (c_\beta \xi^\beta \widehat{u}(\xi)) = P(\xi)\widehat{u}(\xi).$$

In particular, for every $N \in \mathbb{N}$, $|\xi|^N \widehat{u}(\xi) \in L^\infty(\mathbb{R}^n)$ and hence $\widehat{u} \in L^1(\mathbb{R}^n)$, being $\widehat{u} \in C^0(\mathbb{R}^n)$.

By Theorems 5.2.6 and 2.3.6, we conclude that

$$\|u\|_{L^2}^2 = \int_{\mathbb{R}^n} u(x)\overline{u(x)}dx = \frac{1}{(2\pi)^n} \int_{\mathbb{R}^n} \overline{u(x)} \int_{\mathbb{R}^n} e^{i\xi \cdot x} \widehat{u}(\xi) d\xi\ dx$$

$$= \frac{1}{(2\pi)^n} \int_{\mathbb{R}^n} \widehat{u}(\xi) \int_{\mathbb{R}^n} e^{i\xi \cdot x}\overline{u(x)} dx\ d\xi = \frac{1}{(2\pi)^n} \int_{\mathbb{R}^n} \widehat{u}(\xi)\overline{\widehat{u}(\xi)} d\xi = \frac{1}{(2\pi)^n} \|\widehat{u}\|_{L^2}^2\,. \quad \square$$

Thanks to Theorem 5.3.1, it is possible to extend the definition of Fourier transform to the whole space $L^2(\mathbb{R}^n)$. Indeed, given $u \in L^2(\mathbb{R}^n)$, by density we can construct a sequence $\{u_k\} \subset L^1(\mathbb{R}^n) \cap L^2(\mathbb{R}^n)$ such that $u_k \to u$ in $L^2(\mathbb{R}^n)$. Theorem 5.3.1 implies that

$$\|\widehat{u}_{k_1} - \widehat{u}_{k_2}\|_{L^2} = (2\pi)^{n/2}\|u_{k_1} - u_{k_2}\|_{L^2}$$

and since $\{u_k\}$ is a converging sequence in $L^2(\mathbb{R}^n)$, we deduce that $\{\widehat{u}_k\}$ is a Cauchy sequence in $L^2(\mathbb{R}^n)$. Thanks to the completeness of $L^2(\mathbb{R}^n)$, there exists $g \in L^2(\mathbb{R}^n)$ such that $\widehat{u}_k \to g$ in $L^2(\mathbb{R}^n)$. Thus we set

$$\mathcal{F}u := g.$$

We leave to the reader as an exercise the proof that g does not depend on the choice of the approximating sequence $\{u_k\}$ and that

$$\|\mathcal{F}u\|_{L^2} = (2\pi)^{n/2}\|u\|_{L^2} \qquad \forall u \in L^2(\mathbb{R}^n). \tag{5.8}$$

We also observe that, applying the transform $\mathcal{F}$ twice, by (5.7) we deduce that

$$\mathcal{F}^2(u(x), \xi) = (2\pi)^n u(-\xi) \qquad \forall u \in L^2(\mathbb{R}^n). \tag{5.9}$$

Hence the function

$$\mathcal{F} : L^2(\mathbb{R}^n) \to L^2(\mathbb{R}^n) \tag{5.10}$$

is bijective, since every $u \in L^2(\mathbb{R}^n)$ is the image of $v = \mathcal{F}^3((2\pi)^{-2n}u)$: indeed by (5.9) we derive

$$\mathcal{F}(v(x), \xi) = (2\pi)^{-2n}\mathcal{F}^4(u(x), \xi) = u(\xi). \tag{5.11}$$

The definition of Fourier transform can be extended to a framework more general than the one considered so far. The advantages of such a generalization will become clear in the next section, in which we will present the applications of the Fourier transform to differential equations, and in Chapter 6, devoted to Sobolev spaces.

Let us begin with some comments about the Fourier transform of functions rapidly vanishing at infinity (see Definition 2.6.13). First of all, we remark that $\mathcal{S}(\mathbb{R}^n) \subset L^1(\mathbb{R}^n)$ and hence the Fourier transform is well defined in $\mathcal{S}(\mathbb{R}^n)$. Moreover, there holds the following

Theorem 5.3.2. *The Fourier transform $\mathcal{F}$ is a linear function from $\mathcal{S}(\mathbb{R}^n)$ into itself. Moreover $\mathcal{F} : \mathcal{S}(\mathbb{R}^n) \to \mathcal{S}(\mathbb{R}^n)$ is bijective and continuous with respect to the convergence in $\mathcal{S}(\mathbb{R}^n)$.*

Proof. Let $\varphi \in \mathcal{S}(\mathbb{R}^n)$. We have to show that, for every non negative integer N and for every multi-index α, it holds $|\xi|^N D^\alpha \widehat{\varphi}(\xi) \in L^\infty(\mathbb{R}^n)$. It suffices to prove that $P(\xi)D^\alpha\widehat{\varphi}(\xi) \in L^\infty(\mathbb{R}^n)$ for every polynomial P, namely for every $P(x) = \sum_\beta c_\beta x^\beta$

with $\beta = (\beta_1, ..., \beta_n) \in \mathbb{N}^n$ multi-index, $x^\beta := x_1^{\beta_1} \cdot ... \cdot x_n^{\beta_n}$ and $c_\beta \in \mathbb{C}$. Setting $g(x) := x^\alpha \varphi(x)$, by Theorem 5.1.7 we deduce that

$$D^\alpha \widehat{\varphi} = (-i)^{|\alpha|} \widehat{g}. \tag{5.12}$$

On the other hand, once defined the differential operator $P(D) := \sum_\beta i^{-|\beta|} c_\beta D^\beta$ associated to the polynomial P, by Theorem 5.1.8 we infer

$$\mathcal{F}(P(D)g) = \sum_\beta i^{-|\beta|} c_\beta \mathcal{F}(D^\beta g) = \sum_\beta (c_\beta \xi^\beta \widehat{g}(\xi)) = P(\xi)\widehat{g}(\xi). \tag{5.13}$$

Since $\varphi \in \mathcal{S}(\mathbb{R}^n)$, also g and $P(D)g$ belong to $\mathcal{S}(\mathbb{R}^n) \subset L^1(\mathbb{R}^n)$. Hence, by Theorem 5.1.5, we have that $\mathcal{F}(P(D)g) \in L^\infty(\mathbb{R}^n)$ and, thanks to (5.13), also $P(\xi)\widehat{g}(\xi) \in L^\infty(\mathbb{R}^n)$. Finally, by (5.12) we conclude that $P(\xi)D^\alpha\widehat{\varphi}(\xi) \in L^\infty(\mathbb{R}^n)$.

The fact that the Fourier transform is bijective from $\mathcal{S}(\mathbb{R}^n)$ into itself follows from Theorem 5.2.6 and (5.11).

We omit the proof of the continuity, since it would require the usage of a general result of Functional Analysis and this goes beyond our purposes. □

Thanks to Theorem 5.3.2, it makes sense to give the following

Definition 5.3.3. *Let $u \in \mathcal{S}'(\mathbb{R}^n)$. Then it is defined Fourier transform of u, and is denoted by $\widehat{u}$ or $\mathcal{F}u$, the functional*

$$\widehat{u}(\varphi) := u(\widehat{\varphi}) \qquad \forall \varphi \in \mathcal{S}(\mathbb{R}^n). \tag{5.14}$$

Such functional turns out to be a tempered distribution as well, thanks to the continuity of the Fourier transform over the space $\mathcal{S}(\mathbb{R}^n)$.

The previous definition is justified by the next fact, whose proof is left to the reader as an exercise.

Exercise 5.3.4. Let $u, v \in L^1(\mathbb{R}^n)$. Verify that the functions $\widehat{u} \cdot v$ and $u \cdot \widehat{v}$ belong to $L^1(\Omega)$ and prove that

$$\int_{\mathbb{R}^n} \widehat{u} v = \int_{\mathbb{R}^n} u \widehat{v}. \tag{5.15}$$

Hint: for the check see Theorem 5.1.5 and for the proof of (5.15) use Theorem 2.3.6. □

A consequence of Theorem 5.3.2 is the following

Theorem 5.3.5. *The Fourier transform $\mathcal{F} : \mathcal{S}'(\mathbb{R}^n) \to \mathcal{S}'(\mathbb{R}^n)$ is a linear bijective function, continuous with respect to the convergence in $\mathcal{S}'(\mathbb{R}^n)$.*

Proof. Let us first prove the surjectivity. Given $v \in \mathcal{S}'(\mathbb{R}^n)$, we define the tempered distribution

$$\langle u, \varphi \rangle := \langle v, \mathcal{F}^{-1}\varphi \rangle \qquad \forall \varphi \in \mathcal{S}(\mathbb{R}^n).$$

The functional u is actually a tempered distribution, due to the continuity of the Fourier transform in $\mathcal{S}(\mathbb{R}^n)$ proved in Theorem 5.3.2. Hence it holds

$$\langle \widehat{u}, \varphi \rangle = \langle u, \widehat{\varphi} \rangle = \langle v, \mathcal{F}^{-1}\widehat{\varphi} \rangle = \langle v, \varphi \rangle \qquad \forall \varphi \in \mathcal{S}(\mathbb{R}^n)$$

which implies $\widehat{u} = v$ and then the surjectivity. Let us pass to the injectivity: let $u_1, u_2 \in \mathcal{S}'(\mathbb{R}^n)$ be such that $\widehat{u}_1 = \widehat{u}_2$. In view of Theorem 5.3.2, for every $\varphi \in \mathcal{S}(\mathbb{R}^n)$ we can uniquely determine a function $\psi \in \mathcal{S}(\mathbb{R}^n)$ such that $\widehat{\psi} = \varphi$, so that

$$\langle u_1, \varphi \rangle = \langle u_1, \widehat{\psi} \rangle = \langle \widehat{u}_1, \psi \rangle = \langle \widehat{u}_2, \psi \rangle = \langle u_2, \widehat{\psi} \rangle = \langle u_2, \varphi \rangle \qquad \forall \varphi \in \mathcal{S}(\mathbb{R}^n).$$

This proves the injectivity.

The continuity of $\mathcal{F}$ in $\mathcal{S}'(\mathbb{R}^n)$ immediately follows from Definition 2.6.14 (convergence in $\mathcal{S}'(\mathbb{R}^n)$) and from the continuity of $\mathcal{F}$ in $\mathcal{S}(\mathbb{R}^n)$. □

We give the following definition:

Definition 5.3.6. *Let $u \in \mathcal{S}'(\mathbb{R}^n)$, $g \in \mathcal{S}(\mathbb{R}^n)$, P be a polynomial and $\alpha = (\alpha_1, ..., \alpha_n) \in \mathbb{N}^n$ be a multi-index. Then, for every $\varphi \in \mathcal{S}(\mathbb{R}^n)$, we set*

$$\langle gu, \varphi \rangle := \langle u, g\varphi \rangle, \quad \langle P \cdot u, \varphi \rangle := \langle u, P \cdot \varphi \rangle, \quad \langle D^\alpha u, \varphi \rangle := (-1)^{|\alpha|} \langle u, D^\alpha \varphi \rangle.$$

It is possible to verify that $g \cdot u$, $P \cdot u$ e $D^\alpha u$ are actually tempered distributions, see for instance [61, Theorem 7.13].

Exercise 5.3.7. Let p.v.$(\frac{1}{x})$ be the tempered distribution defined in Example 2.6.18. Prove that

$$x \cdot \text{p.v.}\left(\frac{1}{x}\right) = 1 \qquad \text{in } \mathcal{S}'(\mathbb{R}).$$ □

There holds the following

Theorem 5.3.8. *Let $u \in \mathcal{S}'(\mathbb{R}^n)$. Then for every $\alpha = (\alpha_1, ..., \alpha_n) \in \mathbb{N}^n$ there hold*

$$D^\alpha \widehat{u}(\xi) = (-i)^{|\alpha|} \mathcal{F}(x^\alpha u(x), \xi) \qquad \text{and} \qquad \mathcal{F}(D^\alpha u(x), \xi) = i^{|\alpha|} \xi^\alpha \widehat{u}(\xi).$$

We recall the notation $x^\alpha = x_1^{\alpha_1} \cdot ... \cdot x_n^{\alpha_n}$ for every $x = (x_1, ..., x_n) \in \mathbb{R}^n$.

Proof. This result follows from Theorems 5.1.7-5.1.8 and from Definition 5.3.6. □

Let us now see some easy but enlightening examples of computation of the Fourier transform of a tempered distribution.

Example 5.3.9. According to Example 2.6.17, we know that the Dirac delta function δ_0 is a tempered distribution. For every test function $\varphi \in \mathcal{S}(\mathbb{R}^n)$ it holds

$$\langle \widehat{\delta_0}, \varphi \rangle = \langle \delta_0, \widehat{\varphi} \rangle = \widehat{\varphi}(0) = \int_{\mathbb{R}^n} \varphi(x) dx = \langle 1, \varphi \rangle.$$

Thus we can affirm that $\widehat{\delta_0} = 1$ in $\mathcal{S}'(\mathbb{R}^n)$. Moreover, we observe that

$$\langle \widehat{1}, \varphi \rangle = \langle 1, \widehat{\varphi} \rangle = \int_{\mathbb{R}^n} \widehat{\varphi}(\xi)\, d\xi = (2\pi)^n \varphi(0).$$

Therefore, $\widehat{1} = (2\pi)^n \delta_0$ in $\mathcal{S}'(\mathbb{R}^n)$. This result could be also deduced with a generalization of Theorem 5.2.6 to the space $\mathcal{S}'(\mathbb{R}^n)$. □

Example 5.3.10. To compute the Fourier transform of p.v.$(\frac{1}{x})$, we exploit Exercise 5.3.7. In view of Theorem 5.3.8 and Example 5.3.10, we know that

$$i \frac{d}{d\xi} \mathcal{F}\left(\text{p.v.} \left(\frac{1}{x} \right), \xi \right) = \mathcal{F}\left(x \cdot \text{p.v.} \left(\frac{1}{x} \right), \xi \right) = \widehat{1} = 2\pi \delta_0 \qquad \text{in } \mathcal{S}'(\mathbb{R}).$$

Therefore

$$\frac{d}{d\xi} \mathcal{F}\left(\text{p.v.} \left(\frac{1}{x} \right), \xi \right) = -2\pi i\, \delta_0 = -\pi i \frac{d}{d\xi} (2H(\xi)) \qquad \text{in } \mathcal{S}'(\mathbb{R}),$$

where H is the Heaviside function. Thus there exists $c \in \mathbb{R}$ such that

$$\mathcal{F}\left(\text{p.v.} \left(\frac{1}{x} \right), \xi \right) = -\pi i\, (2H(\xi) + c) \qquad \text{in } \mathcal{S}'(\mathbb{R}).$$

By Theorem 5.1.6 (vii) (generalized to tempered distributions!) we deduce that $c = -1$. □

In Example 2.6.16 we have shown that the functions of $L^2(\mathbb{R}^n)$ are also tempered distributions. This means that the elements of $L^2(\mathbb{R}^n)$ admit two distinct definitions of Fourier transform. It is natural to ask whether they coincide in some sense or not. We leave the answer to the reader as an exercise:

Exercise 5.3.11. Prove that if $u \in L^2(\mathbb{R}^n)$ then the definitions of Fourier transform given in $L^2(\mathbb{R}^n)$ and in $\mathcal{S}'(\mathbb{R}^n)$ coincide. In other words, prove that if $v \in L^2(\mathbb{R}^n)$ is the Fourier transform of u as an element of $L^2(\mathbb{R}^n)$ and $\widehat{u} \in \mathcal{S}'(\mathbb{R}^n)$ is the Fourier transform as an element of $\mathcal{S}'(\mathbb{R}^n)$, then

$$\widehat{u}(\varphi) = \int_{\mathbb{R}^n} v\varphi \qquad \forall \varphi \in \mathcal{S}(\mathbb{R}^n).$$

□

5.4 Fourier transform and differential equations

5.4.1 Applications to ordinary differential equations

We start considering linear equations with constant coefficients of the form

$$a_k y^{(k)} + a_{k-1} y^{(k-1)} + ... + a_1 y' + a_0 y = f(x)\,, \tag{5.16}$$

with $a_0, a_1, ..., a_k \in \mathbb{R}$ and $f \in C^0(\mathbb{R})$. It is known that such equations can be solved with elementary techniques, however it is worth to see how to provide a solution to (5.16) by using the Fourier transform. For this purpose, let us assume for simplicity that the function f introduced in (5.16) belongs also to $L^1(\mathbb{R})$, so that we may try to find solutions belonging to $L^1(\mathbb{R})$ as well. By applying in a formal way the Fourier transform to both sides of (5.16) and by applying Theorem 5.1.8, we obtain

$$\widehat{y}(\xi) \cdot \sum_{j=0}^{k} a_j (i\xi)^j = \widehat{f}(\xi). \tag{5.17}$$

After the application of the transform, the differential equation (5.16) reduces to a simple algebraic equation which allows to determine explicitly the function $\widehat{y}$. In general, we cannot expect the so found function $\widehat{y}$ to be the Fourier transform of a function of $L^1(\mathbb{R})$. This is not surprising, since it may happen that equation (5.16) does not admit classical solutions in $L^1(\mathbb{R})$, as occurs in the next

Example 5.4.1. Consider the equation

$$y'' + y = e^{-|x|}. \tag{5.18}$$

Arguing as above, we infer

$$(1 - \xi^2)\widehat{y}(\xi) = \frac{2}{1 + \xi^2}$$

and hence

$$\widehat{y}(\xi) = \frac{2}{(1 - \xi^2)(1 + \xi^2)}.$$

Such function does not belong to $L^1(\mathbb{R})$. On the other hand, using the method of the variation of the arbitrary constants, it is clear that the general solution to (5.18) is given by

$$\begin{aligned} y(x) &= A\cos x + B\sin x + \int_0^x \sin(x - t)e^{-|t|}dt \\ &= A\cos x + B\sin x + \frac{1}{2}(\sin|x| - \cos x + e^{-|x|}) \end{aligned}$$

with $A, B \in \mathbb{R}$ arbitrarily chosen. But $y \notin L^1(\mathbb{R})$ for any choice of $A, B \in \mathbb{R}$. □

Once $\widehat{y}$ is determined by (5.17), it remains to understand how derive from it the originary function y. If we proceed in a formal way, by Theorem 5.2.3, we deduce that

$$y = \mathcal{F}^{-1}(\widehat{y}) = \mathcal{F}^{-1}\left(\frac{1}{\sum_{j=0}^{k} a_j (i\xi)^j}\right) * f. \tag{5.19}$$

Therefore, under suitable assumptions, (5.19) gives a solution to (5.16). Let us see a concrete application of this procedure:

Example 5.4.2. Consider the second order equation

$$y'' - a^2 y = f(x), \tag{5.20}$$

with $a > 0$ and $f \in V$, where V is the space defined in (5.6).

We look for solutions in $L^1(\mathbb{R})$. By (5.19), we have

$$y(x) = \mathcal{F}^{-1}\left(-\frac{1}{a^2+\xi^2}\right) * f.$$

Exploiting Theorem 5.2.6, Proposition 4.11.3 and Remark 4.11.4, we obtain

$$K(x) := \mathcal{F}^{-1}\left(-\frac{1}{a^2+\xi^2}, x\right) = -\frac{1}{2\pi}\int_{\mathbb{R}} e^{ix\xi}\frac{1}{a^2+\xi^2}\, d\xi = -\frac{1}{2a}e^{-a|x|}$$

and finally

$$y(x) = (K * f)(x) = -\frac{1}{2a}\int_{\mathbb{R}} e^{-a|x-y|} f(y)\, dy.$$

This is not the unique solution to (5.20), since other solutions are also the functions of the form

$$Ae^{ax} + Be^{-ax} + (K * f)(x),$$

with $A, B \in \mathbb{R}$ arbitrarily chosen. However, the sole solution which belongs to $L^1(\mathbb{R})$ is the one corresponding to $A = B = 0$. □

Example 5.4.3. Let us consider the airfoil in a fluid flow (see the picture below), with only one degree of freedom, the translation in the direction of the x axis.

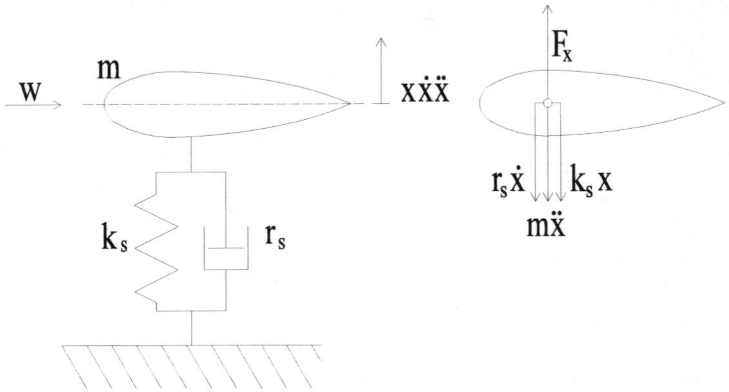

A one degree of freedom airfoil in a fluid flow

The equation that governs the motion of the system under study is the following [17]:

$$m \frac{d^2x}{dt^2} + r_s \frac{dx}{dt} + k_s x = F_x , \qquad (5.21)$$

where m is the mass of the vibrating system, r_s and k_s (constant and positive) are the structural damping and the structural stiffness respectively, and F_x is the component in the vertical direction x of the resultant of the aerodynamic forces acting on the body. For the constraints, both the orthogonal component to F_x and the aerodynamic momentum acting on the system are balanced. The acting forces are the drag force F_r, parallel to the relative velocity v_r of the fluid flow with respect to the airfoil, and the lift force F_p, orthogonal to the former force. There hold the following relations:

$$F_r = \frac{1}{2} \rho C_r S v_r^2 \qquad F_p = \frac{1}{2} \rho C_p S v_r^2 ,$$

where ρ is the fluid density and S is the cross sectional area. The coefficients C_r and C_p depend, for large ranges of values of Reynolds number Re, on the considered airfoil and on the *angle of attack* α between the direction of v_r and a characteristic direction of the body, attached to it. As the next picture shows, if we denote by w the absolute velocity of the fluid flow, to compute the modulus of v_r we use the simple relation $v_r^2 = w^2 + \dot{x}^2$.

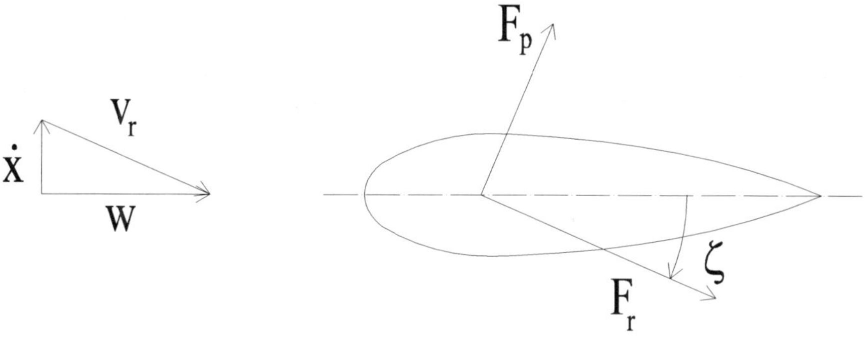

Aerodynamic forces acting on the airfoil

Denoting by ζ the angle between v_r and the horizontal axis, we obtain:

$$\zeta = \arctan\left(\frac{\dot{x}}{w}\right) \approx \frac{\dot{x}}{w}$$

in which the last approximation is due to the small value of $\dot{x}$ compared to w. If we choose as characteristic direction of the airfoil attached to it the same direction (horizontal) of the undisturbed fluid flow, the angle of attack α coincides with ζ. Therefore

$$F_x = -F_r\ \sin(\alpha) + F_p\ \cos(\alpha) = -F_r\ \sin(\zeta) + F_p\ \cos(\zeta)$$

and (5.21) becomes

$$m\,\ddot{x} + r_s\,\dot{x} + k_s\,x = \frac{1}{2}\,\rho\,S\,(w^2 + \dot{x}^2)\,\left[-C_r(\alpha)\ \sin\left(\frac{\dot{x}}{w}\right) + C_p(\alpha)\ \cos\left(\frac{\dot{x}}{w}\right)\right] \tag{5.22}$$

which displays a nonlinearity in the term in the variable $\dot{x}$. To remove such nonlinearity, we linearize F_x in the neighborhood of the equilibrium configuration $\alpha_0 = 0$: in fact, α_0 is a zero of C_p, hence $x_0 = 0$. The linearization gives

$$C_r(\alpha) \approx C_r(0) + C_r'(0)\alpha\ , \qquad C_p(\alpha) \approx C_p(0) + C_p'(0)\alpha.$$

By [17, Figure 5.9, p.445], we derive $C_r'(0) = C_p(0) = 0$ and the linearized form of (5.22) becomes

$$m\,\ddot{x} + \left\{\frac{1}{2}\rho\,S\,w\,\left(C_r(0) - C_p'(0)\right) + r_s\right\}\ \dot{x} + k_s\,x = 0 \tag{5.23}$$

which is a homogeneous ordinary differential equation of order 2 with constant coefficients. Let us assume that the airfoil, in addition to the aerodynamic forces previously described, is subject to a time-dependent external force $f(t) \in V$ parallel to the x axis. Then the equation (5.23) becomes

$$m\,\ddot{x} + \left\{ \frac{1}{2}\,\rho\, S\, w\,\left(C_r(0) - C_p'(0)\right) + r_s \right\}\,\dot{x} + k_s\, x = f(t)\,. \tag{5.24}$$

If we denote by $\widehat{x}$ and by $\widehat{f}$ the Fourier transforms of x and f respectively, by (5.24) we obtain

$$-\xi^2\, m\,\widehat{x} + i\, r\,\xi\,\widehat{x} + k_s\,\widehat{x} = \widehat{f}\,,$$

where

$$r = \left\{ \frac{1}{2}\,\rho\, S\, w\,\left(C_r(0) - C_p'(0)\right) + r_s \right\}.$$

Thus

$$\widehat{x}(\xi) = \frac{\widehat{f}(\xi)}{-\xi^2\, m + i\, r\,\xi + k_s}.$$

Finally,

$$x(t) = \mathcal{F}^{-1}\left(\frac{1}{-\xi^2\, m + i\, r\,\xi + k_s} \right) * f.$$

In view of Theorem 5.2.6 we get

$$K(t) := \mathcal{F}^{-1}\left(\frac{1}{-\xi^2\, m + i\, r\,\xi + k_s}, t \right) = \frac{1}{2\pi}\int_{\mathbb{R}} e^{it\xi}\frac{1}{-\xi^2\, m + i\, r\,\xi + k_s}\, d\xi.$$

In conclusion, we have $x = K * f$. We leave to the reader the last computations (use Proposition 4.11.3 and Remark 4.11.4). □

In some cases, the Fourier transform turns out to be useful also for the solution of linear equations with variable coefficients.

Example 5.4.4. Consider the equation

$$y'' + xy' + y = 0. \tag{5.25}$$

Transforming both terms and exploiting Theorems 5.1.7-5.1.8, we obtain

$$0 = -\xi^2\widehat{y} + \mathcal{F}(xy'(x), \xi) + \widehat{y} = -\xi^2\widehat{y} + i\frac{d}{d\xi}\mathcal{F}(y'(x), \xi) + \widehat{y} = -\xi^2\widehat{y} + i(i\xi\widehat{y}(\xi))' + \widehat{y}$$
$$= -\xi^2\widehat{y} - (\widehat{y} + \xi\widehat{y}') + \widehat{y} = -\xi^2\widehat{y} - \xi\widehat{y}'.$$

Consequently, the function $\widehat{y}$ solves the first order equation $\widehat{y}' + \xi\widehat{y} = 0$, which, after integration, gives

$$\widehat{y}(\xi) = Ae^{-\xi^2/2}, \qquad A \in \mathbb{R}.$$

Finally, by Theorem 5.1.6 and Proposition 5.1.9, we find

$$y(x) = \widetilde{A}e^{-x^2/2},$$

with $\widetilde{A} = A/\sqrt{2\pi}$. The so found solutions are the unique solutions to (5.25) in the space $L^1(\mathbb{R})$. □

5.4.2 Applications to partial differential equations

In this section we examine again some of the partial differential equations which we studied in Chapter 3 and we show that the Fourier transform is a powerful tool for their solution.

Vibrating string equation

Let us assume that $g \in V \cap C^2(\mathbb{R})$ and $h \in V \cap C^1(\mathbb{R})$ (V is the space defined in (5.6)), and let us consider the problem of a vibrating string, with infinite length and zero forcing term:

$$\begin{cases} u_{tt} - c^2 u_{xx} = 0 & (x,t) \in \mathbb{R} \times (0,\infty) \\ u(x,0) = g(x) & x \in \mathbb{R} \\ u_t(x,0) = h(x) & x \in \mathbb{R}\ . \end{cases} \tag{5.26}$$

We proceed formally, by applying to both sides of the equation in (5.26) the Fourier transform with respect to the sole spatial variable x. By this we mean that, given a generic function $v = v(x,t)$, we consider

$$\mathcal{F}(v(x,t),\xi) = \widehat{v}(\xi,t) = \int_{\mathbb{R}} e^{-i\xi x} v(x,t)\ dx.$$

By (5.26) and Theorem 5.1.8, we obtain

$$\begin{aligned} \partial_t^2 \widehat{u}(\xi,t) &= \int_{\mathbb{R}} e^{-i\xi x} \partial_t^2 u(x,t)\ dx = c^2 \int_{\mathbb{R}} e^{-i\xi x} \partial_x^2 u(x,t)\ dx \\ &= c^2 \mathcal{F}(\partial_x^2 u(x,t),\xi) = -c^2 \xi^2 \widehat{u}(\xi,t). \end{aligned}$$

So we arrive to the second order Cauchy problem

$$\begin{cases} \partial_t^2 \widehat{u}(\xi,t) + c^2\xi^2 \widehat{u}(\xi,t) = 0 \\ \widehat{u}(\xi,0) = \widehat{g}(\xi) \\ \partial_t \widehat{u}(\xi,0) = \widehat{h}(\xi) \end{cases}$$

whose solution is

$$\widehat{u}(\xi,t) = \frac{1}{2}\widehat{g}(\xi)(e^{ic\xi t} + e^{-ic\xi t}) + \frac{\widehat{h}(\xi)}{2ic\xi}(e^{ic\xi t} - e^{-ic\xi t}).$$

Transforming back, thanks to Theorem 5.2.6, we get the following expressions of u and g:

$$u(x,t) = \frac{1}{4\pi}\int_{\mathbb{R}} (e^{i(x+ct)\xi} + e^{i(x-ct)\xi})\widehat{g}(\xi)\ d\xi + \frac{1}{4\pi i c}\int_{\mathbb{R}} (e^{i(x+ct)\xi} - e^{i(x-ct)\xi})\frac{\widehat{h}(\xi)}{\xi}\ d\xi, \tag{5.27}$$

$$g(x) = \frac{1}{2\pi} \int_{\mathbb{R}} e^{ix\xi} \widehat{g}(\xi)\, d\xi \qquad \forall x \in \mathbb{R};$$

hence

$$\frac{1}{4\pi} \int_{\mathbb{R}} (e^{i(x+ct)\xi} + e^{i(x-ct)\xi}) \widehat{g}(\xi)\, d\xi = \frac{1}{2}(g(x+ct) + g(x-ct)) \qquad \forall x \in \mathbb{R}. \quad (5.28)$$

On the other hand, we also have

$$h(x) = \frac{1}{2\pi} \int_{\mathbb{R}} e^{ix\xi} \widehat{h}(\xi)\, d\xi \qquad \forall x \in \mathbb{R}$$

and integrating such identity, by Theorem 2.3.6, we infer

$$\frac{1}{2c} \int_{x-ct}^{x+ct} h(y)dy = \frac{1}{4\pi c} \int_{x-ct}^{x+ct} \left(\int_{\mathbb{R}} e^{iy\xi} \widehat{h}(\xi)\, d\xi \right) dy$$

$$= \frac{1}{4\pi c} \int_{\mathbb{R}} \widehat{h}(\xi) \left(\int_{x-ct}^{x+ct} e^{iy\xi} dy \right) d\xi = \frac{1}{4\pi i c} \int_{\mathbb{R}} (e^{i(x+ct)\xi} - e^{i(x-ct)\xi}) \frac{\widehat{h}(\xi)}{\xi}\, d\xi. \quad (5.29)$$

Thus, substituting (5.28) and (5.29) into (5.27), we conclude that

$$u(x,t) = \frac{1}{2}(g(x+ct) + g(x-ct)) + \frac{1}{2c} \int_{x-ct}^{x+ct} h(y)dy$$

which coincides with the D'Alembert formula (3.31).

Wave equation in dimension $n = 2$

Let us assume that $g \in V \cap C^3(\mathbb{R}^2)$ and $h \in V \cap C^2(\mathbb{R}^2)$ (being V defined in (5.6)) and let us consider the wave equation in the plane

$$\begin{cases} u_{tt} - c^2 \Delta u = 0 & (x,t) \in \mathbb{R}^2 \times (0, \infty) \\ u(x,0) = g(x) & x \in \mathbb{R}^2 \\ u_t(x,0) = h(x) & x \in \mathbb{R}^2\ . \end{cases} \quad (5.30)$$

We apply to both terms of the equation in (5.30) the Fourier transform with respect to the sole spatial variable x. Thanks to Theorem 5.1.8, we arrive to the second order Cauchy problem

$$\begin{cases} \partial_t^2 \widehat{u}(\xi,t) + c^2 |\xi|^2 \widehat{u}(\xi,t) = 0 \\ \widehat{u}(\xi,0) = \widehat{g}(\xi) \\ \partial_t \widehat{u}(\xi,0) = \widehat{h}(\xi) \end{cases}$$

whose solution is

$$\widehat{u}(\xi,t) = \widehat{g}(\xi)\cos(c|\xi|t) + \widehat{h}(\xi)\frac{\sin(c|\xi|t)}{c|\xi|} = \widehat{g}(\xi)\frac{\partial}{\partial t}\left(\frac{\sin(c|\xi|t)}{c|\xi|}\right) + \widehat{h}(\xi)\frac{\sin(c|\xi|t)}{c|\xi|}. \tag{5.31}$$

Proceeding formally, in view of Theorems 5.2.6 and 5.2.3, we obtain

$$\begin{aligned} u(x,t) &= \mathcal{F}^{-1}\left(\frac{\partial}{\partial t}\left(\frac{\sin(c|\xi|t)}{c|\xi|}\right)\right) * g(x) + \mathcal{F}^{-1}\left(\frac{\sin(c|\xi|t)}{c|\xi|}, x\right) * h(x) \\ &= \frac{\partial}{\partial t}\left[\mathcal{F}^{-1}\left(\frac{\sin(c|\xi|t)}{c|\xi|}\right) * g(x)\right] + \mathcal{F}^{-1}\left(\frac{\sin(c|\xi|t)}{c|\xi|}, x\right) * h(x) \end{aligned} \tag{5.32}$$

where the inverse transform $\mathcal{F}^{-1}$ is to be meant in the sense of tempered distributions. For every $\varepsilon \geq 0$ we set

$$\Psi_\varepsilon(\xi,t) = e^{-\varepsilon c|\xi|}\frac{\sin(c|\xi|t)}{c|\xi|}$$

and

$$\psi_\varepsilon(x,t) = \mathcal{F}^{-1}(\Psi_\varepsilon(\xi,t),x) = \frac{1}{4\pi^2}\int_{\mathbb{R}^2} e^{ix\cdot\xi-\varepsilon c|\xi|}\frac{\sin(c|\xi|t)}{c|\xi|}\,d\xi\ . \tag{5.33}$$

Since for every fixed $t > 0$ it holds $\Psi_\varepsilon \to \Psi_0$ in $\mathcal{S}'(\mathbb{R}^2)$ as $\varepsilon \to 0^+$, Theorem 5.3.5 implies that

$$\psi_\varepsilon \to \psi_0 = \mathcal{F}^{-1}\left(\frac{\sin(c|\xi|t)}{c|\xi|}, x\right)$$

in $\mathcal{S}'(\mathbb{R}^2)$ as $\varepsilon \to 0^+$. Since the function we want to transform back is radially symmetric, by analogy with assertion (v) in Theorem 5.1.6, we deduce that also its inverse transform is radially symmetric. Rewriting in polar coordinates the integral appearing in (5.33) and assuming that, up to rotations, $x = (r,0) \in \mathbb{R}^2$, we obtain

$$\begin{aligned} \psi_\varepsilon(x,t) &= \frac{1}{8\pi^2 ci}\int_0^{2\pi}\left(\int_0^{\infty}\left(e^{i(r\rho\cos\vartheta+c\rho t)-\varepsilon c\rho} - e^{i(r\rho\cos\vartheta-c\rho t)-\varepsilon c\rho}\right)d\rho\right)d\vartheta \\ &= \frac{1}{8\pi^2 ci}\int_0^{2\pi}\left(-\frac{1}{i(r\cos\vartheta+ct)-\varepsilon c} + \frac{1}{i(r\cos\vartheta-ct)-\varepsilon c}\right)d\vartheta \\ &= \frac{1}{8\pi^2 ci}(I_{1,\varepsilon}(x,t) + I_{2,\varepsilon}(x,t)). \end{aligned} \tag{5.34}$$

Let us now pass to the computation of

$$\lim_{\varepsilon\to 0^+} I_{1,\varepsilon}(x,t).$$

Exploiting the 2π-periodicity of the cosine function, the values of the integrals which lead to (5.34) do not change if we replace the interval of integration $(0, 2\pi)$ by $(-\pi, \pi)$. Making the change of variable $s = \tan\frac{\vartheta}{2}$, we get

$$I_{1,\varepsilon}(x,t) = \int_{-\infty}^{+\infty} \frac{-2\,ds}{(-\varepsilon c + i(ct-r))s^2 - \varepsilon c + i(ct+r)} = \int_{-\infty}^{+\infty} \frac{-2\,ds}{A_\varepsilon s^2 + B_\varepsilon}. \tag{5.35}$$

In the latter equality we have used the relation $\cos\vartheta = (1-s^2)/(1+s^2)$. To compute this integral we will use Proposition 4.11.2. Let $z_{1,\varepsilon}, z_{2,\varepsilon} \in \mathbb{C}$ be the roots of the polynomial $Q_\varepsilon(z) = A_\varepsilon z^2 + B_\varepsilon$, with the convention $\mathrm{Im} z_{1,\varepsilon} > 0$ and $\mathrm{Im} z_{2,\varepsilon} < 0$, and let $f_\varepsilon(z) := \frac{-2}{A_\varepsilon z^2 + B_\varepsilon}$. Thus we may write

$$-2\int_{-\infty}^{+\infty} \frac{ds}{A_\varepsilon s^2 + B_\varepsilon} = 2\pi i\ \mathrm{Res}(f_\varepsilon, z_{1,\varepsilon}). \tag{5.36}$$

Since $z_{1,\varepsilon}$ is a pole of order 1 for f_ε, exploiting the formulas for the computation of residues (see Section 4.8), we deduce that

$$\mathrm{Res}(f_\varepsilon, z_{1,\varepsilon}) = -\frac{1}{A_\varepsilon z_{1,\varepsilon}}. \tag{5.37}$$

At this point it is necessary to distinguish the cases $r < ct$ and $r > ct$. For $r < ct$ we have

$$\lim_{\varepsilon\to 0^+} z_{1,\varepsilon} = i\sqrt{\frac{ct+r}{ct-r}}, \tag{5.38}$$

while for $r > ct$

$$\lim_{\varepsilon\to 0^+} z_{1,\varepsilon} = \sqrt{\frac{r+ct}{r-ct}}. \tag{5.39}$$

Finally, by (5.35)-(5.39), we obtain

$$\lim_{\varepsilon\to 0^+} I_{1,\varepsilon}(x,t) = \begin{cases} \frac{2\pi i}{\sqrt{c^2t^2-r^2}} & \text{if } |x| = r < ct \\ \frac{2\pi}{\sqrt{r^2-c^2t^2}} & \text{if } |x| = r > ct. \end{cases}$$

Proceeding in the same way, one can show that

$$\lim_{\varepsilon\to 0^+} I_{2,\varepsilon}(x,t) = \begin{cases} \frac{2\pi i}{\sqrt{c^2t^2-r^2}} & \text{if } |x| = r < ct \\ -\frac{2\pi}{\sqrt{r^2-c^2t^2}} & \text{if } |x| = r > ct. \end{cases}$$

By combining (5.34) with these two limits, we derive

$$\mathcal{F}^{-1}\left(\frac{\sin(c|\xi|t)}{c|\xi|}, x\right) = \lim_{\varepsilon\to 0^+} \psi_\varepsilon(x,t) = \begin{cases} \frac{1}{2\pi c\sqrt{c^2t^2-r^2}} & \text{if } |x| = r < ct \\ 0 & \text{if } |x| = r > ct. \end{cases} \tag{5.40}$$

Inserting (5.40) into (5.32) and exploiting the fact that $\psi_0(x,t) = 0$ for $|x| > ct$, we get

$$u(x,t) = \frac{\partial}{\partial t}\int_{B_{ct}(x)} \frac{g(y)}{2\pi c\sqrt{c^2t^2 - |x-y|^2}}\,dy + \int_{B_{ct}(x)} \frac{h(y)}{2\pi c\sqrt{c^2t^2 - |x-y|^2}}\,dy.$$

Making the change of variables $z = \frac{1}{ct}(y-x)$, the first term of the above equality can be rewritten as

$$\frac{\partial}{\partial t}\int_{B_{ct}(x)} \frac{g(y)}{2\pi c\sqrt{c^2t^2 - |x-y|^2}}\,dy = \frac{\partial}{\partial t}\int_{B_1(0)} \frac{t\,g(x+ctz)}{2\pi\sqrt{1-|z|^2}}\,dz$$

$$= \int_{B_1(0)} \frac{g(x+ctz) + t\nabla g(x+ctz)\cdot cz}{2\pi\sqrt{1-|z|^2}}\,dz = \int_{B_{ct}(x)} \frac{g(y) + \nabla g(y)\cdot(y-x)}{2\pi ct\sqrt{c^2t^2 - |y-x|^2}}\,dy.$$

Whence we deduce that

$$u(x,t) = \frac{1}{2}⨍_{B_{ct}(x)} \frac{ct\{t\,h(y) + g(y) + \nabla g(y)\cdot(y-x)\}}{\sqrt{c^2t^2 - |y-x|^2}}\,dy.$$

This formula coincides with (3.87).

Heat equation in dimension n

Let $g \in V$ (see (5.6)) and consider the problem

$$\begin{cases} u_t - \Delta u = 0 & (x,t) \in \mathbb{R}^n \times (0,\infty) \\ u(x,0) = g(x) & x \in \mathbb{R}^n. \end{cases} \tag{5.41}$$

We proceed formally and we apply to both terms of the equation the Fourier transform with respect to the sole spatial variable. Differentiating under the integral sign, we obtain

$$\begin{aligned} \partial_t\widehat{u}(\xi,t) &= \int_{\mathbb{R}^n} e^{-i\xi\cdot x}\partial_t u(x,t)\,dx = \int_{\mathbb{R}^n} e^{-i\xi\cdot x}\Delta u(x,t)\,dx \\ &= \mathcal{F}(\Delta u(x,t),\xi) = -|\xi|^2\widehat{u}(\xi,t). \end{aligned}$$

Transforming formally also the initial datum, we deduce that $\widehat{u}(\xi,t)$ solves the Cauchy problem

$$\begin{cases} \partial_t\widehat{u}(\xi,t) + |\xi|^2\widehat{u}(\xi,t) = 0 \\ \widehat{u}(\xi,0) = \widehat{g}(\xi) \end{cases} \tag{5.42}$$

and hence $\widehat{u}(\xi,t) = \widehat{g}(\xi)e^{-|\xi|^2 t}$. Transforming back, in view of Theorems 5.2.6 and 5.2.3, we infer

$$u(x,t) = \mathcal{F}^{-1}(\widehat{u}(\xi,t),x) = \mathcal{F}^{-1}(e^{-|\xi|^2 t},x) * g(x).$$

Thanks to Theorem 5.1.6 (iii) and Proposition 5.1.9, we find

$$\mathcal{F}^{-1}(e^{-|\xi|^2 t}, x) = \frac{1}{(4\pi t)^{n/2}}\, e^{-\frac{|x|^2}{4t}}.$$

This yields

$$u(x,t) = \frac{1}{(4\pi t)^{n/2}} \int_{\mathbb{R}^n} e^{-\frac{|x-y|^2}{4t}} g(y)\, dy = \int_{\mathbb{R}^n} K(x,y,t) g(y)\, dy\,, \tag{5.43}$$

where $K(x,y,t)$ is defined by (3.103). Note that (5.43) coincides with (3.105).

Fundamental solution of the Laplace equation in dimension $n \geq 3$

We look for a solution $u \in \mathcal{S}'(\mathbb{R}^n)$ ($n \geq 3$) to the distributional equation

$$-\Delta u = \delta_0, \tag{5.44}$$

namely (cf. (3.124)) a tempered distribution u such that

$$\langle u, -\Delta\varphi\rangle = \varphi(0) \qquad \forall \varphi \in \mathcal{D}(\mathbb{R}^n).$$

Applying the Fourier transform to both sides of (5.44) and exploiting Theorem 5.3.8 and Example 5.3.10, we obtain $|\xi|^2 \widehat{u}(\xi) = 1$. Thus we have to solve

$$\widehat{u}(\xi) = \frac{1}{|\xi|^2} \qquad \text{in } \mathcal{S}'(\mathbb{R}^n). \tag{5.45}$$

Let us now proceed as in [61, Exercise 6, Chapter 8]. We set

$$K_\lambda(x) = |x|^{-\lambda} \qquad \forall x \in \mathbb{R}^n \backslash \{0\}$$

for every $0 < \lambda < n$, so that $K_\lambda \in \mathcal{S}'(\mathbb{R}^n)$. Thus

$$\widehat{K}_\lambda(x) = c(n,\lambda) K_{n-\lambda}(x) \tag{5.46}$$

with

$$c(n,\lambda) = \frac{\pi^{n/2} 2^{n-\lambda} \Gamma\left(\frac{n-\lambda}{2}\right)}{\Gamma\left(\frac{\lambda}{2}\right)} \tag{5.47}$$

and with $\Gamma(\alpha) = \int_0^\infty t^{\alpha-1} e^{-t} dt$ for every $\alpha > 0$.

We postpone the proof of (5.46) for a moment and we go on by transforming back (5.45). Applying (5.46) for $\lambda = 2$ and exploiting the identity (5.7) (valid also in the framework of tempered distributions), we obtain

$$u(x) = \mathcal{F}^{-1}(K_2(\xi), x) = (2\pi)^{-n} \widehat{K}_2(x) = (2\pi)^{-n} c(n,2) K_{n-2}(x)$$

$$= \frac{\Gamma\left(\frac{n}{2}-1\right)}{4\pi^{n/2}}|x|^{2-n} = \frac{\Gamma\left(\frac{n}{2}+1\right)}{n(n-2)\pi^{n/2}}|x|^{2-n}.$$

The last term coincides, when $y = 0$, with the function $\Phi(x, 0)$ defined in (3.123).

Now it remains to prove (5.46). We have to distinguish 3 cases (we recall that we are interested in the case $\lambda = 2$).

Case $n < 2\lambda < 2n$. In this case the function K_λ can be decomposed as $K_\lambda(x) = K_\lambda(x) \cdot \chi_{B_1(0)} + K_\lambda(x) \cdot \chi_{\mathbb{R}^n \setminus B_1(0)}$, being $\chi_{B_1(0)}$ and $\chi_{\mathbb{R}^n \setminus B_1(0)}$ the characteristic functions of the unit ball $B_1(0)$ and of its complement, respectively. The two functions in which we have decomposed K_λ belong to $L^1(\mathbb{R}^n)$ and $L^2(\mathbb{R}^n)$ respectively, hence, by Theorem 5.1.5 and (5.10), $\widehat{K}_\lambda$ can be written as the sum of a function of $C_0^0(\mathbb{R}^n)$ and a function of $L^2(\mathbb{R}^n)$. We remark that $K_\lambda \notin L^1(\mathbb{R}^n)$; nevertheless, after some computations we succeed in applying Theorem 5.1.6 (v) to the function K_λ and we obtain that $\widehat{K}_\lambda$ is radially symmetric:

$$\widehat{K}_\lambda(x) = \widehat{K}_\lambda(y) \qquad \text{if } |x| = |y|. \tag{5.48}$$

Moreover, one can verify that

$$\widehat{K}_\lambda(\alpha x) = |\alpha|^{\lambda - n}\widehat{K}_\lambda(x) \qquad \text{for a.e. } x \in \mathbb{R}^n.$$

We may conclude that for $x \neq 0$ it holds

$$\widehat{K}_\lambda(x) = |x|^{\lambda-n}\widehat{K}_\lambda\left(\frac{x}{|x|}\right) = c(n,\lambda)K_{n-\lambda}(x)\,, \tag{5.49}$$

with $c(n, \lambda) = \widehat{K}_\lambda\left(\frac{x}{|x|}\right)$ to be determined: indeed, by (5.48), we deduce that c does not depend on x. By (5.49) we infer

$$\int_{\mathbb{R}^n} |x|^{-\lambda}\widehat{\varphi}(x)\, dx = \int_{\mathbb{R}^n} K_\lambda(x)\widehat{\varphi}(x)\, dx$$

$$= \int_{\mathbb{R}^n} \widehat{K}_\lambda(x)\varphi(x)\, dx = c(n,\lambda)\int_{\mathbb{R}^n} |x|^{\lambda-n}\varphi(x)\, dx.$$

Choosing $\varphi(x) = e^{-|x|^2}$ and using Proposition 5.1.9, from this chain of equalities we derive the value of $c(n, \lambda)$ as in (5.47).

Case $0 < 2\lambda < n$. Set $\mu = n - \lambda$ so that $n < 2\mu < 2n$. In view of the previous case, we derive

$$\widehat{K}_{n-\lambda}(x) = \widehat{K}_\mu(x) = c(n,\mu)K_{n-\mu}(x) = c(n, n-\lambda)K_\lambda(x)$$

and, transforming back both terms, we get

$$K_{n-\lambda}(x) = c(n, n-\lambda)\mathcal{F}^{-1}(K_\lambda(\xi), x)$$

$$= c(n, n-\lambda)\mathcal{F}^{-1}(K_\lambda(-\xi), x) = c(n, n-\lambda)(2\pi)^{-n}\widehat{K}_\lambda(x)$$

which, in turn, gives

$$\widehat{K}_\lambda(x) = \frac{(2\pi)^n}{c(n, n-\lambda)} K_{n-\lambda}(x) = \frac{\pi^{n/2}2^{n-\lambda}\Gamma\left(\frac{n-\lambda}{2}\right)}{\Gamma\left(\frac{\lambda}{2}\right)} K_{n-\lambda}(x).$$

Case $2\lambda = n$**.** We can proceed by approximation, by choosing a suitable sequence $\lambda_k \to \lambda^+$, so that we can apply the first case. Since $K_{\lambda_k} \to K_\lambda$ in $\mathcal{S}'(\mathbb{R}^n)$, by Theorem 5.3.5 we have that also $\widehat{K}_{\lambda_k} \to \widehat{K}_\lambda$ in $\mathcal{S}'(\mathbb{R}^n)$. Similarly, we deduce that $K_{n-\lambda_k} \to K_{n-\lambda}$ and $\widehat{K}_{n-\lambda_k} \to \widehat{K}_{n-\lambda}$ in $\mathcal{S}'(\mathbb{R}^n)$. On the other hand, $c(n, \lambda_k) \to c(n, \lambda)$. Therefore, passing to the limit in $\widehat{K}_{\lambda_k} = c(n, \lambda_k)\widehat{K}_{n-\lambda_k}$, we obtain the desired result also in this case.

5.5 Laplace transform

Also in this section we deal only with complex functional spaces. We begin with the

Definition 5.5.1. *Let* $u \in L^1_{loc}(\mathbb{R})$ *be such that* $\operatorname{supp} u \subset [0, \infty)$ *(the support is to be meant in the sense of distributions, see Definition 2.6.5). Assume that there exists* $\lambda \in \mathbb{R}$ *such that the map* $t \mapsto e^{-\lambda t}u(t)$ *belongs to* $L^1(\mathbb{R})$ *and define* $\lambda_a(u)$ *as the infimum of such* λ*. Then, such function* u *is said* **transformable function** *and for every* $s \in \mathbb{C}$ *satisfying* $\operatorname{Re} s > \lambda_a(u)$ *it is set*

$$\mathcal{L}(u(t), s) = \int_0^\infty e^{-st}u(t)\,dt. \tag{5.50}$$

The function $\mathcal{L}u$ *is called* **Laplace transform** *of* u*.*

Clearly $\mathcal{L}$ is a linear operator. A sufficient condition to ensure the transformability of a function is given in the following

Proposition 5.5.2. *Let* $u \in L^1_{loc}(\mathbb{R})$ *be such that* $\operatorname{supp} u \subset [0, \infty)$*. If there exist* $M > 0$ *and* $\gamma \in \mathbb{R}$ *such that*

$$|u(t)| \le Me^{\gamma t} \qquad \forall t \in \mathbb{R}\ ,$$

then u *is transformable and* $\lambda_a(u) \le \gamma$*.*

Proof. The proof is left to the reader as an exercise. □

Note that if u is transformable, in view of Theorem 2.3.7, it holds

$$\lim_{\operatorname{Re} s \to +\infty} \mathcal{L}(u(t), s) = 0\,. \tag{5.51}$$

For the computation of the Laplace transform, the following properties turn out to be useful:

Theorem 5.5.3. *Let u be a transformable function. Then:*

(i) for every $t_0 > 0$ the function $t \mapsto u(t - t_0)$ is transformable and moreover $\mathcal{L}(u(t-t_0), s) = e^{-t_0 s}\mathcal{L}(u(t), s)$ for $\operatorname{Re} s > \lambda_a(u)$;

(ii) for every $s_0 \in \mathbb{C}$ the function $t \mapsto e^{s_0 t}u(t)$ is transformable and moreover $\mathcal{L}(e^{s_0 t}u(t), s) = \mathcal{L}(u(t), s - s_0)$ for $\operatorname{Re} s > \operatorname{Re} s_0 + \lambda_a(u)$;

(iii) for every $c > 0$ the function $t \mapsto u(ct)$ is transformable and moreover $\mathcal{L}(u(ct), s) = \frac{1}{c}\mathcal{L}(u(t), s/c)$ for $\operatorname{Re} s > c\lambda_a(u)$.

Proof. The proof of this result is an easy exercise and it is left to the reader. □

Let us now present some significant examples of computation of the Laplace transform.

Example 5.5.4. The Heaviside function H introduced in (2.40) is transformable, moreover $\lambda_a(H) = 0$. For $s \in \mathbb{C}$ with $\operatorname{Re} s > 0$ it holds

$$\mathcal{L}(H(t), s) = \int_0^{\infty} e^{-st} H(t)\, dt = \int_0^{\infty} e^{-st}\, dt = \left[-\frac{e^{-st}}{s}\right]_0^{\infty} = \frac{1}{s} - \lim_{t\to+\infty} \frac{e^{-st}}{s} = \frac{1}{s},$$

where we have used $|e^{-st}| = e^{-(\operatorname{Re} s)t} \to 0$ as $t \to +\infty$, being $\operatorname{Re} s > 0$. In conclusion,

$$\mathcal{L}(H(t), s) = \frac{1}{s} \qquad \forall\ \operatorname{Re} s > 0.$$

Furthermore, applying Theorem 5.5.3 (ii), for every $s_0 \in \mathbb{C}$ we obtain

$$\mathcal{L}(e^{s_0 t}H(t), s) = \frac{1}{s - s_0} \qquad \forall\ \operatorname{Re}\ s > \operatorname{Re}\ s_0.$$

□

Example 5.5.5. Consider the function $u(t) = \cos(\omega t)\ H(t)$, with $\omega > 0$. Since

$$\cos(\omega t) = \frac{e^{i\omega t} + e^{-i\omega t}}{2},$$

by Example 5.5.4 we infer

$$\mathcal{L}(\cos(\omega t)\ H(t), s) = \frac{1}{2}\ \mathcal{L}(e^{i\omega t}\ H(t), s) + \frac{1}{2}\ \mathcal{L}(e^{-i\omega t}\ H(t), s) =$$

$$= \frac{1}{2}\,\frac{1}{s - i\omega} + \frac{1}{2}\,\frac{1}{s + i\omega} = \frac{s}{s^2 + \omega^2} \qquad \forall\ \operatorname{Re} s > 0.$$

□

For the applications to ordinary differential equations, it is useful to define in some sense the Laplace transform of the Dirac delta function (even if it is not an element of $L^1_{loc}(\mathbb{R})$!):

Example 5.5.6. As we have shown in Example 2.6.9, the Dirac delta function δ_0 can be seen as the distributional limit as $\varepsilon \to 0^+$ of

$$u_\varepsilon(t) = \begin{cases} \dfrac{1}{\varepsilon} & \text{if } t \in [0,\varepsilon] \\ 0 & \text{if } t \in \mathbb{R} \setminus [0,\varepsilon]. \end{cases} \tag{5.52}$$

We define the Laplace transform of δ_0 as the limit as $\varepsilon \to 0^+$ of $\mathcal{L}u_\varepsilon$:

$$\mathcal{L}(\delta_0, s) = \lim_{\varepsilon \to 0^+} \int_0^\infty e^{-st} u_\varepsilon(t)\, dt = \lim_{\varepsilon \to 0^+} \int_0^\varepsilon \frac{1}{\varepsilon} e^{-st}\, dt = \lim_{\varepsilon \to 0^+} \frac{e^{-s\varepsilon} - 1}{-s\varepsilon} = 1.$$

We conclude that $\mathcal{L}\delta_0 = 1$. □

Now we see how the Laplace transform behaves with respect to differentiation.

Theorem 5.5.7. *Let u be a transformable function. Then $\mathcal{L}u$ is a holomorphic function on the complex half-plane* $\operatorname{Re} s > \lambda_a(u)$. *Moreover, the function $t \mapsto t\,u(t)$ is transformable and there holds*

$$\frac{d}{ds}\mathcal{L}(u(t), s) = -\mathcal{L}(t\,u(t), s) \qquad \forall \operatorname{Re} s > \lambda_a(u).$$

Proof. For $\operatorname{Re} s > \lambda_a(u)$ fixed, we evaluate the difference quotient

$$\frac{\mathcal{L}u(s+h) - \mathcal{L}u(s)}{h} = \int_0^\infty e^{-st} \frac{e^{-ht} - 1}{h} u(t)\, dt$$

as $h \to 0$ in $\mathbb{C}$. Distinguishing the cases $|ht| \le 1$ and $|ht| > 1$, we deduce that there exists a constant $C_1 > 0$ independent of h and t, such that

$$\frac{|e^{-ht} - 1|}{|ht|} \le \begin{cases} C_1 & \text{if } |ht| \le 1 \\ e^{|\operatorname{Re} h| t} + 1 & \text{if } |ht| \ge 1 \end{cases}$$

and consequently

$$\frac{|e^{-ht} - 1|}{|h|} \le (C_1 + 2) t e^{|\operatorname{Re} h| t} \qquad \forall t > 0,\ \forall h \in \mathbb{C} \setminus \{0\}.$$

On the other hand, for every $\varepsilon > 0$ there exists a constant $C_2 > 0$ such that

$$t \le C_2 e^{\varepsilon t} \qquad \forall t > 0\ .$$

Therefore we can estimate the integrand:

$$\left| e^{-st} \frac{e^{-ht} - 1}{h} u(t) \right| \le C_2(C_1 + 2) e^{-[\operatorname{Re} s - \varepsilon - |\operatorname{Re} h|]t} |u(t)|;$$

once fixed $0 < \varepsilon < \operatorname{Re} s - \lambda_a(u)$, defining $\delta = \operatorname{Re} s - \varepsilon - \lambda_a(u)$, we may assume that $|\operatorname{Re} h| < \delta/2$ since $h \to 0$ in $\mathbb{C}$. In this way, we deduce that

$$\left| e^{-st} \frac{e^{-ht} - 1}{h} u(t) \right| \leq C_2(C_1 + 2) e^{-[\lambda_a(u)+\delta/2]t} |u(t)| \in L^1(\mathbb{R}).$$

Taking into account that

$$\lim_{h \to 0} e^{-st} \frac{e^{-ht} - 1}{h} u(t) = -e^{-st} t\, u(t)$$

and applying Theorem 2.3.7, we immediately get the desired result. □

Example 5.5.8. Consider the function

$$u(t) = \begin{cases} 0 & \text{if } t \leq 0 \\ t & \text{if } t > 0 \end{cases} \tag{5.53}$$

which is known as “ramp” function, for obvious reasons. Once we have observed that $u(t) = t\, H(t)$ with H the Heaviside function, applying Theorem 5.5.7 and taking into account Example 5.5.4, we obtain

$$\frac{d}{ds}\left(\frac{1}{s}\right) = \frac{d}{ds} \mathcal{L}(H(t), s) = -\mathcal{L}(u(t), s).$$

Thus we may conclude that

$$\mathcal{L}(u(t), s) = \frac{1}{s^2} \qquad \forall \operatorname{Re} s > 0.$$ □

By iterating Theorem 5.5.7, it is possible to compute the Laplace transform of any function which coincides with a polynomial over $[0, \infty)$:

Example 5.5.9. Consider the function

$$u(t) = \sum_{k=0}^{n} a_k t^k H(t).$$

By Theorem 5.5.7 we deduce that for every integer $k \geq 1$

$$\frac{d^k}{ds^k} \mathcal{L}(H(t), s) = (-1)^k \mathcal{L}(t^k H(t), s)\ .$$

Therefore, recalling that $\mathcal{L}(H(t), s) = 1/s$, we derive

$$\mathcal{L}(t^k H(t), s) = \frac{k!}{s^{k+1}} \qquad \forall \operatorname{Re} s > 0.$$

Clearly, the last equality implies that

$$\mathcal{L}(u(t), s) = \sum_{k=0}^{n} \frac{a_k k!}{s^{k+1}} \qquad \forall \operatorname{Re} s > 0.$$ □

In the next result we explain what happens when we apply the Laplace transform to the derivative of a function.

Theorem 5.5.10. *Let u be a transformable function such that $u \in C^1([0,\infty))$ and assume that also the function v defined by*

$$v(t) = \begin{cases} u'(t) & \text{if } t \geq 0 \\ 0 & \text{if } t < 0 \end{cases}$$

is transformable. Then there holds

$$\mathcal{L}(v(t),s) = s\mathcal{L}(u(t),s) - u(0^+) \qquad \forall s \in \mathbb{C},\ \operatorname{Re} s > \max\{\lambda_a(u), \lambda_a(v)\} \tag{5.54}$$

where $u(0^+) = \lim_{t\to 0^+} u(t)$.

Moreover, the following formulas hold true:

$$\lim_{\operatorname{Re} s \to +\infty} s\,\mathcal{L}(u(t),s) = u(0^+) \tag{5.55}$$

and, if $v \in L^1(\mathbb{R})$, then $\lambda_a(u) \leq 0$ and

$$\lim_{\substack{s\to 0 \\ \operatorname{Re} s > 0}} s\mathcal{L}(u(t),s) = \lim_{t\to+\infty} u(t)\,. \tag{5.56}$$

Proof. We begin by proving (5.54). Integrating by parts, we obtain

$$\mathcal{L}(v(t),s) = \lim_{r\to+\infty} \int_0^r e^{-st} u'(t)\,dt = \lim_{r\to+\infty} \left\{ [e^{-st}u(t)]_0^r + \int_0^r s\,e^{-st}u(t)\,dt \right\}$$

$$= \int_0^\infty s\,e^{-st}u(t)\,dt - u(0^+) + \lim_{r\to+\infty} e^{-sr}u(r).$$

Then it remains to show that

$$\lim_{r\to+\infty} e^{-sr}u(r) = 0. \tag{5.57}$$

For every $t \geq 0$ it holds

$$\frac{d}{dt}(e^{-st}u(t)) = -se^{-st}u(t) + e^{-st}u'(t)\ ,$$

moreover, since u and v are transformable we deduce that the function $\frac{d}{dt}(e^{-st}u(t))$ belongs to $L^1(0,\infty)$ for $\operatorname{Re} s > \max\{\lambda_a(u), \lambda_a(v)\}$, thus

$$\lim_{r\to+\infty} e^{-sr}u(r) = u(0) + \int_0^\infty \frac{d}{dt}(e^{-st}u(t))\,dt.$$

This proves the existence of the limit in (5.57). On the other hand, being $t \mapsto e^{-st}u(t)$ an element of $L^1(0,\infty)$ for $\operatorname{Re} s > \max\{\lambda_a(u), \lambda_a(v)\}$, such limit can only be zero. This concludes the proof of (5.57) and (5.54).

Let us pass to the proof of (5.55). Let v be as in the statement of the theorem. Thanks to (5.54), the proof of (5.55) will be achieved once proved that

$$\lim_{\operatorname{Re} s\to+\infty} \int_0^\infty e^{-st}u'(t)\,dt = 0.$$

It suffices to fix $s_0 > \lambda_a(v)$ and to consider the following upper bound for the integrand:

$$|e^{-st}u'(t)| = e^{-(\operatorname{Re} s)t}|u'(t)| \leq e^{-s_0 t}|u'(t)| \in L^1(\mathbb{R}) \qquad \forall t > 0, \forall s \in \mathbb{C}, \operatorname{Re} s > s_0.$$

At this point, it is enough to observe that $\lim_{\operatorname{Re} s\to+\infty} e^{-st}u'(t) = 0$ for every $t > 0$ and to apply Theorem 2.3.7.

We conclude with the proof of (5.56). Since $v \in L^1(\mathbb{R})$, the limit $L := \lim_{t\to+\infty} u(t)$ is finite and hence $t \mapsto e^{-st}u(t) \in L^1(\mathbb{R})$ whenever $\operatorname{Re} s > 0$; therefore $\lambda_a(u) \leq 0$. Since $\operatorname{Re} s > 0$ we know that for all $t > 0$

$$|e^{-st}v(t)| \leq |v(t)| \in L^1(\mathbb{R}).$$

Whence, by Theorem 2.3.7 we have that

$$\lim_{\substack{s\to 0,\\ \operatorname{Re} s>0}} \int_0^\infty e^{-st}u'(t)\,dt = \int_0^\infty u'(t)\,dt = L - u(0^+)$$

and by applying formula (5.54) we obtain the result. □

Note that (5.56) is not reversible: if the limit on the left exists, then the limit on the right may not exist. To see this, consider the function $u(t) = H(t)\sin t$.

Remark 5.5.11. We can reinterpret formula (5.54) as follows. Let u satisfy the assumptions of Theorem 5.5.10 and denote again by $u : \mathbb{R} \to \mathbb{C}$ its trivial extension to $(-\infty, 0)$. Arguing as in Example 2.6.12, we can show that the distributional derivative of u, here denoted by u', can be written in the form $u' = v + u(0^+)\delta_0$ with v as in the statement of Theorem 5.5.10. Thus we may formally recast (5.54) as

$$\mathcal{L}(u'(t), s) = s\mathcal{L}(u(t), s) \qquad \forall s \in \mathbb{C},\ \operatorname{Re} s > \max\{\lambda_a(u), \lambda_a(v)\}\,. \tag{5.58}$$

The previous identity is not just formal: it has a rigorous validity, since the theory of Laplace transform can be extended to distributions. Here we will not treat such extension, except for the Dirac delta, see Example 5.5.6.

If we choose $u(t) = H(t)$ in (5.58), with H the Heaviside function, we obtain

$$\mathcal{L}(\delta_0, s) = s\mathcal{L}(H(t), s)$$

which is perfectly consistent with the Examples 5.5.4-5.5.6. □

By Iterating formula (5.54) we obtain the

Corollary 5.5.12. *Let u be a transformable function such that $u \in C^k([0,\infty))$ and assume that for every $j \in \{1, ..., k\}$ also the functions*

$$v_j(t) = \begin{cases} u^{(j)}(t) & \text{if } t \geq 0 \\ 0 & \text{if } t < 0 \end{cases}$$

are transformable. Then, there holds the formula

$$\mathcal{L}(v_k(t), s) = s^k \mathcal{L}(u(t), s) - \sum_{j=1}^{k} s^{k-j} u^{(j-1)}(0^+)$$

for $\operatorname{Re} s > \max\{\lambda_a(u), \lambda_a(v_1), ..., \lambda_a(v_k)\}$, *where* $u^{(j-1)}(0^+) = \lim_{t\to 0^+} u^{(j-1)}(t)$ *for every* $j \in \{1, ..., k\}$.

Let us see an application of Theorem 5.5.10:

Example 5.5.13. Given $\omega > 0$, consider the function

$$u(t) = \sin(\omega t)\ H(t).$$

In view of Theorem 5.5.10 and Example 5.5.5, we obtain

$$\mathcal{L}(\sin(\omega t)\,H(t), s) = \mathcal{L}\left(-\frac{1}{\omega}\frac{d}{dt}(\cos(\omega t))\,H(t), s\right)$$

$$= -\frac{s}{\omega}\mathcal{L}(\cos(\omega t)\,H(t), s) + \frac{1}{\omega} = \frac{\omega}{s^2+\omega^2}. \qquad \square$$

We now state two results which relate the Laplace transform to integral functions.

Theorem 5.5.14. *Let u be a transformable function and assume that also $t \mapsto u(t)/t$ is a transformable function. Then there holds the formula*

$$\mathcal{L}(u(t)/t, s) = \int_s^\infty \mathcal{L}(u(t), \sigma)\, d\sigma \qquad \forall s \in \mathbb{R},\ s > \lambda_a(u).$$

Proof. This follows from Theorem 5.5.7 and (5.51). □

Theorem 5.5.15. *Let u be a transformable function. Then also the function*

$$U(t) = \int_0^t u(r)\, dr$$

is transformable and there holds the formula

$$\mathcal{L}(U(t), s) = \frac{1}{s}\mathcal{L}(u(t), s) \qquad \forall s \in \mathbb{C},\ \operatorname{Re} s > \max\{\lambda_a(u), 0\}.$$

Proof. It is enough to go over the proof of Theorem 5.5.10, replacing u with U. Actually, under the considered hypotheses, we cannot affirm that $U \in C^1([0,\infty))$. However U is differentiable at almost every point of $[0,\infty)$ and its derivative coincides with u. Furthermore, it is possible to verify that U is transformable: indeed, if $\operatorname{Re} s > \max\{\lambda_a(u), 0\}$, it holds

$$\begin{aligned}
\lim_{r\to+\infty} \int_0^r \left| e^{-st} \left(\int_0^t u(\sigma)\, d\sigma \right) \right| dt &\leq \lim_{r\to+\infty} \int_0^r \left(\int_0^t e^{-(\operatorname{Re} s)t} |u(\sigma)| d\sigma \right) dt \\
&= \lim_{r\to+\infty} \int_0^r \left(\int_\sigma^r e^{-(\operatorname{Re} s)t} |u(\sigma)| dt \right) d\sigma \\
&= -\frac{1}{\operatorname{Re} s} \lim_{r\to+\infty} \int_0^r \left(e^{-(\operatorname{Re} s)r} - e^{-(\operatorname{Re} s)\sigma} \right) |u(\sigma)| d\sigma \\
&= \frac{1}{\operatorname{Re} s} \int_0^\infty e^{-(\operatorname{Re} s)\sigma} |u(\sigma)| d\sigma < +\infty\,.
\end{aligned}$$

Moreover, all the steps in the proof of Theorem 5.5.10 can be formally repeated replacing u with U and u' with U'. □

We conclude this section with a result which relates the Laplace transform to the convolution.

Theorem 5.5.16. *Let u, v be transformable functions. Then the convolution $u * v$ is a transformable function and there holds the formula*

$$\mathcal{L}((u*v)(t), s) = \mathcal{L}(u(t), s) \cdot \mathcal{L}(v(t), s) \qquad \forall \operatorname{Re} s > \max\{\lambda_a(u), \lambda_a(v)\}.$$

Proof. We begin by observing that $\operatorname{supp}(u*v) \subset [0,\infty)$. In fact, being the supports of u and v contained into $[0,\infty)$, we deduce that

$$(u*v)(t) = \int_{\mathbb{R}} u(t-y)v(y)\, dy = \int_0^t u(t-y)v(y)\, dy$$

and the latter integral vanishes for $t < 0$ since $v \equiv 0$ in $(-\infty, 0)$. If $\operatorname{Re} s > \max\{\lambda_a(u), \lambda_a(v)\}$, then

$$e^{-st}(u*v)(t) = \int_0^t e^{-s(t-y)} u(t-y) e^{-sy} v(y)\, dy = (e^{-st}u(t)) * (e^{-st}v(t)) \in L^1(\mathbb{R})$$

by Theorem 5.2.1. This proves than $u * v$ is transformable. Finally, exploiting Theorem 2.3.6 and making a change of variables, we get

$$\mathcal{L}((u*v)(t), s) = \int_{\mathbb{R}} \left(\int_{\mathbb{R}} e^{-st} u(t-y) v(y)\, dy \right) dt = \int_{\mathbb{R}} v(y) \left(\int_{\mathbb{R}} e^{-st} u(t-y)\, dt \right) dy$$

$$= \int_{\mathbb{R}} e^{-sy} v(y) \left(\int_{\mathbb{R}} e^{-s(t-y)} u(t-y)\, dt \right) dy = \int_{\mathbb{R}} e^{-sy} v(y) \left(\int_{\mathbb{R}} e^{-st} u(t)\, dt \right) dy$$

$$= \int_{\mathbb{R}} s^{-st} u(t)\, dt \cdot \int_{\mathbb{R}} s^{-st} v(t)\, dt = \mathcal{L}(u(t), s) \cdot \mathcal{L}(v(t), s)\,,$$

where we have used the fact that the integral (5.50), which defines the Laplace transform, can be extended to all $\mathbb{R}$ since the integrands vanish in $(-\infty, 0)$. □

5.6 Inverse Laplace transform

Here we present the problem of the inversion of the Laplace transform, namely, given a function f of a complex variable, we want to establish which assumptions ensure the existence and uniqueness of a transformable function u such that $\mathcal{L}u = f$. If such a function u exists, we shall say that u is the **inverse Laplace transform** of f and we shall set $u = \mathcal{L}^{-1} f$.

A complete presentation of the *inversion* topic would require the introduction of the Laplace transform in the more general framework of Laplace transformable distributions (see [38, Capitolo VII]). For the sake of simplicity, here we limit ourselves to determine a sufficient condition on the function f of a complex variable that ensures the existence of $\mathcal{L}^{-1} f$ in the framework of transformable functions.

In order to obtain a formula which allows to derive explicitly the inverse transform of f, we proceed formally assuming that u is a transformable function satisfying $\mathcal{L}u = f$. Taking into account that $\operatorname{supp} u \subset [0, \infty)$, setting $s = x + iy$ with $x, y \in \mathbb{R}$, we obtain

$$f(s) = \mathcal{L}(u(t), s) = \int_{\mathbb{R}} e^{-iyt} e^{-xt} u(t)\, dt = \mathcal{F}(e^{-xt} u(t), y).$$

Assuming that the map $y \mapsto f(x + iy)$ belongs to the space V defined in (5.6), by Theorem 5.2.6 we deduce that

$$e^{-xt} u(t) = \mathcal{F}^{-1}(f(x + iy), t) = \frac{1}{2\pi} \int_{\mathbb{R}} e^{ity} f(x + iy)\, dy$$

and this implies

$$u(t) = \frac{1}{2\pi} \int_{\mathbb{R}} e^{(x+iy)t} f(x + iy)\, dy.$$

Eventually, setting $s = x + iy$ and $\Gamma_x = \{s = x + iy;\ y \in \mathbb{R}\}$, we obtain the formula

$$u(t) = \frac{1}{2\pi i} \int_{\Gamma_x} e^{st} f(s)\, ds. \tag{5.59}$$

Thus it is proved the

Theorem 5.6.1. *Let $r_0 \in \mathbb{R}$ and let f be a holomorphic complex function over the half-plane $\{s \in \mathbb{C};\ \operatorname{Re} s > r_0\}$. Assume that for every $r > r_0$ there exist $\alpha > 1$ and $C > 0$ such that*

$$|f(s)| \leq \frac{C}{1+|s|^\alpha} \qquad \forall \operatorname{Re} s > r.$$

Then there exists a unique transformable function u such that $\mathcal{L}u = f$. Moreover, u is given by formula (5.59) for every $x > r$.

5.7 Laplace transform and differential equations

5.7.1 Applications to ordinary differential equations

The Laplace transform proves efficient in the solution of Cauchy problems associated to ordinary differential equations. In fact, according to Corollary 5.5.12, it is possible to reduce to the study of an algebraic equation. For this reason, it will be useful the easy (and immediate)

Proposition 5.7.1. *Let*

$$f(s) = \frac{a_m\ s^m + a_{m-1}\ s^{m-1} + \cdots + a_0}{b_n\ s^n + b_{n-1}\ s^{n-1} + \cdots + b_0} \qquad (s \in \mathbb{C}),$$

with $n > m$. Let $p_1, ..., p_N \in \mathbb{C}$ (with $N \leq n$) be the distinct roots of the denominator with algebraic multiplicities $r_1, ..., r_N$ respectively. Then there exist some coefficients $c_{lj} \in \mathbb{C}$ (with $l = 1, ..., N$ and $j = 1, ..., r_l$) such that

$$f(s) = \sum_{j=1}^{r_1} \frac{c_{1j}}{(s+p_1)^j} + \sum_{j=1}^{r_2} \frac{c_{2j}}{(s+p_2)^j} + \ldots + \sum_{j=1}^{r_N} \frac{c_{Nj}}{(s+p_N)^j}. \tag{5.60}$$

Let us analyze more in depth linear equations with constant coefficients of the form

$$a_k y^{(k)} + a_{k-1} y^{(k-1)} + \ldots + a_1 y' + a_0 y = f(t), \tag{5.61}$$

with $a_0, a_1, ..., a_k \in \mathbb{R}$ and $f \in C^0(\mathbb{R})$, with the initial conditions

$$y(0) = y_0,\ y'(0) = y_1, ..., y^{(k-1)}(0) = y_{k-1}. \tag{5.62}$$

Assuming that the datum f is a transformable function, applying formally the Laplace transform to both sides of (5.61) and taking into account (5.62), by Corollary 5.5.12 we obtain

$$a_k \left(s^k \mathcal{L}(y(t), s) - \sum_{j=1}^{k} s^{k-j} y_{j-1} \right) + a_{k-1} \left(s^{k-1} \mathcal{L}(y(t), s) - \sum_{j=1}^{k-1} s^{k-1-j} y_{j-1} \right)$$

$$+ ... + a_0\mathcal{L}(y(t), s) = \mathcal{L}(f(t), s).$$

From the last equality we infer

$$\mathcal{L}(y(t), s) = \frac{\mathcal{L}(f(t), s) + F_1(s)}{F_2(s)}$$

for some suitable polynomials F_1 and F_2, which depend on the initial conditions $y_0, ..., y_{k-1}$ and on the coefficients of the equation. Applying formally Theorem 5.5.16 to both sides of the last identity, we derive

$$y(t) = \mathcal{L}^{-1}\left(\frac{1}{F_2}\right) * f + \mathcal{L}^{-1}\left(\frac{F_1}{F_2}\right).$$

Thus, finally, we are led to compute the inverse transform of a rational function. To do that, it is convenient to apply Proposition 5.7.1. Once we have written the functions $1/F_1$ and F_1/F_2 in the form (5.60), we can use Example 5.5.9 and assertion (ii) in Theorem 5.5.3.

In order to better understand how this procedure works, let us see some concrete examples which model engineering and physical phenomena.

Example 5.7.2. The hyperstatic beam represented in the picture, hinged at A ($x = 0$) and B ($x = L$), is subject to a transverse load per unit length P uniformly distributed; we want to determine the deformed configuration of the beam.

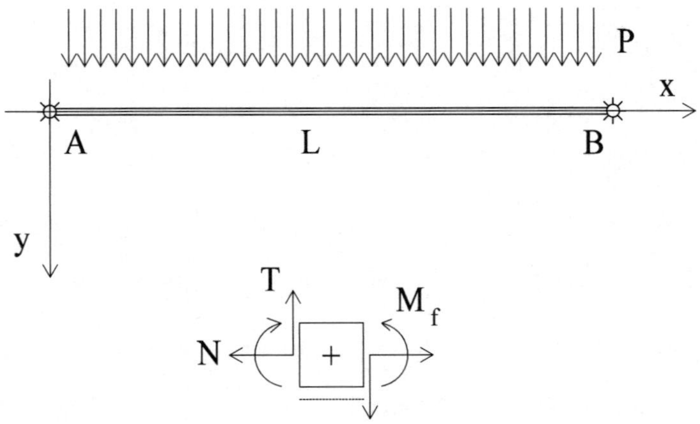

Beam subject to a distributed load

We will consider the following assumptions:
- the beam is homogeneous and slender;

• the undeformed beam is straight and deforms remaining in the linearly elastic range; moreover, the bending moment is normal to the xy plane;
• due to the action of the loads and the constraints, the cross sections of the beam rotate, remaining plane;
• the applied load produce only small curvatures in the beam.

We denote by $y = y(x)$ the function that describes the beam in the undeformed configuration. Calling $q(x)$ the load, by [8, 13, 67] we infer that $y(x)$ satisfies the equation

$$\frac{d^4 y}{dx^4} = \frac{q(x)}{EJ}, \tag{5.63}$$

with the boundary conditions

$$y(0) = 0 \qquad y''(0) = 0 \qquad y(L) = 0 \qquad y''(L) = 0. \tag{5.64}$$

Let us now solve problem (5.63)-(5.64) when the load satisfies $q(x) \equiv P$ for some constant P. It is well known that such problem can be solved directly by means of four consecutive integrations, obtaining a polynomial of fourth degree as solution. Nevertheless, we will proceed in an alternative way, by solving (5.63)-(5.64) with the Laplace transform. Applying $\mathcal{L}$ to both sides of (5.63) and using Corollary 5.5.12, we obtain

$$s^4\, Y(s) - s^3\, y(0) - s^2\, y'(0) - s\, y''(0) - y'''(0) = \frac{P}{EJs}, \tag{5.65}$$

with $Y(s) = \mathcal{L}(y(x), s)$. Imposing the first two conditions in (5.64) and setting $k_1 = y'(0)$, $k_2 = y'''(0)$ we get

$$Y(s) = \frac{k_1}{s^2} + \frac{k_2}{s^4} + \frac{P}{EJs^5}.$$

In view of Example 5.5.9 and by the conditions in (5.64) at $x = L$, we deduce that

$$y(x) = k_1 x + \frac{k_2 x^3}{3!} + \frac{P}{EJ}\frac{x^4}{4!} = k_1 x + \frac{k_2 x^3}{6} + \frac{P}{EJ}\frac{x^4}{24} \qquad k_1 = \frac{P\,L^3}{24EJ} \qquad k_2 = -\frac{P\,L}{2EJ}\,.$$

Thus, the expression of the deformed beam becomes

$$y(x) = \frac{P}{24EJ}(L^3 x - 2Lx^3 + x^4) \qquad \forall x \in [0, L].$$

We have considered the deformations of the beam when it is subject to a constant vertical load P; now we want to determine how the beam deforms when it is subject to a load concentrated at a single point. Let P_0 be such load and let $x = a$ be its point of application, see the picture below.

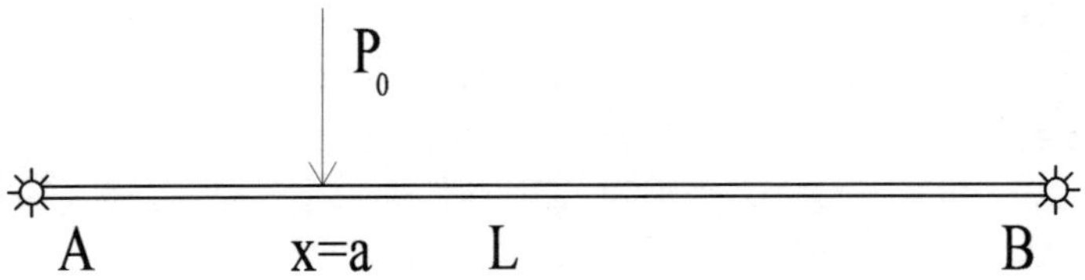

Beam subject to a concentrated load

This problem can be solved by considering a uniformly distributed load that acts on the interval $[a, a+\varepsilon]$, whose magnitude per unit length is given by W_0 (see the picture below), and then passing to the limit as $\varepsilon \to 0^+$.

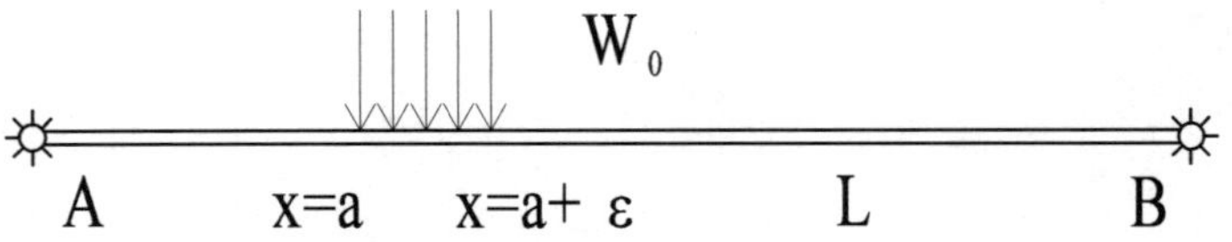

Beam subject to a uniformly distributed load W_0

The total load P_0 acting on $[a, a+\varepsilon]$ is given by $W_0 \cdot (a+\varepsilon-a) = W_0\,\varepsilon$. Therefore,

$$W_\varepsilon(x) = \begin{cases} \dfrac{P_0}{\varepsilon} & a \le x \le a+\varepsilon \\ 0 & \text{elsewhere.} \end{cases} \tag{5.66}$$

Proceeding as in Example 2.6.9, by (5.66) we deduce that $W_\varepsilon \to P_0\delta_a$ in $\mathcal{D}'(0, L)$ as $\varepsilon \to 0^+$. Hence, by (5.63) we obtain the distributional equation

$$\frac{d^4y}{dx^4} = \frac{P_0}{EJ}\,\delta_a$$

with conditions (5.64). Applying the Laplace transform, by Theorem 5.5.3 (i), Corollary 5.5.12 and by Example 5.5.6, we derive

$$s^4\;Y(s) - s^3\;y(0) - s^2\;y'(0) - s\;y''(0) - y'''(0) = \frac{P_0}{EJ}e^{-as}. \tag{5.67}$$

Transforming back and imposing the validity of conditions (5.64) at $x = 0$, we get

$$y(x) = k_1x + \frac{k_2x^3}{6} + \frac{P_0}{6EJ}(x-a)^3H(x-a)\,,$$

where we have set $k_1 = y'(0)$ and $k_2 = y'''(0)$. In view of conditions (5.64) at $x = L$, we derive

$$y(x) = \left(\frac{P_0L(L-a)}{6EJ} - \frac{P_0(L-a)^3}{6EJL}\right)x - \frac{P_0(L-a)}{6EJL}x^3 + \frac{P_0}{6EJ}(x-a)^3H(x-a). \;\square$$

Example 5.7.3. We want to compute the response $x(t)$ of the one degree of freedom vibrating linear system (represented in the picture), with mass m, rigidity modulus of the elastic element k, damping coefficient r, further affected by an external force $f(t)$.

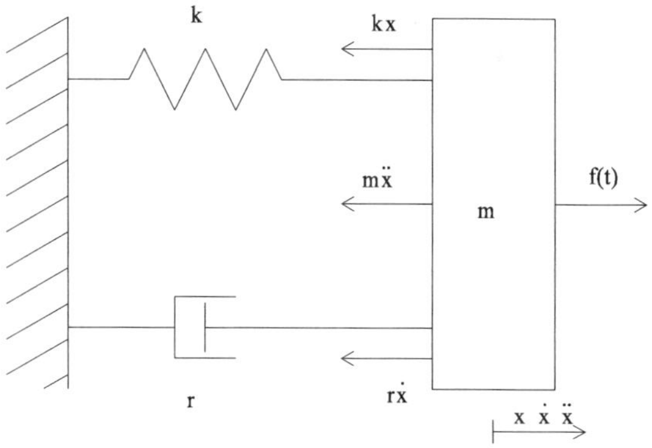

A one degree of freedom system

By the second law of dynamics, the equation of motion of this system is given by the following differential equation [15, 16, 33]:

$$m\,\frac{d^2x}{dt^2} + r\,\frac{dx}{dt} + kx = f(t). \tag{5.68}$$

Prescribing zero initial conditions $x(0) = x'(0) = 0$ and applying the Laplace transform to both sides of (5.68), we obtain:

$$m\,s^2\,X(s) + r\,s\,X(s) + kX(s) = F(s)\,, \tag{5.69}$$

with $X(s) = \mathcal{L}(x(t), s)$ and $F(s) = \mathcal{L}(f(t), s)$. Equation (5.69) can be rewritten as:

$$X(s) = G(s)\;F(s) \qquad \text{with} \qquad G(s) = \frac{1}{m\,s^2 + r\,s + k}\,. \tag{5.70}$$

If the external force is an impulse $f(t) = \delta_0$ (Dirac delta function centered at $t = 0$), by Example 2.6.9 we infer $F(s) = \mathcal{L}(\delta_0, s) = 1$. By (5.70) we deduce

$$\mathcal{L}(x(t), s) = X(s) = G(s)\;F(s) = G(s) = \frac{1}{m\,s^2 + r\,s + k}.$$

If $r^2 - 4mk < 0$, the denominator of G admits two complex conjugate roots $p_1 = \alpha + i\omega$ and $p_2 = \alpha - i\omega$, with $\alpha = -\frac{r}{2m} < 0$. Proposition 5.7.1 gives

$$G(s) = \frac{1}{2m\omega i(s - p_1)} - \frac{1}{2m\omega i(s - p_2)}\,,$$

therefore, by Example 5.5.4, we get

$$x(t) = \mathcal{L}^{-1}(G(s), t) = \frac{1}{2m\omega i}\, e^{p_1 t} - \frac{1}{2m\omega i}\, e^{p_2 t} = \frac{1}{\omega m}\, e^{\alpha t} \sin(\omega t) \qquad \forall t > 0. \quad \square$$

Example 5.7.4. In [3] it is considered the circuit pictured below, with a resistance R, a capacitance C, an inductance L connected in series. Let $v(t)$, $i(t)$, $q(t)$ be respectively the voltage, the current and the charge on the capacitor at the generic time t.

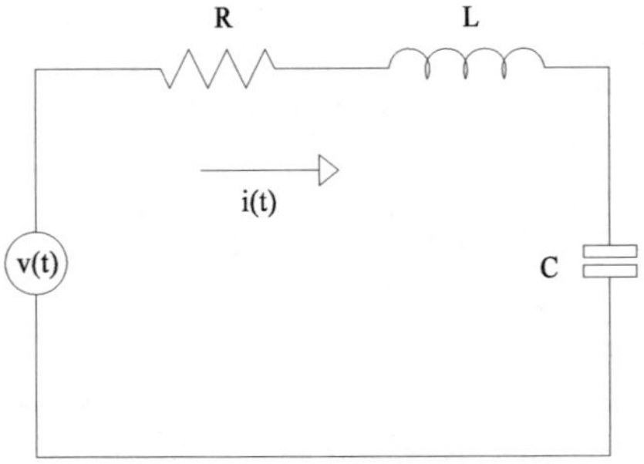

Circuit RLC

Assuming to know the initial values of the current and of the charge, $i(0) = i_0$ and $q(0) = q_0$, we want to determine the behavior of $i(t)$. By the relation $i = \frac{dq}{dt}$ and by applying the first Kirchhoff law to the circuit under study, we arrive to the following integro-differential equation in the unknown $i(t)$:

$$v(t) = R\, i(t) + L\, \frac{di}{dt} + \frac{q_0}{C} + \frac{1}{C} \int_0^t i(t)\, dt. \tag{5.71}$$

Differentiating (5.71) with respect to t, we obtain a second order differential equation. But this requires that the voltage $v(t)$ is differentiable, while in practice it may have discontinuities. On the contrary, if we apply the Laplace transform directly to (5.71), such difficulty is avoided.

Denoting by $I(s)$ and $V(s)$ the Laplace transforms of $i(t)$ and $v(t)$ respectively, by Theorem 5.5.15 we obtain

$$\mathcal{L}\left(\int_0^t i(\tau)\, d\tau, s\right) = \frac{I(s)}{s}.$$

Applying the Laplace transform to both sides of (5.71) and exploiting Theorem 5.5.10, we get

$$V(s) = R\, I(s) + L\, (s\, I(s) - i_0) + \frac{q_0}{Cs} + \frac{I(s)}{Cs},$$

whence we deduce that

$$I(s) = \frac{L i_0 C s - q_0}{LCs^2 + RCs + 1} + \frac{Cs}{LCs^2 + RCs + 1} V(s). \tag{5.72}$$

In the right hand side, the first term and the first factor of the second term are rational functions in the variable s, with the degree of the numerator lower than the degree of the denominator. Let us set

$$LCs^2 + RCs + 1 = N(s) \qquad Li_0Cs - q_0 = M(s) \qquad Cs = Q(s)$$

and let us consider the most common case in which the zeros of the denominator, the same for both terms, are distinct: $s_1 \neq s_2$. Applying Proposition 5.7.1 to the first term and transforming back, we obtain

$$\mathcal{L}^{-1}\left(\frac{M(s)}{N(s)}, t\right) = A_1\, e^{s_1 t} + A_2\, e^{s_2 t} = A(t).$$

Applying Proposition 5.7.1 to the second factor of the second term and transforming back, we get

$$\mathcal{L}^{-1}\left(\frac{Q(s)}{N(s)}, t\right) = B_1\, e^{s_1 t} + B_2\, e^{s_2 t} = B(t).$$

Moreover, by Theorem 5.5.16,

$$\frac{Q(s)}{N(s)}\, V(s) = \mathcal{L}(B(t) * v(t), s)$$

whence we derive the formula

$$i(t) = A(t) + B(t) * v(t) = A(t) + \int_0^t v(\tau)\; B(t-\tau)\; d\tau \qquad \forall t > 0. \qquad \square$$

The Laplace transform is useful also for solving ordinary equations with variable coefficients characterized by terms of the kind $t^m\; y^{(n)}(t)$, whose Laplace transform is $(-1)^m \frac{d^m}{ds^m}\mathcal{L}(y^{(n)}(t), s)$.

Example 5.7.5. Consider the following differential equation

$$t\,\frac{d^2y}{dt^2} + 2\frac{dy}{dt} + t\; y = 0\,,$$

with $y(0^+) = 1$ and $y(\pi) = 0$. Setting $y'(0^+) = C$ and applying the Laplace transform to both sides of the equation, by Theorems 5.5.7-5.5.10, we infer

$$-\frac{d}{ds}[s^2\; Y(s) - s\,y(0^+) - y'(0^+)] + 2[s\,Y(s) - y(0^+)] - \frac{d}{ds}Y(s) = 0$$

which implies $-(s^2+1)\; Y'(s) - 1 = 0$, namely

$$Y'(s) = -\frac{1}{s^2+1}\,,$$

where Y' denotes the derivative of Y with respect to s. Integrating, we obtain $Y(s) = -\arctan s + A$. Since by (5.55) it necessarily holds $\lim_{s\to+\infty} Y(s) = 0$, then $A = \frac{\pi}{2}$. Therefore we obtain

$$Y(s) = \frac{\pi}{2} - \arctan s \qquad \forall s > 0.$$

Taking into account Example 5.5.13 and Theorem 5.5.14, we derive

$$y(t) = \mathcal{L}^{-1}\left(\frac{\pi}{2} - \arctan s, t\right) = \frac{\sin t}{t} H(t) \qquad \forall t \in \mathbb{R},$$

which satisfies the condition $y(\pi) = 0$. Therefore, it is the required solution. □

5.7.2 Applications to partial differential equations

In this section we show how the Laplace transform method can be successfully applied also for solving linear partial differential equations. In particular, we illustrate the usage of the Laplace transform for the study of some diffusion problems and of the wave equation in one spatial dimension.

Heat equation on a bounded interval

Let us consider the following problem with Dirichlet boundary condition:

$$\begin{cases} u_t - u_{xx} = f & x \in (0,1),\ t \in (0,\infty) \\ u(x,0) = u_0(x) & x \in (0,1) \\ u(0,t) = g_0(t),\ u(1,t) = g_1(t) & t \in [0,\infty), \end{cases} \tag{5.73}$$

with f defined in $(0,1) \times (0,\infty)$, the initial datum u_0 defined in $(0,1)$ and the boundary data g_0 and g_1 defined in $(0,\infty)$.
Denoting by U, F, G_0, G_1 the "partial" Laplace transforms (with respect to t) of the solution u, the datum f and the boundary data g_0 and g_1 respectively, by (5.73) we obtain the following transformed problem:

$$\begin{cases} s\,U(x,s) - u_0(x) - U_{xx}(x,s) = F(x,s) & 0 < x < 1 \\ U(0,s) = G_0(s) \\ U(1,s) = G_1(s). \end{cases} \tag{5.74}$$

If we consider the particular case in which

$$f = 0 \qquad u_0 = 0 \qquad g_0 = 0 \qquad g_1 = 1,$$

then the solution of the transformed problem reads

$$U(x,s) = \frac{1}{s}\,\frac{\sinh(x\sqrt{s})}{\sinh(\sqrt{s})}\,.$$

Transforming back (see Section 7.1, Capitolo VII in [38]), we obtain the solution to the original problem:

$$u(x,t) = x + \sum_{n=1}^{\infty}(-1)^n\,\frac{2}{n\pi}\,e^{-n^2\pi^2 t}\,\sin(n\pi x).$$

Heat equation on a half-line

Let us consider the following Cauchy-Dirichlet problem for the heat equation on a half-line:

$$\begin{cases} u_t - u_{xx} = 0 & (x,t)\in(0,\infty)\times(0,\infty) \\ u(x,0) = u_0(x) & x\in[0,\infty) \\ u(0,t) = 0 & t\in[0,\infty)\,. \end{cases} \tag{5.75}$$

Denoting by U the partial Laplace transform with respect to t of the solution u, by (5.75) we obtain the following transformed problem:

$$\begin{cases} s\,U(x,s) - u_0(x) - U_{xx}(x,s) = 0 & x>0 \\ U(0,s) = 0\,. \end{cases} \tag{5.76}$$

By solving the second order equation (5.76), we obtain the general solution

$$U(x,s) = A(s)\,e^{-x\sqrt{s}} + B(s)\,e^{x\sqrt{s}} + \frac{1}{\sqrt{s}}\int_0^x \sinh(\sqrt{s}(y-x))u_0(y)\,dy\,, \tag{5.77}$$

with $A(s)$ and $B(s)$ independent of x. Since the computation of the inverse transform of the integral term appearing in (5.77) is difficult, we assume the initial condition to be constant, for instance $u_0 \equiv 1$. Clearly, with such a choice of u_0, the general solution to the equation in (5.76) is given by

$$U(x,s) = A(s)e^{-x\sqrt{s}} + B(s)e^{x\sqrt{s}} + \frac{1}{s}.$$

For the sake of simplicity, let us assume that $B(s) = 0$, so that $U(x,s) \to 0$ as $s \to +\infty$ (see (5.55)) and let us find $A(s)$ by exploiting the condition $U(0,s) = 0$. We get

$$U(x,s) = \frac{1}{s} - \frac{1}{s}\,e^{-x\sqrt{s}}.$$

Now we have to compute the inverse Laplace transform of the map $s \mapsto e^{-x\sqrt{s}}/s$. We need to specify that for every $s = \rho e^{i\vartheta} \in \mathbb{C}$ with $\rho > 0$ and $\vartheta \in (-\pi, \pi)$ we set

$$\sqrt{s} = \sqrt{\rho} e^{i\vartheta/2}.$$

In this way the function $s \mapsto \sqrt{s}$ is holomorphic on $\mathbb{C}$ minus the set of real non positive numbers. Applying formula (5.59) we infer that

$$\mathcal{L}^{-1}\left(\frac{e^{-x\sqrt{s}}}{s}, t\right) = \frac{1}{2\pi i} \int_{\Gamma_\alpha} e^{st} \frac{e^{-x\sqrt{s}}}{s} ds$$

for an arbitrary $\alpha > 0$. At this point, we define for $\varepsilon > 0$ small enough and $R > 0$ large enough, the closed path C as in the picture below,

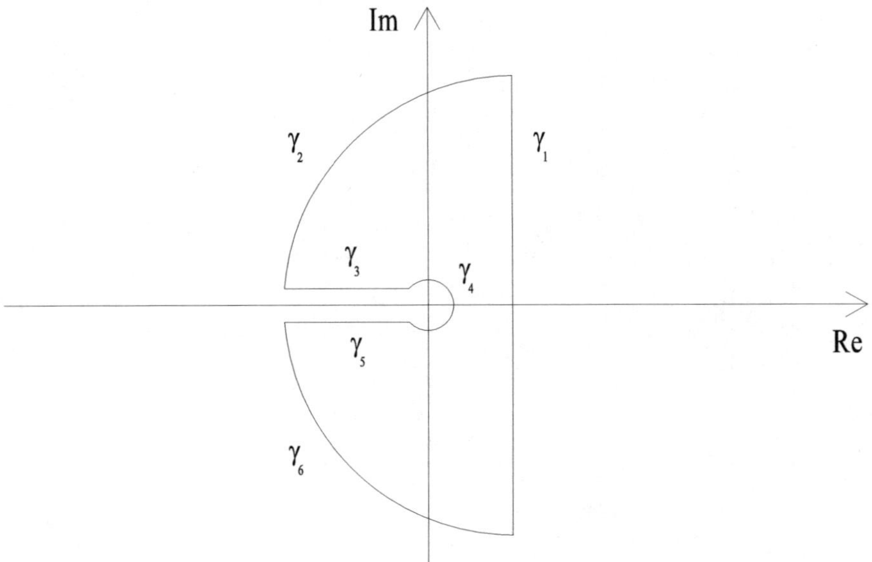

Integration path

where ε is the radius of the arc of circle γ_4, R is the radius of the two arcs of circle γ_2 and γ_6, and finally α is the abscissa of the vertical segment.

Exercise 5.7.6. Prove that the final solution is given by

$$u(x,t) = \frac{1}{\pi} \int_0^{\infty} e^{-\rho t} \frac{\sin(x\sqrt{\rho})}{\rho} d\rho.$$

Hint: pass to the limit as $\varepsilon \to 0^+$ and $R \to +\infty$. See also [38, Paragrafo 7.3, Capitolo VII]. □

Vibrating string equation on a bounded interval

Let us consider the following problem:

$$\begin{cases} u_{tt} - u_{xx} = 0 & (x,t) \in (0,\pi) \times (0,\infty) \\ u(x,0) = 1 & x \in (0,\pi) \\ u_t(x,0) = 0 & x \in (0,\pi) \\ u(0,t) = u(\pi,t) = 0 & t \in [0,\infty). \end{cases} \tag{5.78}$$

Denoting by U the partial Laplace transform with respect to t of the solution u, by (5.78) we obtain the following transformed problem:

$$\begin{cases} s^2\, U(x,s) - s - \dfrac{\partial^2 U}{\partial x^2}(x,s) = 0 & x \in (0,\pi) \\ U(0,s) = U(\pi,s) = 0. \end{cases} \tag{5.79}$$

The solution of the transformed problem reads

$$U(x,s) = \frac{1}{s} - \frac{1}{s}\frac{\sinh(xs)}{\sinh(\pi s)} - \frac{1}{s}\frac{\sinh[(\pi - x)s]}{\sinh(\pi s)}.$$

Transforming back (see [38, Paragrafo 8.1, Capitolo VII]), we obtain the solution of the original problem:

$$u(x,t) = \frac{2}{\pi}\sum_{n=1}^{\infty} \frac{1-(-1)^n}{n}\cos(nt)\sin(nx).$$

Vibrating string equation on a half-line

Let us consider the following problem:

$$\begin{cases} u_{tt} - u_{xx} = 0 & (x,t) \in (0,\infty) \times (0,\infty) \\ u(x,0) = 0 \quad u_t(x,0) = 1 & x \in [0,\infty) \\ u(0,t) = 0 & t \in [0,\infty). \end{cases} \tag{5.80}$$

Denoting by U the partial Laplace transform with respect to t of the solution u, by (5.80) we obtain the following transformed problem:

$$\begin{cases} s^2\, U(x,s) - 1 - \dfrac{\partial^2 U}{\partial x^2}(x,s) = 0 & x > 0 \\ U(0,s) = 0. \end{cases} \tag{5.81}$$

The solution of the transformed problem reads

$$U(x,s) = \frac{1}{s^2} + A(s)\, e^{-xs} + B(s)\, e^{xs}\,,$$

with $A(s)$ and $B(s)$ independent of x. At this point, setting $B(s) = 0$ (see (5.55)), we can determine $A(s)$ exploiting the Dirichlet condition $U(0,s) = 0$. We get:

$$U(x,s) = \frac{1}{s^2} - \frac{1}{s^2}\, e^{-xs}$$

Transforming back (see [38, Paragrafo 8.3, Capitolo VII]), we obtain the solution

$$u(x,t) = (t)^+ - (t-x)^+ = \min\{x,t\}, \quad x > 0,\ t > 0.$$

Chapter 6

Hilbertian Sobolev spaces

The Sobolev spaces H^k with k positive integer, are those vector subspaces of L^2 of functions having k weak derivatives in L^2. The main reason which leads to the study of these spaces is illustrated in the first section of this chapter. In the rest of the chapter, the presentation of Sobolev spaces follows step by step, so that the reader can decide where to stop the study. We shall begin with first order Sobolev spaces in dimension $n = 1$, then we shall move on to higher dimensions $n \geq 2$. We proceed with Sobolev spaces of integer order $k \geq 2$ and then of real positive order $s \geq 0$. Along with the notion of traces, we will introduce the subspaces of functions having zero trace. By duality, this allows to define Sobolev spaces with negative exponent.

6.1 Why study Sobolev spaces?

Sobolev spaces have their main motivation in the study of differential equations, both ordinary and partial ones. In order to illustrate their use, let us consider a simple example. Given a function $f \in C^0([a,b])$, we want to determine a function $u(x)$ verifying

$$\begin{cases} -u'' + u = f & \text{in } (a,b) \\ u(a) = u(b) = 0. \end{cases} \tag{6.1}$$

A **classic** (or strong) solution to problem (6.1) is a solution $u \in C^2([a,b])$ which verifies (6.1) in the usual sense, namely pointwise. However, a good theory of existence and uniqueness of solutions is possible only in the framework of Hilbert spaces; while the spaces C^k, even if complete, are not Hilbert spaces. To explain how Sobolev spaces appear in a natural way, we multiply (6.1) by $\varphi \in \mathcal{D}(a,b)$ and we integrate by parts: we get

$$\int_a^b u'\varphi' + \int_a^b u\varphi = \int_a^b f\varphi \quad \forall \varphi \in \mathcal{D}(a,b). \tag{6.2}$$

We observe that in order that (6.2) makes sense, it suffices that $u, \varphi, u', \varphi' \in L^2(a,b)$. So we say that a function u such that $u, u' \in L^2(a,b)$ is a **weak** solution to (6.1) if it satisfies (6.2). The space of functions u such that $u \in L^2(\Omega)$ and $u' \in L^2(\Omega)$ is the right space in which looking for solutions to (6.2). Such space is exactly what will be called Sobolev space. We had to pay in regularity of the solution; but this matter will be overcome later. The following scheme briefly describes the *variational method* in the theory of (ordinary or partial) differential equations:

1. specify the notion of weak solution in a suitable Sobolev space;

2. establish *existence* and *uniqueness* of the weak solution by applying Lax-Milgram Theorem (Corollary 1.7.4) which, we recall, requires a Hilbertian structure;

3. show that the weak solution found enjoys more regularity, for instance C^2: here a good theory of *regularity* is necessary (a short account of this theory can be found in Section 7.1.5);

4. prove that a regular (of class C^2) weak solution is a classic solution.

Step 4. is easy. Indeed, in the case of problem (6.2), let us assume that $u \in C^2([a,b])$, $u(a) = u(b) = 0$ and that u verifies (6.2). Integrating by parts we get:

$$\int_a^b (-u'' + u - f)\varphi = 0 \qquad \forall \varphi \in \mathcal{D}(a,b).$$

Thus, by Proposition 2.4.22, we deduce that $-u'' + u - f = 0$ a.e. in (a,b). Since all the functions involved are continuous, we can remove "a.e." and obtain $-u''+u = f$, in pointwise sense in (a,b).

6.2 First order Hilbertian Sobolev spaces

6.2.1 First order Sobolev spaces in dimension $n = 1$

In this section, we shall denote by $I \subseteq \mathbb{R}$ the generic interval $I = (a,b)$, possibly unbounded. If necessary, further assumptions will be specified. We begin with the following

Definition 6.2.1. *The* **Sobolev space** $H^1(I)$ *is defined by*

$$H^1(I) = \left\{ u \in L^2(I);\ \exists g \in L^2(I) \ \ s.t. \ \int_I u\, \varphi' = -\int_I g\, \varphi,\ \forall \varphi \in \mathcal{D}(I) \right\}.$$

For every $u \in H^1(I)$, the function g introduced in Definition 6.2.1 will be called **weak derivative** of u and will be denoted by u'. If $u \in C^1(I) \cap L^2(I)$ and if $u' \in L^2(I)$ (where u' is the classical derivative of u), then $u \in H^1(I)$ and the classical derivative of u coincides with the weak derivative of u. In particular, if I is bounded, then $C^1(\overline{I}) \subset H^1(I)$.

Exercise 6.2.2. Let $I = (-1, 1)$ and let $u(x) = |x|$. Prove that $u \in H^1(I)$ and determine the weak derivative of u. □

We endow the space $H^1(I)$ with the scalar product defined by

$$(u, v)_{H^1} = (u, v)_{L^2} + (u', v')_{L^2} \qquad \forall u, v \in H^1(I)$$

and with the induced norm

$$\|u\|_{H^1} = (\|u\|_{L^2}^2 + \|u'\|_{L^2}^2)^{\frac{1}{2}} \qquad \forall u \in H^1(I).$$

Thus we obtain the

Theorem 6.2.3. *The space $H^1(I)$ is a separable Hilbert space.*

Proof. We start showing that $H^1(I)$ is complete. Let then $\{u_n\} \subset H^1(I)$ be a Cauchy sequence. Thus for every $\varepsilon > 0$ there exists $\bar{n} \in \mathbb{N}$ such that

$$\|u_n - u_m\|_{L^2}^2 + \|u_n' - u_m'\|_{L^2}^2 = \|u_n - u_m\|_{H^1}^2 < \varepsilon^2 \qquad \forall n, m > \bar{n}.$$

Consequently, $\{u_n\}$ and $\{u_n'\}$ are both Cauchy sequences in $L^2(I)$ and, since $L^2(I)$ is complete (see Theorem 2.4.10), we deduce that there exist two functions $u, v \in L^2(I)$ such that

$$u_n \to u \quad \text{and} \quad u_n' \to v \quad \text{in } L^2(I) \tag{6.3}$$

as $n \to \infty$. At this point we have to prove that $u \in H^1(I)$ and that $u' = v$. By (6.3) we deduce

$$\int_I u_n \varphi' \to \int_I u \varphi' \qquad \text{and} \qquad \int_I u_n' \varphi \to \int_I v \varphi \qquad \forall \varphi \in \mathcal{D}(I). \tag{6.4}$$

By definition of weak derivative we also have that

$$\int_I u_n' \varphi = -\int_I u_n \varphi' \quad \forall \varphi \in \mathcal{D}(I)$$

for every $n \in \mathbb{N}$; by combining the last identity with (6.4) we obtain

$$\int_I v \varphi = -\int_I u \varphi' \quad \forall \varphi \in \mathcal{D}(I).$$

This proves that u admits a weak derivative in $L^2(I)$ and that $u' = v$. Therefore $u \in H^1(I)$ and, in view of (6.3), $u_n \to u$ in $H^1(I)$ as $n \to \infty$.

In order to show the separability of $H^1(I)$, we introduce the linear operator $T : H^1(I) \to L^2(I) \times L^2(I)$ defined by

$$Tu := (u, u') \quad \forall u \in H^1(I).$$

Thanks to Theorem 2.4.10, the product space $L^2(I) \times L^2(I)$ turns out to be a separable Hilbert space if endowed with the scalar product

$$[(u_1, u_2), (v_1, v_2)] := (u_1, v_1)_{L^2} + (u_2, v_2)_{L^2} \qquad \forall (u_1, u_2), (v_1, v_2) \in L^2(I) \times L^2(I).$$

Therefore the operator T is an isometry from $H^1(I)$ into $L^2(I) \times L^2(I)$. In conclusion, the Sobolev space $H^1(I)$ is separable, being isometric to a subspace of the separable space $L^2(I) \times L^2(I)$. □

The next result shows that a function of $H^1(I)$ can be identified with a continuous function, up to negligible sets. Before stating the result, let us introduce the following

Lemma 6.2.4. *Let $f \in L^1_{loc}(I)$ satisfy*

$$\int_I f\varphi' = 0 \quad \forall \varphi \in \mathcal{D}(I).$$

Then there exists a constant C such that $f = C$ a.e. in I.

Proof. See [10, Lemma VIII.1]. □

Now we can state the

Theorem 6.2.5. *Let $u \in H^1(I)$. Then there exists a function $\widetilde{u} \in C^0(\overline{I})$ such that $u = \widetilde{u}$ a.e. in I and*

$$\widetilde{u}(x) - \widetilde{u}(y) = \int_x^y u'(t)\, dt \quad \forall x, y \in \overline{I}.$$

Proof. Let us fix $y_0 \in I$ and let us set

$$\widetilde{v}(x) = \int_{y_0}^x u'(t)\, dt.$$

We leave to the reader to check that $\widetilde{v} \in C^0(\overline{I})$ and that

$$\int_I \widetilde{v}\varphi' = -\int_I u'\varphi \qquad \forall \varphi \in \mathcal{D}(I).$$

On the other hand, by definition of weak derivative, it holds

$$\int_I u\varphi' = -\int_I u'\varphi \qquad \forall \varphi \in \mathcal{D}(I)$$

and hence

$$\int_I (u - \widetilde{v})\varphi' = 0 \qquad \forall \varphi \in \mathcal{D}(I).$$

In view of Definition 2.6.11, we infer that $(u - \widetilde{v})' = 0$ in the sense of distributions; but then, applying Lemma 6.2.4, we deduce that $u - \widetilde{v} = C$ a.e. in I, for some constant $C \in \mathbb{R}$. Therefore, the function $\widetilde{u}(x) = \widetilde{v}(x) + C$ fulfills the required properties. □

Let us clarify the importance of Theorem 6.2.5. We recall that if $u \in H^1(I)$, then every function v such that $v = u$ a.e. in I still belongs to $H^1(I)$. Theorem 6.2.5 states that every function $u \in H^1(I)$ admits a unique continuous representative, namely there exists a continuous function which belongs to the equivalence class of u, where the equivalence is meant with respect to the relation of equality a.e. We recall that also the property "u admits a continuous representative" is distinct to the property "u is continuous a.e." (see after Theorem 2.7.1).

Finally, by Theorem 6.2.5, we infer that if $u \in H^1(I)$ and if $u' \in C^0(\overline{I})$, then $u \in C^1(\overline{I})$ (more precisely, $\widetilde{u} \in C^1(\overline{I})$).

Exercise 6.2.6. (i) Let $u, v \in H^1(I)$; prove that $u \cdot v \in H^1(I)$ and that $(u \cdot v)' = u' \cdot v + u \cdot v'$.

(ii) Let $u \in H^1(I)$ and let $G \in C^1(\mathbb{R})$ satisfy $G(0) = 0$; prove that $G \circ u \in H^1(I)$ and that $(G \circ u)' = (G' \circ u) \cdot u'$. □

We now introduce an important closed subspace of H^1.

Definition 6.2.7. *We denote by $H^1_0(I)$ the closure of $\mathcal{D}(I)$ in $H^1(I)$.*

The space $H^1_0(I)$ is a separable Hilbert space if endowed with the product induced by $H^1(I)$. If $I = \mathbb{R}$, we know that $\mathcal{D}(\mathbb{R})$ is dense in $H^1(\mathbb{R})$ (see [10, Teorema VIII.6]) and hence $H^1_0(\mathbb{R}) = H^1(\mathbb{R})$. In the case $I \neq \mathbb{R}$, the following result gives a characterization of the elements of $H^1_0(I)$:

Theorem 6.2.8. *Let $I \neq \mathbb{R}$ and let $u \in H^1(I)$. Then $u \in H^1_0(I)$ if and only if $u = 0$ on ∂I.*

Proof. See [10, Teorema VIII.11]. □

This theorem fits well the model problem (6.1), in which the differential equation is coupled with homogeneous Dirichlet conditions. Since u vanishes on ∂I, the solution to (6.2) has to be searched in $H^1_0(I)$. Let us state an important result about the functions of $H^1_0(I)$.

Theorem 6.2.9. (Poincaré inequality). *Assume that $I = (a, b)$ is bounded. Then*

$$\|u\|_{L^2} \leq (b - a) \cdot \|u'\|_{L^2} \quad \forall u \in H^1_0(I).$$

As a consequence, the map $u \mapsto \|u'\|_{L^2}$ defines a norm in $H^1_0(I)$ which is equivalent to the norm of $H^1(I)$.

Proof. For every $u \in H_0^1(I)$ it holds

$$\|u\|_{L^2}^2 = \int_a^b |u(t)|^2 dt \le (b-a) \cdot \|u\|_{L^\infty}^2. \tag{6.5}$$

On the other hand, in view of Theorem 6.2.5 and of the Hölder inequality (2.17), for every $x \in I$ we have

$$|u(x)| = \left| \int_a^x u'(t)dt \right| \le \int_a^b |u'(t)|dt \le (b-a)^{1/2} \cdot \|u'\|_{L^2}$$

and hence

$$\|u\|_{L^\infty} \le (b-a)^{1/2} \cdot \|u'\|_{L^2}. \tag{6.6}$$

By (6.5)-(6.6) we obtain

$$\|u\|_{L^2} \le (b-a) \cdot \|u'\|_{L^2}.$$

Finally, the inequalities

$$\|u'\|_{L^2} \le \|u\|_{H^1} = (\|u\|_{L^2}^2 + \|u'\|_{L^2}^2)^{1/2} \le ((b-a)^2+1)^{1/2} \|u'\|_{L^2} \qquad \forall u \in H^1(I)$$

imply that the map $u \mapsto \|u'\|_{L^2}$ defines a norm in $H_0^1(I)$, which is equivalent to the norm of $H^1(I)$. □

We denote by $H^{-1}(I)$ the dual space of $H_0^1(I)$. Thanks to Theorem 1.5.5, it is possible to identify $L^2(I)$ with its dual, but we cannot identify $H_0^1(I)$ with its dual at the same time. In this way we obtain the following Hilbert triplet

$$H_0^1(I) \subset L^2(I) \subset H^{-1}(I)$$

(see Section 1.8). The continuous embedding of $L^2(I)$ into $H^{-1}(I)$ is to be meant as follows: we associate with each $u \in L^2(I)$ the linear functional $I_u : H_0^1(I) \mapsto \mathbb{R}$ defined by

$$I_u(v) := \int_I uv \qquad \forall v \in H_0^1(I).$$

Clearly, $I_u \in H^{-1}(I)$ and hence we can identify the function u with the functional I_u, obtaining the desired embedding. The elements of H^{-1} can be represented as "derivatives" of L^2-functions; more precisely, it holds the following

Proposition 6.2.10. *Let $F \in H^{-1}(I)$. Then there exist $f_0, f_1 \in L^2(I)$ such that*

$$\langle F, v \rangle = \int_I f_0 v + \int_I f_1 v' \qquad \forall v \in H_0^1(I)$$

and $\|F\|_{H^{-1}} = \max\{\|f_0\|_{L^2}, \|f_1\|_{L^2}\}$. If I is bounded, f_0 can be taken equal to zero.

Proof. See [10, Proposizione VIII.13]. □

The functions f_0 and f_1 determined in Proposition 6.2.10 are not unique. In the case in which I is bounded, Proposition 6.2.10 states that

$$\langle -F, \varphi \rangle = -\int_I f_1 \varphi' = _{\mathcal{D}'} \langle f_1', \varphi \rangle_{\mathcal{D}} \qquad \forall \varphi \in \mathcal{D}'(I) \ ,$$

where $\langle \cdot, \cdot \rangle$ denotes the duality between $H^{-1}(I)$ and $H_0^1(I)$, while $_{\mathcal{D}'}\langle \cdot, \cdot \rangle_{\mathcal{D}}$ denotes the duality product between the space of distributions $\mathcal{D}'(I)$ and the space of test functions $\mathcal{D}(I)$. In view of what we have just said and by Definition 2.6.11, we deduce that $-F$ is the distributional derivative of $f_1 \in L^2(I)$. Basically, the functionals of $H^{-1}(I)$ have "-1 derivatives" in $L^2(I)$.

6.2.2 First order Sobolev spaces in dimension $n \geq 2$

To better understand this subsection, it is important to remind Definition 3.1.2, which characterizes the regularity of the boundary of an open set $\Omega \subset \mathbb{R}^n$.

We begin with the following

Definition 6.2.11. *Let $\Omega \subseteq \mathbb{R}^n$ be an open set and let $u \in L^2(\Omega)$. We will say that u admits a weak derivative in $L^2(\Omega)$ with respect to x_i if there exists $g \in L^2(\Omega)$ such that*

$$\int_\Omega u \, \frac{\partial \varphi}{\partial x_i} \, dx = -\int_\Omega g \, \varphi \, dx, \ \forall \varphi \in \mathcal{D}(\Omega) \, .$$

We shall then call such g the **partial weak derivative** *of u with respect to x_i and we will write $\frac{\partial u}{\partial x_i} = g_i$.*

We are now in position to define the Sobolev space $H^1(\Omega)$:

Definition 6.2.12. *Let $\Omega \subseteq \mathbb{R}^n$ be an open set. The* **Sobolev space** *$H^1(\Omega)$ is defined by*

$$H^1(\Omega) = \left\{ u \in L^2(\Omega); \ \frac{\partial u}{\partial x_i} \in L^2(\Omega), \ \forall i = 1, ..., n \right\} \ ,$$

where $\frac{\partial u}{\partial x_i}$ denotes the weak derivative of u introduced in Definition 6.2.11.

Given $u \in H^1(\Omega)$, we will set

$$\nabla u = \left(\frac{\partial u}{\partial x_1}, \frac{\partial u}{\partial x_2}, ..., \frac{\partial u}{\partial x_n} \right) .$$

Clearly, if $u \in C^1(\Omega) \cap L^2(\Omega)$ and if $\frac{\partial u}{\partial x_i} \in L^2(\Omega)$ for every $i = 1, ..., n$ (where $\frac{\partial u}{\partial x_i}$ stays for the partial derivative of u in the classical sense), then $u \in H^1(\Omega)$. Moreover, the classical partial derivatives coincide with the weak ones. In particular, if Ω is bounded, then

$$C^1(\overline{\Omega}) \subset H^1(\Omega).$$

Conversely, it is proved that if $u \in H^1(\Omega) \cap C^0(\Omega)$ and if $\frac{\partial u}{\partial x_i} \in C^0(\Omega)$ for every $i = 1, ..., n$ (where $\frac{\partial u}{\partial x_i}$ stays for the weak partial derivative), then $u \in C^1(\Omega)$.

The space $H^1(\Omega)$ is endowed with the scalar product

$$(u,v)_{H^1} = (u,v)_{L^2} + \sum_{i=1}^{n} \left(\frac{\partial u}{\partial x_i}, \frac{\partial v}{\partial x_i} \right)_{L^2} = \int_\Omega (uv + \nabla u \nabla v) \qquad \forall u, v \in H^1(\Omega)$$

with induced norm

$$\|u\|_{H^1} = \left(\|u\|_{L^2}^2 + \sum_{i=1}^{n} \left\| \frac{\partial u}{\partial x_i} \right\|_{L^2}^2 \right)^{\frac{1}{2}} = (\|u\|_{L^2}^2 + \|\nabla u\|_{L^2}^2)^{\frac{1}{2}} \qquad \forall u \in H^1(\Omega).$$

Exercise 6.2.13. Prove that the map $u \mapsto \|u\|_{L^2} + \|\nabla u\|_{L^2}$ defines an equivalent norm in $H^1(\Omega)$. □

From the separability and the completeness of $L^2(\Omega)$, we derive the

Theorem 6.2.14. *The space $H^1(\Omega)$ is a separable Hilbert space.*

Proof. We refer to a more general case (see Theorem 6.3.5). □

Definition 6.2.15. *We denote by $H_0^1(\Omega)$ the closure of $\mathcal{D}(\Omega)$ in $H^1(\Omega)$.*

If equipped with the scalar product of $H^1(\Omega)$, $H_0^1(\Omega)$ is a Hilbert space. The functions of $H_0^1(\Omega)$ are essentially the functions of $H^1(\Omega)$ which "vanish on $\partial\Omega$". However, to give a precise meaning to this sentence is a delicate matter, since a function $u \in H^1(\Omega)$ is defined only a.e. and $\partial\Omega$ is negligible, having Lebesgue measure zero (see Section 2.4). The function u may not have a continuous representative, since in dimension $n \geq 2$ Theorem 6.2.5 is not valid, as the following example shows.

Example 6.2.16. Let $n = 2$ and let $\Omega = \{x \in \mathbb{R}^2;\ |x| < \frac{1}{2}\}$. Consider over Ω the function $f(x) = \log|\log|x||$. Then $f \in H^1(\Omega)$ but $f \notin L^\infty(\Omega)$. □

However, we can affirm that among the functions of $H_0^1(\Omega)$, continuous functions are the ones that vanish on $\partial\Omega$:

Theorem 6.2.17. *Let $\Omega \subset \mathbb{R}^n$ be an open set with C^1 boundary. Let $u \in H^1(\Omega) \cap C^0(\overline{\Omega})$. Then*

$$u = 0 \text{ on } \partial\Omega \qquad \Longleftrightarrow \qquad u \in H_0^1(\Omega).$$

Proof. See [10, Teorema IX.17]. □

Theorem 6.2.9 admits a natural extension to the higher dimensional case $n \geq 2$:

Theorem 6.2.18. (Poincaré inequality). *Assume that Ω is bounded. Then there exists a constant C (depending on Ω) such that*

$$\|u\|_{L^2} \le C\|\nabla u\|_{L^2} \qquad \forall u \in H_0^1(\Omega).$$

In particular, the map $u \mapsto \|\nabla u\|_{L^2}$ is a norm in $H_0^1(\Omega)$, which is equivalent to the norm $\|u\|_{H^1}$.

Proof. We start by showing the inequality in the case $u \in \mathcal{D}(\Omega)$. The function u can be extended trivially outside Ω, obtaining $u \in \mathcal{D}(\mathbb{R}^n)$. Since Ω is bounded, there exist $a, b \in \mathbb{R}$ such that $\overline{\Omega} \subset (a,b) \times \mathbb{R}^{n-1}$. Then it holds

$$u(x) = \int_a^{x_1} \frac{\partial u}{\partial x_1}(t, x_2, ..., x_n)dt \qquad\qquad \forall x = (x_1, ..., x_n) \in \Omega$$

and by the Hölder inequality (2.17) we have

$$|u(x)| \le (x_1 - a)^{1/2} \left(\int_a^{x_1} \left| \frac{\partial u}{\partial x_1}(t, x_2, ..., x_n) \right|^2 dt \right)^{1/2}$$

$$\le (b-a)^{1/2} \left(\int_a^{b} \left| \frac{\partial u}{\partial x_1}(t, x_2, ..., x_n) \right|^2 dt \right)^{1/2} .$$

Integrating with respect to $x_2, ..., x_n$ over $\mathbb{R}^{n-1}$ and recalling that supp$u \subset \Omega$, we obtain

$$\int_{\mathbb{R}^{n-1}} |u(x)|^2 dx_2 \cdots dx_n \le (b-a) \int_{(a,b)\times\mathbb{R}^{n-1}} \left| \frac{\partial u}{\partial x_1}(t, x_2, ..., x_n) \right|^2 dt dx_2 \cdots dx_n$$

$$= (b-a) \int_\Omega \left| \frac{\partial u}{\partial x_1} \right|^2 dx \le (b-a)\|\nabla u\|_{L^2}^2.$$

Integrating with respect to x_1 and recalling that supp$u \subset \Omega$, the last chain of inequalities yields

$$\|u\|_{L^2}^2 = \int_\Omega |u(x)|^2 dx = \int_a^b \left(\int_{\mathbb{R}^{n-1}} |u(x)|^2 dx_2 \cdots dx_n \right) dx_1 \le (b-a)^2 \|\nabla u\|_{L^2}^2.$$

The previous inequality con be extended by density to the functions of $H_0^1(\Omega)$, being by definition $\mathcal{D}(\Omega)$ dense in $H_0^1(\Omega)$. □

We denote by $H^{-1}(\Omega)$ the dual space of $H_0^1(\Omega)$. In analogy with the one-dimensional case, we have the Hilbert triplet

$$H_0^1(\Omega) \subset L^2(\Omega) \subset H^{-1}(\Omega) \ ,$$

where the continuous embedding $L^2(\Omega) \subset H^{-1}(\Omega)$ is to be meant as follows: each function $u \in L^2(\Omega)$ is identified with the linear functional $I_u \in H^{-1}(\Omega)$ defined by

$$\langle I_u, v\rangle := \int_\Omega uv \qquad \forall v \in H_0^1(\Omega).$$

The elements of H^{-1} can be represented by means of L^2-functions; more precisely, it holds the following

Proposition 6.2.19. *Let $F \in H^{-1}(\Omega)$. Then there exist $f_0, f_1, ..., f_n \in L^2(\Omega)$ such that*

$$\langle F, v\rangle = \int_\Omega f_0 v + \sum_{i=1}^n \int_\Omega f_i \frac{\partial v}{\partial x_i} \qquad \forall v \in H_0^1(\Omega)$$

and

$$\|F\|_{H^{-1}} = \max_{0\le i\le n} \{\|f_i\|_{L^2}\}.$$

If Ω is bounded, f_0 can be taken equal to zero.

Proof. See [10, Proposizione IX.20]. □

Let $\Omega \subset \mathbb{R}^n$ be a bounded open set and assume that $f = (f_1, ..., f_n) \in [C^1(\overline{\Omega})]^n$. Then, by Gauss-Green formula (whose general formulation can be found in (6.16)), we have

$$\int_\Omega v \,\mathrm{div} f = -\int_\Omega f \cdot \nabla v \qquad \forall v \in H_0^1(\Omega).$$

Thus, Proposition 6.2.19 states that a function $F \in H^{-1}(\Omega)$ is the weak divergence of a suitable field of $[L^2(\Omega)]^n$.

Recalling Definition 1.8.5, we pass to the statement of the following embedding theorem:

Theorem 6.2.20. (Sobolev Theorem). *Let Ω be $\mathbb{R}^n$ or an open set with Lipschitz boundary. Then there hold the following continuous embeddings:*
(i) if $n = 2$ then $H^1(\Omega) \subset L^q(\Omega)$, for every $2 \le q < \infty$;
(ii) if $n \ge 3$ then $H^1(\Omega) \subset L^q(\Omega)$, for every $2 \le q \le 2^* = \frac{2n}{n-2}$.

Proof. See [1, Theorem 5.4]. □

If in addition Ω is bounded, then the embeddings stated in Theorem 6.2.20 become compact, as it is shown in the following

Theorem 6.2.21. (Theorem of Rellich, Sobolev, Gagliardo, Nirenberg). *Let $\Omega \subset \mathbb{R}^n$ be a bounded open set with Lipschitz boundary. Then there hold the following continuous embeddings:*
(i) if $n = 2$ then $H^1(\Omega) \subset\subset L^q(\Omega)$, for every $1 \le q < \infty$;
(ii) if $n \ge 3$ then $H^1(\Omega) \subset\subset L^q(\Omega)$, for every $1 \le q < 2^* = \frac{2n}{n-2}$.

Proof. See [1, Theorem 6.2]. □

6.3 The spaces H^k $(k \in \mathbb{N})$

By iterating Definition 6.2.11 we can define higher order weak derivatives:

Definition 6.3.1. *Let $\Omega \subseteq \mathbb{R}^n$ be an open set and let $u \in L^1_{loc}(\Omega)$. Given a multi-index α we will say that u admits a weak derivative $D^\alpha u$ if there exists $g \in L^1_{loc}(\Omega)$ such that*

$$\int_\Omega u D^\alpha \varphi = (-1)^{|\alpha|} \int_\Omega g\varphi \qquad \forall \varphi \in \mathcal{D}(\Omega).$$

If so, we set $D^\alpha u := g$.

Once the meaning of higher order weak derivatives is specified, we can give the following

Definition 6.3.2. *Let $k \in \mathbb{N}$ and let $\Omega \subseteq \mathbb{R}^n$ be an open set. The Sobolev space $H^k(\Omega)$ is defined by*

$$H^k(\Omega) = \{u \in L^2(\Omega);\ D^\alpha u \in L^2(\Omega) \quad \forall \alpha \in \mathbb{N}^n \text{ s.t. } |\alpha| \leq k\}.$$

It is endowed with the scalar product

$$(u,v)_{H^k} = \sum_{0\leq|\alpha|\leq k} \int_\Omega D^\alpha u D^\alpha v \qquad \forall u, v \in H^k(\Omega) \tag{6.7}$$

and with induced norm

$$\|u\|_{H^k} = \left(\sum_{0\leq|\alpha|\leq k} \|D^\alpha u\|^2_{L^2} \right)^{\frac{1}{2}}. \tag{6.8}$$

Exercise 6.3.3. Prove that (6.8) defines a norm. □

Remark 6.3.4. Formally, we can set $H^0(\Omega) = L^2(\Omega)$; this allows to define inductively the Sobolev spaces

$$H^{k+1}(\Omega) = \{u \in H^k(\Omega);\ \nabla u \in (H^k(\Omega))^n\} \qquad \forall k \geq 0.$$

In particular, it holds $H^m(\Omega) \subset H^k(\Omega)$ if $m > k \geq 0$, moreover $\|u\|_{H^k} \leq \|u\|_{H^m}$ for every $u \in H^m(\Omega)$. □

Among the L^p spaces only L^2 has the structure of Hilbert space; this important property extends to all the H^k spaces, as states the

Theorem 6.3.5. *The spaces $H^k(\Omega)$ are separable Hilbert spaces with the scalar product* (6.7).

Proof. The key point is the completeness of $H^k(\Omega)$ with respect to the norm (6.8): let $\{u_m\}$ be a Cauchy sequence in $H^k(\Omega)$, then $\{D^\alpha u_m\}_{m=1}^{\infty}$ is a Cauchy sequence in $L^2(\Omega)$ for every α such that $0 \le |\alpha| \le k$.

But $L^2(\Omega)$ is complete, therefore there exists $u_\alpha \in L^2(\Omega)$ such that the convergence

$$D^\alpha u_m \to u_\alpha \qquad \forall \alpha \in \mathbb{N}^n \;\; s.t. \;\; 0 \le |\alpha| \le k \tag{6.9}$$

holds true in $L^2(\Omega)$. Now we have to prove that $u_\alpha = D^\alpha u$. By $u_m \xrightarrow{L^2} u$ we deduce $u_m \xrightarrow{\mathcal{D}'} u$ and, since the differentiation is a continuous operator in $\mathcal{D}'(\Omega)$, we also infer that

$$\left.\begin{array}{r} D^\alpha u_m \xrightarrow{\mathcal{D}'} D^\alpha u \\ (6.9) \Rightarrow D^\alpha u_m \xrightarrow{\mathcal{D}'} u_\alpha \end{array}\right\} \Rightarrow u_\alpha = D^\alpha u.$$

In view of (6.9) and since $u_\alpha = D^\alpha u$, we immediately obtain $u_m \to u$ in $H^k(\Omega)$. The separability of $H^k(\Omega)$ follows from the separability of $L^2(\Omega)$, being $H^k(\Omega)$ isometric to a subspace of $(L^2(\Omega))^N$, where N denotes the number of all the partial derivatives D^α for α varying in $\mathbb{N}^n$ such that $0 \le |\alpha| \le k$. □

In the particular case $\Omega = \mathbb{R}^n$ we can give an equivalent definition for the space $H^k(\mathbb{R}^n)$ by means of the Fourier transform of a tempered distribution (see Definitions 2.6.14 and 5.14). Since the Fourier transform acts in complex settings, we need to define the complex Sobolev space

$$H^k(\mathbb{R}^n, \mathbb{C}) = \{u_1 + iu_2;\; u_1, u_2 \in H^k(\mathbb{R}^n)\}, \tag{6.10}$$

endowed with the complex scalar product

$$(u, v)_{H^k} = \sum_{0 \le |\alpha| \le k} \int_{\mathbb{R}^n} D^\alpha u \overline{D^\alpha v}. \tag{6.11}$$

In the next statement we will denote by $\mathcal{S}'(\mathbb{R}^n, \mathbb{C})$ the complex space of tempered distributions.

Theorem 6.3.6. *For every $k \in \mathbb{N}$ there holds*

$$H^k(\mathbb{R}^n, \mathbb{C}) = \{u \in \mathcal{S}'(\mathbb{R}^n, \mathbb{C});\; (1 + |\xi|^2)^{\frac{k}{2}} \cdot \widehat{u}(\xi) \in L^2(\mathbb{R}^n, \mathbb{C})\} \tag{6.12}$$

and the norm

$$\|u\|_* = \|(1 + |\xi|^2)^{\frac{k}{2}} \cdot \widehat{u}(\xi)\|_{L^2}$$

turns out to be equivalent to the norm induced by the scalar product (6.11).

Proof. Thanks to the definition of Fourier transform for L^2 functions described in Section 5.3, we can affirm that $u \in H^k(\mathbb{R}^n, \mathbb{C})$ if and only if $\mathcal{F}(D^\alpha u) \in L^2(\mathbb{R}^n, \mathbb{C})$ for every multi-index α such that $|\alpha| \leq k$. In view of Theorem 5.3.8, the last condition is equivalent to say that $\xi^\alpha \widehat{u}(\xi) \in L^2(\mathbb{R}^n, \mathbb{C})$. Moreover, by (5.8), we infer

$$(2\pi)^{n/2}\|D^\alpha u\|_{L^2} = \|\mathcal{F}(D^\alpha u)\|_{L^2} = \|\xi^\alpha \widehat{u}\|_{L^2}. \tag{6.13}$$

On the other hand, it is easy to verify that there exist two positive constants C_1, C_2 depending only on n and k such that

$$C_1(1+|\xi|^2)^k \leq \sum_{0\leq|\alpha|\leq k} (\xi^\alpha)^2 \leq C_2(1+|\xi|^2)^k \qquad \forall \xi \in \mathbb{R}^n.$$

Therefore, by (6.13), we obtain

$$(2\pi)^n\|u\|_{H^k}^2 = (2\pi)^n \sum_{0\leq|\alpha|\leq k} \|D^\alpha u\|_{L^2}^2 = \sum_{0\leq|\alpha|\leq k} \int_{\mathbb{R}^n} (\xi^\alpha)^2 |\widehat{u}(\xi)|^2 d\xi$$

and also

$$C_1 \int_{\mathbb{R}^n} (1+|\xi|^2)^k |\widehat{u}(\xi)|^2 d\xi \leq (2\pi)^n \|u\|_{H^k}^2 \leq C_2 \int_{\mathbb{R}^n} (1+|\xi|^2)^k |\widehat{u}(\xi)|^2 d\xi.$$

This proves that $u \in H^k(\mathbb{R}^n, \mathbb{C})$ if and only if $(1+|\xi|^2)^{k/2}\widehat{u} \in L^2(\mathbb{R}^n, \mathbb{C})$ and that the norm $\|\cdot\|_*$ is equivalent to the norm induced by (6.11). □

Let us now see some properties of the spaces H^k.

Proposition 6.3.7. (property of the intermediate derivatives). *Let $\Omega \subset \mathbb{R}^n$ be a bounded open set with Lipschitz boundary. If $u \in L^2(\Omega)$ and $D^\alpha u \in L^2(\Omega)$ for $|\alpha| = k$, then $u \in H^k(\Omega)$.*

Proof. See [1, Theorem 4.17]. □

There holds also the following result:

Theorem 6.3.8. *(i) If Ω is an open set with finite measure and $u \in C^k(\overline{\Omega})$, then $u \in H^k(\Omega)$ and*

$$\|u\|_{H^k}^2 \leq |\Omega| \cdot \sum_{|\alpha|\leq k} \|D^\alpha u\|_{L^\infty}^2,$$

where $|\Omega|$ denotes the Lebesgue measure of Ω.

(ii) If $\Omega \subset \mathbb{R}^n$ is a bounded open set with Lipschitz boundary, then $C^\infty(\overline{\Omega})$ is dense in $H^k(\Omega)$ for every $k \in \mathbb{N}$. Moreover $\mathcal{D}(\mathbb{R}^n)$ is dense in $H^k(\mathbb{R}^n)$ for every $k \in \mathbb{N}$.

Proof. Assertion (i) immediately follows from Proposition 2.4.7. For the proof of part (ii) we refer to [1, Theorem 3.18, Corollary 3.19]. □

We conclude this section by pointing out that the spaces $H^k(\Omega)$ can be generalized in a natural way defining

$$W^{k,p}(\Omega) = \{u \in L^p(\Omega);\ D^\alpha u \in L^p(\Omega)\ \forall \alpha \in \mathbb{N}^n;\ |\alpha| \le k\}.$$

The Sobolev space $W^{k,p}(\Omega)$ is a Banach space, but for $p \neq 2$ it is not a Hilbert space any more. Nevertheless, also these spaces turn out to be extremely useful, see for instance Proposition 7.1.26 and the application to the regularity theory for the Navier-Stokes equation (8.43). For a deeper analysis about these spaces we refer to [1].

6.4 The spaces H^s $(s \geq 0)$

We now wish to define the spaces H^s with $s \geq 0$ real number, in such a way to give a sort of continuity in the regularity of functions $u \in H^s$ as s increases: to increase regularity of a function $u \in H^k$ it will not be necessary to require $u \in H^{k+1}$.

In view of Theorem 6.3.6, it is natural the

Definition 6.4.1. *For every $s \geq 0$ we define*

$$H^s(\mathbb{R}^n, \mathbb{C}) = \{u \in \mathcal{S}'(\mathbb{R}^n, \mathbb{C});\ (1 + |\xi|^2)^{\frac{s}{2}} \cdot \widehat{u}(\xi) \in L^2(\mathbb{R}^n, \mathbb{C})\}\ ,$$

endowed with the scalar product

$$(u, v)_{H^s} = \int_{\mathbb{R}^n} (1 + |\xi|^2)^s\ \widehat{u}(\xi)\overline{\widehat{v}(\xi)}\ d\xi$$

for every $u, v \in H^s(\mathbb{R}^n, \mathbb{C})$.

Since so far we have dealt mostly with real Sobolev spaces, it appears reasonable the following

Definition 6.4.2. *For every $s \geq 0$ we define*

$$H^s(\mathbb{R}^n) = \{u \in H^s(\mathbb{R}^n, \mathbb{C});\ \mathrm{Im}\ u = 0\}\ ,$$

endowed with the norm

$$\|u\|_{H^s} = \left(\int_{\mathbb{R}^n} (1 + |\xi|^2)^s |\widehat{u}(\xi)|^2\ d\xi\right)^{\frac{1}{2}} \tag{6.14}$$

for every $u \in H^s(\mathbb{R}^n)$.

Remark 6.4.3. The space $H^s(\mathbb{R}^n)$ is complete with respect to the norm (6.14). □

If $\Omega \subset \mathbb{R}^n$ is a bounded open set with Lipschitz boundary, then $H^s(\Omega)$ can be simply defined as the space of the restrictions to Ω of the functions of $H^s(\mathbb{R}^n)$. Also for $s \geq 0$ there holds the following result, which extends Theorem 6.3.8:

Theorem 6.4.4. *If $\Omega \subset \mathbb{R}^n$ is a bounded open set with Lipschitz boundary, then $C^\infty(\overline{\Omega})$ is dense in $H^s(\Omega)$ for every $s \geq 0$, moreover $\mathcal{D}(\mathbb{R}^n)$ is dense in $H^s(\mathbb{R}^n)$.*

To define $H^s(\Omega)$ for an arbitrary open set Ω, it is not necessary to exploit the Fourier transform (see [1, Theorem 7.48]):

Definition 6.4.5. *Let $\Omega \subseteq \mathbb{R}^n$ be an open set and let $s > 0$; if $s \in (k, k+1)$ we set $s = k + t$ with $t \in (0, 1)$ and we define the space*

$$H^s(\Omega) = \left\{ u \in H^k(\Omega);\ \int_\Omega \int_\Omega \frac{|D^\alpha u(x) - D^\alpha u(y)|^2}{|x-y|^{n+2t}}\, dx\, dy < +\infty; \quad |\alpha| = k \right\}$$

which, endowed with the scalar product

$$(u, v)_{H^s} = (u, v)_{H^k} + \sum_{|\alpha|=k} \int_\Omega \int_\Omega \frac{\{D^\alpha u(x) - D^\alpha u(y)\} \cdot \{D^\alpha v(x) - D^\alpha v(y)\}}{|x-y|^{n+2t}}\, dx\, dy$$

and equipped with the induced norm, becomes a Hilbert space.

Exercise 6.4.6. Verify that if $s = k$ with k a non negative integer, then the Sobolev space introduced in Definition 6.4.5 coincides with the one defined in Section 6.3. □

6.5 The spaces H_0^s and the trace operators

We begin with the

Definition 6.5.1. *For every $s \geq 0$ it is defined the space $H_0^s(\Omega)$ as the closure of $\mathcal{D}(\Omega)$ in $H^s(\Omega)$.*

Thanks to [46, Théorème 11.1, Chapitre 1] we infer that if Ω is an open set with boundary regular enough, then the space $\mathcal{D}(\Omega)$ is dense in $H^s(\Omega)$ if and only if $0 \leq s \leq \frac{1}{2}$. Thus, if $s \leq \frac{1}{2}$ it holds $H_0^s(\Omega) = H^s(\Omega)$, while if $s > \frac{1}{2}$ it simply holds $H_0^s(\Omega) \subset H^s(\Omega)$. Note that also $H_0^s(\mathbb{R}^n) = H^s(\mathbb{R}^n)$ for every $s \geq 0$ (see [46, Remarque 7.4, Chapitre 1]). In analogy to what we have said for the space $H_0^1(\Omega)$ (see Theorem 6.2.17), the elements of $H_0^s(\Omega)$ can be considered as functions which vanish on $\partial\Omega$ in a suitable sense. Actually, in the case $s = k \in \mathbb{N}$ we can affirm that, along with a function $u \in H_0^k(\Omega)$, also its derivatives up to order $k - 1$ vanish on $\partial\Omega$.

We now face the problem of defining the value of a function $u \in H^s(\Omega)$ on the boundary. As already said in Section 6.2.2 for the case $s = 1$, since $H^s(\Omega)$ is a space of functions which are defined up to a negligible set, a priori, to talk about the boundary value of a function $u \in H^s(\Omega)$ does not make sense. To obviate this difficulty, we try to construct an extension γ_0 of the operator γ defined by

$$\begin{aligned} \gamma : C^0(\overline{\Omega}) &\to C^0(\partial\Omega) \\ u &\mapsto u_{|\partial\Omega} \end{aligned}$$

to the widest possible classes of functions. Obviously, we will also have to enlarge the range and define a suitable function space over $\partial\Omega$. If Ω is a bounded open set with C^∞ boundary, we know that for every $x_0 \in \partial\Omega$ there exists $\varphi \in C^\infty(\mathbb{R}^{n-1})$ such that $x_n = \varphi(x_1, ..., x_{n-1})$ describes $\partial\Omega$ in a neighborhood $\mathcal{U}$ of x_0 (being $x_1, ..., x_n$ the coordinates of a generic point $x \in \mathcal{U} \cap \partial\Omega$; the explicit coordinate is not necessarily x_n, but we can always reduce to this situation, with a suitable change of frame). Therefore for every $x \in \mathcal{U} \cap \partial\Omega$ it holds

$$u(x) = u(x_1, ..., x_n) = u(x_1, ..., x_{n-1}, \varphi(x_1, ..., x_{n-1})) = u_0(x_1, ..., x_{n-1}).$$

Let $\mathcal{U}^*$ be the projection of $\mathcal{U} \cap \partial\Omega$ on $\mathbb{R}^{n-1}$ (with variables $x_1, ..., x_{n-1}$). Using the previous notations, we can give the

Definition 6.5.2. *We shall say that $u \in H^s(\partial\Omega)$ if, for every $x_0 \in \partial\Omega$, it holds $u_0 \in H^s(\mathcal{U}^*)$.*

In this way, we have defined the space $H^s(\partial\Omega)$ using the definition of H^s in $(n-1)$-dimensional spaces.

If the boundary Ω is not of class C^∞, in general it is not possible to define $H^s(\partial\Omega)$ for every value of $s > 0$, but only for the "compatible" ones (in a sense that we will not specify) with the irregularity of $\partial\Omega$ (see [39]). Actually, if Ω is a polygon in $\mathbb{R}^2$ (or more generally $\partial\Omega$ is piecewise C^∞) there is a way to define $H^s(\partial\Omega)$ for every $s > 0$, but this goes beyond the scopes of this book.

By Theorem 6.4.4 we know that $C^\infty(\overline{\Omega})$ is dense in $H^s(\Omega)$ for every $s \geq 0$, therefore the operator γ can be extended by continuity to the whole $H^s(\Omega)$. Thus we can state the

Theorem 6.5.3. *Let $s > \frac{1}{2}$ and let Ω be a bounded open set with C^{m+1} boundary, where m denotes the integral part of s. Then there exists a unique linear continuous surjective operator γ_0 such that*

$$\gamma_0 : H^s(\Omega) \to H^{s-\frac{1}{2}}(\partial\Omega)$$

and that for every $u \in C^\infty(\overline{\Omega})$, $\gamma_0 u = u_{|\partial\Omega}$. Moreover, there exists at least one linear continuous operator

$$\mathcal{R} : H^{s-\frac{1}{2}}(\partial\Omega) \to H^s(\Omega)$$

such that $\gamma_0(\mathcal{R}g) = g$, for every $g \in H^{s-\frac{1}{2}}(\partial\Omega)$.

Proof. See [46, Théorème 9.4, Chapitre 1] and [76, Theorem 8.7]. □

Definition 6.5.4. *The operator γ_0 is called* **trace operator**, *while $\mathcal{R}$ is called* **lifting operator** .

We observe that the existence of the lifting operator $\mathcal{R}$, appearing in second part of the theorem, ensures that the first part cannot be improved, namely the range of γ_0 cannot be, for instance, $H^s(\partial\Omega)$.

Exercise 6.5.5. Let $\Omega = \{x \in \mathbb{R}^2,\ |x| < 1,\ x_2 > 0\}$ and let $f(x) = |x|^t$ $(t \in (0,\infty) \setminus \mathbb{N})$. Prove that
(i) $f \in H^s(\Omega)$ if and only if $s < t+1$;
(ii) $\gamma_0 f = |x_1|^t$ on the straight part of $\partial\Omega$;
(iii) $\gamma_0 f \in H^r(\partial\Omega)$ if and only if $r < t + \frac{1}{2}$. □

Also the idea to define the trace of a function on the boundary is linked to boundary value problems for differential equations: if we want to solve a differential equation in distributional sense (namely without continuity assumptions on the unknown function), we have to think of boundary conditions in the sense of traces.

Example 6.5.6. Let $\Omega = \{x \in \mathbb{R}^2,\ |x| < 1\}$ and consider the problem

$$\begin{cases} \Delta u = 0 & \text{in } \Omega \\ \gamma_0 u = g. \end{cases}$$

Prove that $u \in H^2(\Omega)$ if and only if $g \in H^{\frac{3}{2}}(\partial\Omega)$ (see [39] for this results and its generalizations). □

For boundary value problems involving normal derivatives of the unknown function, we have to consider higher order traces:

Theorem 6.5.7. *Let $k \in \mathbb{N}$, $s > k + \frac{1}{2}$ and let $\Omega \subset \mathbb{R}^n$ be a bounded open set with C^{m+2} boundary, where m denotes the integral part of s. Then there exist and are uniquely determined $k+1$ linear, continuous and surjective operators such that*

$$\gamma_j : H^s(\Omega) \to H^{s-j-\frac{1}{2}}(\partial\Omega) \qquad j = 0, ..., k$$

and that

$$\gamma_j u = \frac{\partial^j u}{\partial \nu^j}_{|\partial\Omega} \qquad \forall u \in C^\infty(\overline{\Omega}).$$

Moreover, there exists at least one linear continuous operator

$$\mathcal{R} : H^{s-\frac{1}{2}}(\partial\Omega) \times ... \times H^{s-k-\frac{1}{2}}(\partial\Omega) \to H^s(\Omega)$$

such that $\gamma_j(\mathcal{R}(g_0, ..., g_k)) = g_j$, for $j = 0, ..., k$, for every $g_j \in H^{s-j-\frac{1}{2}}(\partial\Omega)$.

Proof. See [46, Théorème 9.4, Chapitre 1] and [76, Theorem 8.7]. □

In the statement of the theorem, ν represents the outer unit normal to $\partial\Omega$, with the convention $\frac{\partial^0 u}{\partial \nu^0} = u$. For $j = 1$ we have the usual normal derivative $\frac{\partial u}{\partial \nu} = (\nabla u \cdot \nu)_{|\partial\Omega}$, while the higher order normal derivatives are repeated normal derivatives in the direction ν.

We can now characterize the H_0^s spaces in terms of traces.

Theorem 6.5.8. *Let Ω be a bounded open set with C^∞ boundary and let $s > \frac{1}{2}$. Then the following conditions are equivalent:*

(i) $u \in H_0^s(\Omega)$;

(ii) $\begin{cases} u \in H^s(\Omega) \\ \gamma_j u = 0 \qquad \forall j \in \mathbb{N} \cap [0, s - \frac{1}{2}). \end{cases}$

Proof. See [46, Théorème 11.5, Chapitre 1]. □

6.6 Embedding theorems

In this section we consider the problem of evaluating the regularity of the elements of $H^s(\Omega)$ in terms of continuity and integrability. Recalling Definition 1.8.5, we can extend Theorem 6.2.20 to the general case $s \geq 0$:

Theorem 6.6.1. (Sobolev Theorem). *Let $\Omega = \mathbb{R}^n$ or an open set with Lipschitz boundary. Let $s \geq 0$, then there hold the following continuous embeddings:*

(i) if $n > 2s$, then $H^s(\Omega) \subset L^q(\Omega)$ for $2 \leq q \leq \frac{2n}{n-2s}$;

(ii) if $n = 2s$, then $H^s(\Omega) \subset L^q(\Omega)$ for $2 \leq q < \infty$;

(iii) if $\frac{n}{2} < s - m$, then $H^s(\Omega) \subset C_B^m(\Omega)$, where $m \in \mathbb{N}$ and $C_B^m(\Omega) = \{u \in C^m(\Omega),\ D^\alpha u \in L^\infty(\Omega),\ 0 \leq |\alpha| \leq m\}$ is equipped with the norm

$$\|u\|_{m,B} = \max_{0 \leq |\alpha| \leq m} \|D^\alpha u\|_{L^\infty}.$$

Proof. See [1, Theorem 7.57]. □

Assertion (iii) in the previous theorem cannot be improved with the equality sign: indeed, if $n = 2$, $H^1(\Omega)$ is not contained into $C^0(\Omega)$ (see Example 6.2.16).

Theorem 6.6.1 ensures the continuity of the elements of H^1 when $n = 1$, while if $n = 2$, the fact that a function belongs to H^1 is not any more sufficient to guarantee its continuity. In the case $n = 2$, in order to have continuity of a function u we have to assume that $u \in H^s$ with $s > 1$. We deduce that, as the space dimension increases, the regularity of the functions of H^s decreases.

Remark 6.6.2. Actually, to obtain the embedding results of Theorem 6.6.1 we can slightly weaken the assumption of Lipschitz regularity for the boundary $\partial\Omega$. On the contrary, in the case of H_0^s spaces, it is possible to obtain the same results of Theorem 6.6.1 without any regularity assumption on $\partial\Omega$, see again [1]. □

The embedding operators appearing in Theorem 6.6.1 are compact operators if the open set Ω is bounded. More precisely, there holds the following theorem, which extends Theorem 6.2.21 to the general case $s \geq 0$:

Theorem 6.6.3. (Theorem of Rellich, Sobolev, Gagliardo, Nirenberg). *Let $\Omega \subset \mathbb{R}^n$ be a bounded open set with Lipschitz boundary and let $s \geq 0$. Then there hold the following compact embeddings:*

(i) if $n > 2s$, then $H^s(\Omega) \subset\subset L^q(\Omega)$ for $1 \leq q < \frac{2n}{n-2s}$;
(ii) if $n = 2s$, then $H^s(\Omega) \subset\subset L^q(\Omega)$ for $1 \leq q < \infty$;
(iii) if $\frac{n}{2} < s - m$ for a certain $m \in \mathbb{N}$, then $H^s(\Omega) \subset\subset C^m(\overline{\Omega})$.

Proof. See [1, Theorem 6.2]. □

Remark 6.6.4. The boundedness assumption for Ω can be slightly weakened, but not omitted: indeed it is possible to construct counterexamples of unbounded open sets Ω for which the embeddings are not compact. Concerning the embedding (i), the upper bound of q is a strict inequality, unlike what happens in Theorem 6.6.1; also for this condition it is possible to provide counterexamples in which the upper bound for q holds with the equality sign, but the embedding is not compact. In (i) and (ii), the lower bound for q is simply $q \geq 1$; indeed, since Ω is bounded we have the continuous embedding $L^q(\Omega) \subset L^p(\Omega)$ for every $1 \leq p \leq q$ (see Proposition 2.4.7) and it can be proved that the composition of a continuous embedding and a compact embedding is a compact embedding. Assuming the boundedness of Ω, the embeddings of the spaces $H_0^s(\Omega)$ into the corresponding spaces are compact without any regularity assumption on $\partial\Omega$. □

Let us now see an example of application of Theorem 6.6.3.

Example 6.6.5. Let Ω be a bounded open set with Lipschitz boundary and assume to have a sequence of functions $\{u_k\} \subset H^s(\Omega)$ which is bounded in $H^s(\Omega)$, namely

$$\exists C > 0, \quad \forall k \in \mathbb{N} \quad \|u_k\|_{H^s} \leq C\,.$$

Thanks to Theorem 1.4.13, we know that there exists a subsequence (which we shall denote again by $\{u_k\}$) which converges weakly in $H^s(\Omega)$:

$$\exists u \in H^s(\Omega), \quad s.t. \quad u_k \stackrel{H^s}{\rightharpoonup} u.$$

If $s < \frac{n}{2}$, then Theorem 6.6.3 implies that $u_k, u \in L^q(\Omega)$ for $q < \frac{2n}{n-2s}$ and that $u_k \to u$ in $L^q(\Omega)$. □

The procedure presented above finds natural applications in the Galerkin method (see Section 6.8 and [50]).

6.7 The spaces H^s $(s < 0)$

For the sake of completeness, we define also the spaces $H^s(\Omega)$ with $s < 0$ (see [46]). To define the space $H^s(\mathbb{R}^n)$ $(s < 0)$ it is still possible to use Definition 6.4.1: we introduce the complex space

$$H^s(\mathbb{R}^n, \mathbb{C}) = \{u \in \mathcal{S}'(\mathbb{R}^n, \mathbb{C}),\ (1 + |\xi|^2)^{\frac{s}{2}} \cdot \widehat{u}(\xi) \in L^2(\mathbb{R}^n, \mathbb{C})\} \qquad \forall s \in \mathbb{R}$$

and the corresponding real space $H^s(\mathbb{R}^n) = \{u \in H^s(\mathbb{R}^n, \mathbb{C});\ \mathrm{Im}\ u = 0\}$. For every $s > 0$, the dual of $H^s(\mathbb{R}^n) = H_0^s(\mathbb{R}^n)$ can be identified with $H^{-s}(\mathbb{R}^n)$ (see [46, Théorème 7.2, Chapitre 1]). Then it reasonable the following

Definition 6.7.1. *For every $s > 0$, we define $H^{-s}(\Omega)$ as the dual space of $H_0^s(\Omega)$.*

In order to have an idea of the structure of the elements of $H^{-s}(\Omega)$ $(s > 0)$, we point out that $\mathcal{D}(\Omega)$ is dense in $H_0^s(\Omega)$ for every $s > 0$ (by definition) and hence, passing to the dual spaces:

$$H^{-s}(\Omega) \subset \mathcal{D}'(\Omega).$$

In other words, the space $H^{-s}(\Omega)$ $(s > 0)$ is a space of distributions.

In the case $s = m \in \mathbb{N}$, there holds the following representation theorem for the elements of $H^{-m}(\Omega)$:

Theorem 6.7.2. *Let $m \in \mathbb{N}$; then every functional $f \in H^{-m}(\Omega)$ can be represented (and the representation is not unique) as follows:*

$$f = \sum_{|\alpha| \le m} D^\alpha f_\alpha\,,$$

where $f_\alpha \in L^2(\Omega)$ for every $|\alpha| \le m$.

Proof. See [1, Theorem 3.8]. □

It is possible to define the space $H^s(\partial\Omega)$ also for negative values of s:

Definition 6.7.3. *For every $s > 0$, it is defined $H^{-s}(\partial\Omega)$ as the dual space of $H^s(\partial\Omega)$.*

As we will see in Section 7.1.4, the space $H^{-1/2}(\partial\Omega)$ turns out to be of fundamental importance in applications. Set

$$\mathbf{E} := \{u \in (L^2(\Omega))^n\ ;\ \mathrm{div} u \in L^2(\Omega)\}\,, \tag{6.15}$$

where the divergence operator is to be meant in the sense of distributions, namely $\mathrm{div} u \in L^2(\Omega)$ if and only if there exists $w \in L^2(\Omega)$ such that

$$\int_\Omega w\varphi = -\int_\Omega u \cdot \nabla\varphi \qquad \forall \varphi \in \mathcal{D}(\Omega).$$

There holds the following

Theorem 6.7.4. *Let Ω be a bounded open set with Lipschitz boundary and denote by ν the outer unit normal to the boundary (which is defined a.e. on $\partial\Omega$). Then there exists a unique linear continuous surjective operator γ_ν such that*

$$\gamma_\nu : \mathbf{E} \to H^{-1/2}(\partial\Omega)$$

and such that $\gamma_\nu u = u \cdot \nu$ for every $u \in (C^0(\overline{\Omega}))^n$.

Proof. See [68, Theorem I.1.2]. □

Definition 6.7.5. *The operator γ_ν is called* **normal trace operator**. *We denote by $\mathbf{E}_0 \subset \mathbf{E}$ the kernel of γ_ν.*

We conclude this section with the following generalization of the formula of integration by parts:

Theorem 6.7.6. *Let Ω be a bounded open set with Lipschitz boundary. Then there holds the following generalized Gauss-Green formula (integration by parts):*

$$\int_\Omega u \nabla w + \int_\Omega w \text{ div } u = \langle \gamma_\nu u, \gamma_0 w \rangle \qquad \forall u \in \mathbf{E}, \quad \forall w \in H^1(\Omega). \tag{6.16}$$

Proof. The validity of this formula follows from the validity of the same formula in the regular case, in which $u \in (C^1(\overline{\Omega}))^n$ and $w \in C^1(\overline{\Omega})$, from the density of such spaces in $\mathbf{E}$ and $H^1(\Omega)$ respectively, and from the continuity of the operators γ_0 and γ_ν. □

6.8 Galerkin method

Assume that we have to prove the existence of solutions to a problem which is set in a separable infinite dimensional Hilbert space V. The first step of Galerkin method consists in approximating V with a sequence of finite dimensional subspaces. The second step consists in finding solutions to the problem "projected" to the finite dimensional subspaces of the sequence. Finally, with a suitable passage to the limit, we show that the so found solutions tend to a solution of the original problem.

Since we deal with problems involving differential equations, the separable Hilbert spaces that we want to approximate are Sobolev spaces. In this section we make the just described mechanism formal and we explain how to apply the Galerkin method.

Definition 6.8.1. *Let V be a separable Hilbert space. A sequence of vector subspaces $\{V_m\}$ is said to be a Galerkin sequence (and it is denoted by $V_m \to V$) if*
(i) $V_m \subset V$, $\quad \dim V_m < +\infty \quad \forall m \in \mathbb{N}$;
(ii) $\forall v \in V \ \exists \{v_m\} \subset V$, $v_m \in V_m$, $v_m \to v$ in V.
In this case we will say that V_m is the Galerkin approximation of order m of V.

Remark 6.8.2. Frequently, it is supposed $\dim V_m = m$ or $V_m \subset V_{m+1}$, but these are not necessary assumptions. Part (ii) of Definition 6.8.1 means that there exists a constructive algorithm to determine the sequence $\{v_m\}$. As a consequence of Definition 6.8.1, we have that $\overline{\bigcup_{m\in\mathbb{N}} V_m} = V$. □

Let us present some significant examples of Galerkin sequences. Our first example holds true for every separable Hilbert space.

Example 6.8.3. Let $\{e_n\}$ be a Hilbertian basis of V and, for every $m \in \mathbb{N}$, set $V_m = \operatorname{span}\{e_1, \ldots, e_m\}$. For every $v \in V$, take

$$v_m = \sum_{k=1}^{m} (v, e_k)_V e_k.$$

By Theorem 1.6.8, $\{V_m\}$ is a Galerkin sequence for V. □

Let us now see how a Sobolev space can be approximated in the one-dimensional case.

Example 6.8.4. Let $V = H_0^1(0,1)$. For every $m \in \mathbb{N}$, let us define the functions $w_1^m, \ldots, w_{m-1}^m \in V$ as follows:

$$\begin{cases} w_j^m\left(\frac{j}{m}\right) = 1\,, \quad w_j^m\left(\frac{j\pm 1}{m}\right) = 0 \\ w_j^m(x) = 0 \quad \text{if } x \notin \left(\frac{j-1}{m}, \frac{j+1}{m}\right) \\ w_j^m \text{ piecewise linear.} \end{cases}$$

We choose $V_m = \operatorname{span}\{w_1^m, \ldots, w_{m-1}^m\}$; then $\dim V_m = m - 1$ and $V_m \not\subset V_{m+1}$.

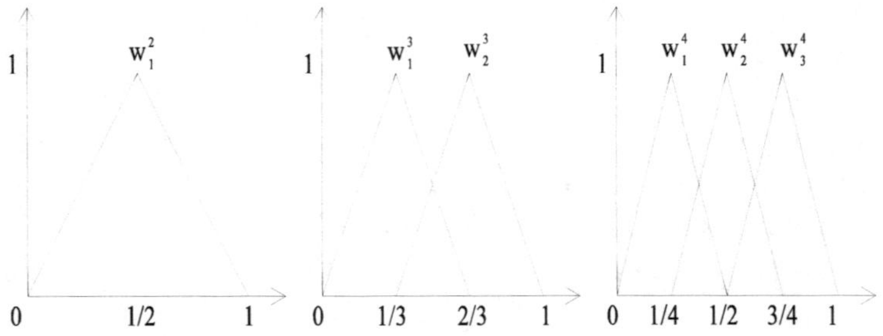

The functions of the spaces V_2, V_3, V_4

For every $v \in V$ we define the piecewise linear function

$$v_m(x) := \sum_{j=1}^{m-1} v\left(\frac{j}{m}\right) w_j^m(x).$$

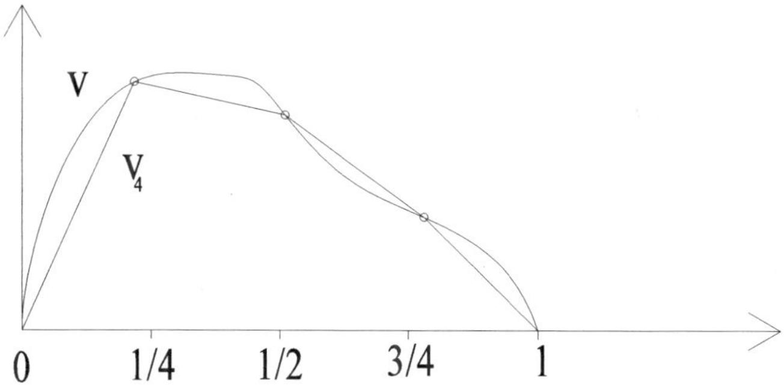

The function v_4

It is immediate to verify that

$$v_m'(x) = m\left[v\left(\frac{j}{m}\right) - v\left(\frac{j-1}{m}\right)\right] \qquad \text{if } x \in \left(\frac{j-1}{m}, \frac{j}{m}\right).$$

We leave to the reader the conclusion, following these steps:

- prove that for every $v \in V$, $v_m \to v$ uniformly;
- prove that for every $v \in V \cap C^1(0,1)$, $v_m \to v$ in V;
- by density, prove that for every $v \in V$, $v_m \to v$ in V. □

Example 6.8.5. Finite elements (n=2). Let $\Omega \subset \mathbb{R}^2$ be a bounded open set and consider the space $V = H_0^1(\Omega)$. For every $m \in \mathbb{N}$ we introduce the family of (closed) triangles $F_m = \{T_k^m\}_{k=1}^m$ such that:

- $\displaystyle\bigcup_{k=1}^m T_k^m \subset \Omega$;
- for every T_i^m, $T_j^m \in F_m$ there holds one of the following:

(a) $T_i^m = T_j^m$	(b) $T_i^m \cap T_j^m = \emptyset$
(c) $T_i^m \cap T_j^m$ coincides with a vertex	(d) $T_i^m \cap T_j^m$ coincides with a side;

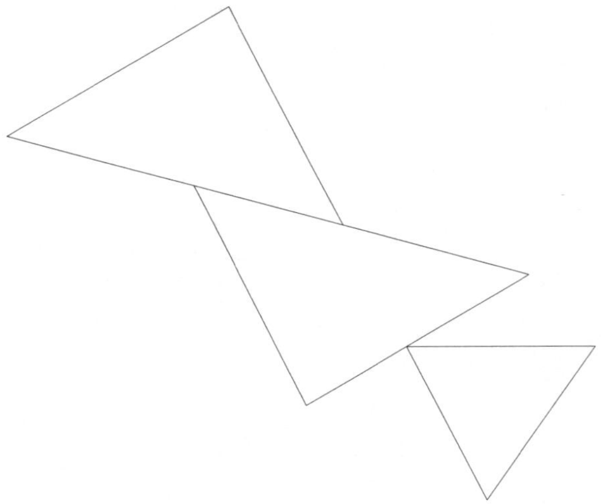

Non admissible triangulation

– for every $T \in F_m$, $\mathrm{diam}(T) \le \varepsilon_m$ with $\varepsilon_m \to 0$ as $m \to \infty$;

– for every compact subset $K \subset \Omega$ there exists $\overline{m} \in \mathbb{N}$ such that $\displaystyle\bigcup_{k=1}^{m} T_k^m \supset K$ for every $m \ge \overline{m}$.

For every $m \in \mathbb{N}$ and for every interior vertex M of the triangulation F_m, we define the function W_M^m so that: it is affine on each triangle of F_m; $W_M^m(M) = 1$ and $W_M^m(A) = 0$ for every vertex A of the triangulation F_m different from M. Finally, we set $W_M \equiv 0$ in $\Omega \setminus \cup_{k=1}^m T_k^m$. Clearly it holds $W_M^m \in V$.

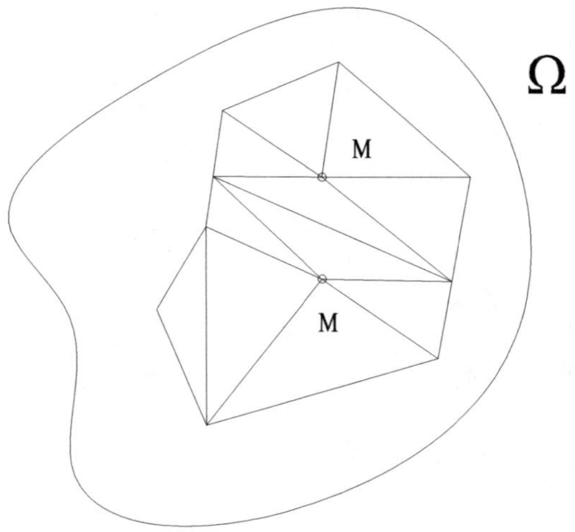

Vertices interior to the triangulation of Ω

We set $V_m = \text{span}\{W^m_M\}$, hence the dimension of V_m equals the number of interior vertices. For every $v \in V \cap C^0(\overline{\Omega})$ we define

$$v_m(x) = \sum_M v(M)\, W^m_M(x). \tag{6.17}$$

It can be shown that $v_m \to v$ in V. Unlike the case $n = 1$, in dimension $n = 2$ the functions of $V = H^1_0(\Omega)$ are not, in general, continuous. Exploiting the fact that the subspace $V \cap C^0(\overline{\Omega})$ is dense in V, given an arbitrary function $w \in V$ it is possible to construct a sequence $\{w_k\}$ such that $w_k \in V_k$ and $w_k \to w$ in V.

To this purpose, we proceed as follows: for every $k \in \mathbb{N}$ we choose $v_k \in V \cap C^0(\overline{\Omega})$ such that $\|w - v_k\| < 1/(2k)$ and we consider $m_k \in \mathbb{N}$ such that $\|(v_k)_{m_k} - v_k\| < 1/(2k)$ with $(v_k)_{m_k}$ as in (6.17). Finally, we set $w_k = (v_k)_{m_k}$. Thus we can construct by induction an increasing sequence of indices m_k and a sequence $\{w_k\}$ such that $w_k \in V_{m_k}$ and $w_k \to w$ in V. □

The Galerkin method is used for proving the existence of solutions for certain equations. Assume that we have to solve a problem (P) in a separable Hilbert space V. We consider a Galerkin sequence $\{V_m\}$ and we introduce a sequence of problems (P_m) which approximate (P) and such that, for every m, (P_m) has a unique solution $u_m \in V_m$. Below we list schematically the steps which lead to determine a solution for problem (P):

Step 1. Introduction of $\{V_m\}$ e (P_m).

Step 2. Derivation of a priori estimates for $\|u_m\|_V$, to prove the boundedness of $\{u_m\}$ in V.

Step 3. Construction of a function u such that $u_m \rightharpoonup u$.

Step 4. Proof that u is a solution to (P).

Chapter 7

Weak solutions of linear partial differential equations

In this chapter we analyze the variational approach for the linear problems studied in Chapter 3. This method allows to prove existence and uniqueness results for partial differential equations, reducing such equations to a variational problem over a suitable functional space. The functional spaces are L^p spaces (cf. Section 2.4) or Sobolev spaces (cf. Chapter 6). In these spaces we apply the abstract results presented in Chapter 1.

7.1 Variational formulation of some boundary value elliptic problems

In this section we want to implement the procedure described in Section 6.1, to solve elliptic partial differential equations. By **elliptic equation** we mean a stationary equation (time-independent) for which we do not solve a Cauchy problem but just a boundary value problem. We shall examine into details some model examples which explain well how the procedure works.

7.1.1 Homogeneous Dirichlet problem

Let $\Omega \subset \mathbb{R}^n$ be a bounded open set of class C^1; we look for a function $u : \Omega \to \mathbb{R}$ verifying

$$\begin{cases} -\Delta u + \alpha u = f & \text{in } \Omega \\ u = 0 & \text{on } \partial\Omega\,, \end{cases} \tag{7.1}$$

with $\alpha \in \mathbb{R}$ and $f \in L^2(\Omega)$. The boundary condition $u = 0$ on $\partial\Omega$ is called **homogeneous Dirichlet condition**.

Definition 7.1.1. *A* **classical** *solution of* (7.1) *is a function* $u \in C^2(\overline{\Omega})$ *satisfying* (7.1). *A* **weak** *solution of* (7.1) *is a function* $u \in H_0^1(\Omega)$ *satisfying*

$$\int_\Omega \nabla u \nabla v + \alpha \int_\Omega uv = \int_\Omega fv \qquad \forall v \in H_0^1(\Omega). \tag{7.2}$$

If u is a classical solution of (7.1), by Theorem 6.2.17 we infer that also $u \in H_0^1(\Omega)$. On the other hand, the weak formulation (7.2) follows from (7.1), multiplying both sides of the equation by a test function $v \in \mathcal{D}(\Omega)$ and integrating by parts:

$$\begin{aligned} \int_\Omega fv = \int_\Omega (-\Delta u + \alpha u)v &= \int_\Omega (\nabla u \nabla v + \alpha uv) - \int_{\partial\Omega} \frac{\partial u}{\partial \nu} v \\ &= \int_\Omega \nabla u \nabla v + \alpha \int_\Omega uv \qquad \forall v \in \mathcal{D}(\Omega)\,, \end{aligned} \tag{7.3}$$

where the boundary term cancels out, being v zero on the boundary. Since $\mathcal{D}(\Omega)$ is dense in $H_0^1(\Omega)$, the identity (7.3) is still valid for every $v \in H_0^1(\Omega)$.

Conversely, let us assume that a weak solution u to (7.1) belongs to $C^2(\overline{\Omega})$. By Theorem 6.2.17, we immediately deduce that $u = 0$ on $\partial\Omega$. Starting from (7.2), in view of (6.16), we obtain

$$\int_\Omega (-\Delta u + \alpha u)v = \int_\Omega fv \qquad \forall v \in \mathcal{D}(\Omega)$$

and hence, in view of Proposition 2.4.22, $-\Delta u + \alpha u = f$ a.e. in Ω. Actually, $-\Delta u + \alpha u = f$ holds true everywhere in Ω, being $u \in C^2(\Omega)$; therefore u is a classical solution of (7.1).

Remark 7.1.2. If u is a classical solution of (7.1), then necessarily $f \in C^0(\overline{\Omega})$. Therefore, if f is a generic function of $L^2(\Omega)$, we cannot expect a weak solution of (7.1) to be also classical. □

There holds the following

Theorem 7.1.3. *Assume that* $\alpha \geq 0$. *Then for every* $f \in L^2(\Omega)$ *there exists a unique weak solution* $u \in H_0^1(\Omega)$ *of* (7.1). *Moreover,* u *can be obtained as the solution of the minimum problem*

$$\min_{v \in H_0^1(\Omega)} \left\{ \frac{1}{2} \int_\Omega (|\nabla v|^2 + \alpha v^2) - \int_\Omega fv \right\} .$$

Proof. We endow the Hilbert space $H_0^1(\Omega)$ with the scalar product (see Theorem 6.2.18)

$$(u, v) = \int_\Omega \nabla u \nabla v \qquad \forall u, v \in H_0^1(\Omega). \tag{7.4}$$

Since by assumption $\alpha \geq 0$, the bilinear form

$$a(u,v) = \int_\Omega (\nabla u \nabla v + \alpha uv) \qquad \forall u,v \in H_0^1(\Omega) \tag{7.5}$$

is coercive over $H_0^1(\Omega)$. Moreover, the linear form $v \mapsto \int_\Omega fv$ is continuous over $H_0^1(\Omega)$. Then, the assertion follows from the Lax-Milgram Theorem (Corollary 1.7.4). □

Remark 7.1.4. As an immediate consequence of the proof, we infer that the results of Theorem 7.1.3 can be generalized to the case $f \in H^{-1}(\Omega)$, provided that we replace $\int_\Omega fv$ with the duality product $\langle f,v\rangle$. □

We now want to establish what happens without the positivity assumption on α. To this purpose, we recall that by *eigenvalue* of $-\Delta$ we mean a number $\lambda \in \mathbb{R}$ for which the variational problem

$$\int_\Omega \nabla u \nabla v = \lambda \int_\Omega uv \qquad \forall v \in H_0^1(\Omega) \tag{7.6}$$

admits a non trivial solution $u \in H_0^1(\Omega)$. Problem (7.6) is the weak version of the Dirichlet problem

$$\begin{cases} -\Delta u = \lambda u & \text{in } \Omega \\ u = 0 & \text{on } \partial\Omega. \end{cases}$$

Setting $H = L^2(\Omega)$ and $V = H_0^1(\Omega)$ equipped with the scalar product (7.4), by the compactness of the embedding $V \subset H$ and by Theorem 1.8.10, it turns out that the eigenvalues of $-\Delta$ can be ordered in an increasing sequence diverging to $+\infty$, and each eigenvalue has finite dimensional eigenspace. Let us denote by $0 < \lambda_1 \leq \lambda_2 \leq \ldots$ the eigenvalues of $-\Delta$, where each eigenvalue appears in the sequence as many times as its geometric multiplicity. Then we denote by $\sigma(-\Delta)$ the spectrum.

Exercise 7.1.5. Prove that the smallest eigenvalue $\lambda_1 > 0$ of $-\Delta$ (in the sense of (7.6)) can be characterized as

$$\lambda_1 = \inf_{u \in H_0^1 \setminus \{0\}} \frac{\|\nabla u\|_{L^2}^2}{\|u\|_{L^2}^2}.$$

Then, show that λ_1 coincides with C^{-2}, where C is the constant defined in the Poincaré inequality, cf. Theorem 6.2.18. □

The conclusions of Theorem 7.1.3 are still valid also for some negative values of α: in fact, as an immediate consequence of Theorem 1.8.10, we obtain the following

Theorem 7.1.6. *The following assertions hold true:*

(i) if $-\alpha \notin \sigma(-\Delta)$ then for every $f \in L^2(\Omega)$ problem (7.2) *admits a unique solution $u \in H_0^1(\Omega)$;*

(ii) if $-\alpha = \lambda \in \sigma(-\Delta)$ then (7.2) *admits a solution if and only if $f \in L^2(\Omega)$ satisfies the condition*

$$\int_\Omega fw = 0$$

for every eigenfunction w associated to λ. Such solution is not unique: if $u \in H_0^1(\Omega)$ is a solution, then the set of solutions can be written as

$$\{u + w \in H_0^1(\Omega) \ ; \ w \ eigenfunction \ associated \ to \ \lambda\}.$$

7.1.2 Non homogeneous Dirichlet problem

Let $\Omega \subset \mathbb{R}^n$ be a bounded open set of class C^1; we look for a function $u : \overline{\Omega} \to \mathbb{R}$ verifying

$$\begin{cases} -\Delta u + \alpha u = f & \text{in } \Omega \\ u = g & \text{on } \partial\Omega\,, \end{cases} \tag{7.7}$$

with f defined over Ω, g on $\partial\Omega$ and $\alpha \in \mathbb{R}$. If $g \in H^{1/2}(\partial\Omega)$, in view of Theorem 6.5.3, we infer that there exists $u_0 \in H^1(\Omega)$ such that $\gamma_0 u_0 = g$, where γ_0 denotes the trace operator. We define the (convex non empty) set

$$K = \{v \in H^1(\Omega);\ v - u_0 \in H_0^1(\Omega)\}$$

which *does not depend* on the particular choice of the function u_0. We are now in position to give the following

Definition 7.1.7. *A* **classical** *solution of* (7.7) *is a function $u \in C^2(\overline{\Omega})$ satisfying* (7.7)*. A* **weak** *solution of* (7.7) *is a function $u \in K$ satisfying*

$$\int_\Omega \nabla u \nabla v + \alpha \int_\Omega uv = \int_\Omega fv \qquad \forall v \in H_0^1(\Omega). \tag{7.8}$$

Reasoning as in the case of homogeneous Dirichlet conditions, it is easy to check that a classical solution is also a weak solution and, conversely, that every weak solution $u \in C^2(\overline{\Omega})$ is also a classical solution.

Let us now analyze the existence of weak solutions.

Theorem 7.1.8. *Let $\lambda_1 > 0$ be the smallest eigenvalue of $-\Delta$ under homogeneous Dirichlet conditions (in the sense of* (7.6)*). Let $f \in L^2(\Omega)$, $g \in H^{1/2}(\partial\Omega)$ and*

$\alpha > -\lambda_1$. Then (7.7) admits a unique weak solution $u \in K$. Moreover, u can be obtained as the solution to the minimum problem

$$\min_{v\in K}\left\{\frac{1}{2}\int_\Omega(|\nabla v|^2+\alpha v^2)-\int_\Omega fv\right\} .$$

Proof. Since $\alpha > -\lambda_1$, the bilinear form (7.5) is coercive and continuous over $H_0^1(\Omega)$, see Exercise 7.1.5. Fixed $u_0 \in H^1(\Omega)$ such that $\gamma_0 u_0 = g$, we look for $z \in H_0^1(\Omega)$ such that

$$\int_\Omega \nabla z\nabla v+\alpha\int_\Omega zv=-\int_\Omega \nabla u_0\nabla v-\alpha\int_\Omega u_0v+\int_\Omega fv \qquad \forall v\in H_0^1(\Omega). \tag{7.9}$$

We observe that the linear functional

$$v\mapsto -\int_\Omega \nabla u_0\nabla v-\alpha\int_\Omega u_0v+\int_\Omega fv$$

is continuous over $H_0^1(\Omega)$. But then, it coincides with a functional Λ which belongs to the dual space $H^{-1}(\Omega)$ and (7.9) can be rewritten as

$$a(z,v)=\langle\Lambda,v\rangle \qquad \forall v\in H_0^1(\Omega)\ ,$$

where $\langle\cdot,\cdot\rangle$ represents the duality between $H^{-1}(\Omega)$ and $H_0^1(\Omega)$. By the Lax-Milgram Theorem (Corollary 1.7.4), the last variational problem admits a unique solution $z \in H_0^1(\Omega)$.

Setting $u = z + u_0$, we obtain that $u \in K$ is a weak solution of (7.7). Hence the existence is proved. Concerning the uniqueness, we argue by contradiction assuming that there exist two weak solutions $u_1, u_2 \in K$: then $w := u_1 - u_2 \in H_0^1(\Omega)$. Subtracting the equations in (7.8) with $u = u_1$ and $u = u_2$, and considering w as test function, we obtain $a(w, w) = 0$ which implies $w \equiv 0$ by the coercivity of $a(\cdot,\cdot)$. Therefore $u_1 \equiv u_2$ and also the uniqueness is proved.

It remains to show the equivalence with the minimum problem; this is the Dirichlet principle that we have already proved for classical solutions, see Theorem 3.6.6. We start showing that the unique weak solution u of (7.7) minimizes the functional

$$J(v)=\frac{1}{2}\int_\Omega|\nabla v|^2+\frac{\alpha}{2}\int_\Omega v^2-\int_\Omega fv \qquad \forall v\in K.$$

Let $w \in K$ then $v = w - u \in H_0^1(\Omega)$ and hence, since u satisfies (7.8),

$$\begin{aligned}J(w)=J(u+v)&=J(u)+\frac{1}{2}\int_\Omega|\nabla v|^2+\frac{\alpha}{2}\int_\Omega v^2+\int_\Omega(\nabla u\nabla v+\alpha uv-fv)\\&=J(u)+\frac{1}{2}\int_\Omega|\nabla v|^2+\frac{\alpha}{2}\int_\Omega v^2\geq J(u).\end{aligned}$$

Viceversa, let us assume that u minimizes J over K. Then for every $v \in H_0^1(\Omega)$ and for every $t \in \mathbb{R}$ it holds $u + tv \in K$ and, setting $\varphi(t) = J(u + tv)$, we deduce that φ admits a global minimum at $t = 0$. Since φ is differentiable we get $\varphi'(0) = 0$. Hence, being

$$\varphi'(t) = t\int_\Omega |\nabla v|^2 + \int_\Omega \nabla u \nabla v + \alpha t\int_\Omega v^2 + \alpha\int_\Omega uv - \int_\Omega fv,$$

we deduce that

$$0 = \varphi'(0) = \int_\Omega \nabla u \nabla v + \alpha\int_\Omega uv - \int_\Omega fv.$$

The arbitrariness of $v \in H_0^1(\Omega)$ allows to conclude that u satisfies (7.8). □

Exercise 7.1.9. In the case $\alpha > 0$, prove Theorem 7.1.8 using Stampacchia Theorem 1.7.3. □

Remark 7.1.10. As already noticed for the homogeneous Dirichlet problem, the result of Theorem 7.1.8 can be extended to the case in which the function $f \in L^2(\Omega)$ is replaced by an element of $(H^1(\Omega))'$. □

7.1.3 Neumann problem

Let $\Omega \subset \mathbb{R}^n$ be a bounded open set of class C^1; we look for a function u verifying

$$\begin{cases} -\Delta u + \alpha u = f & \text{in } \Omega \\ \dfrac{\partial u}{\partial \nu} = g & \text{on } \partial\Omega\,, \end{cases} \tag{7.10}$$

with $f \in L^2(\Omega)$, $g \in H^{-1/2}(\partial\Omega)$, $\alpha \in \mathbb{R}$. The boundary condition $\frac{\partial u}{\partial \nu} = 0$ over $\partial\Omega$ is called **homogeneous Neumann condition**.

Definition 7.1.11. *A* **classical** *solution of* (7.10) *is a function* $u \in C^2(\overline{\Omega})$ *satisfying* (7.10). *A* **weak** *solution of* (7.10) *is a function* $u \in H^1(\Omega)$ *satisfying*

$$\int_\Omega \nabla u \nabla v + \alpha\int_\Omega uv = \int_\Omega fv + \langle g, \gamma_0 v\rangle \qquad \forall v \in H^1(\Omega)\,, \tag{7.11}$$

where $\langle\cdot,\cdot\rangle$ *denotes the duality between* $H^{-1/2}(\partial\Omega)$ *and* $H^{1/2}(\partial\Omega)$ *(by Theorem 6.5.3,* $\gamma_0 v \in H^{1/2}(\partial\Omega)$ *for every* $v \in H^1(\Omega)$*).*

Let us now justify such definition. Given a classical solution u of (7.10), we multiply both sides of the equation by a test function $v \in C^1(\overline{\Omega})$. Using (6.16) we obtain

$$\int_\Omega fv = \int_\Omega(-\Delta u + \alpha u)v = \int_\Omega \nabla u \nabla v - \int_{\partial\Omega} \frac{\partial u}{\partial \nu} v + \int_\Omega \alpha uv$$

$$= \int_\Omega \nabla u \nabla v + \int_\Omega \alpha uv - \int_{\partial\Omega} gv.$$

Since $C^1(\overline{\Omega})$ is dense in $H^1(\Omega)$ (see Theorem 6.3.8), the previous identity holds true also for every $v \in H^1(\Omega)$, provided that v is replaced by $\gamma_0 v$ on $\partial\Omega$.

Conversely, let us assume that $u \in C^2(\overline{\Omega})$ is a weak solution of (7.10); thus, again by (6.16), we infer

$$\int_\Omega (-\Delta u + \alpha u)v + \int_{\partial\Omega} \frac{\partial u}{\partial \nu} v = \langle g, v\rangle + \int_\Omega fv \qquad \forall v \in C^1(\overline{\Omega}). \tag{7.12}$$

Taking $v \in \mathcal{D}(\Omega)$ in (7.12), we obtain

$$\int_\Omega (-\Delta u + \alpha u)v = \int_\Omega fv \qquad \forall v \in \mathcal{D}(\Omega)$$

which, in view of Proposition 2.4.22, implies $-\Delta u + \alpha u = f$ in Ω. Inserting the latter condition into (7.12) we deduce that

$$\int_{\partial\Omega} \frac{\partial u}{\partial \nu} v = \langle g, v\rangle \qquad \forall v \in C^1(\overline{\Omega}).$$

The previous identity holds true for every $v \in H^1(\Omega)$, since $C^1(\overline{\Omega})$ is dense in $H^1(\Omega)$ and $\gamma_0 : H^1(\Omega) \to H^{1/2}(\partial\Omega)$ is a continuous surjective function. Hence we may conclude that $\frac{\partial u}{\partial \nu} = g$ on $\partial\Omega$.

The next step will consist again in proving the existence of a weak solution.

Theorem 7.1.12. *Let $f \in L^2(\Omega)$, $g \in H^{-1/2}(\partial\Omega)$ and $\alpha > 0$. Then there exists a unique weak solution $u \in H^1(\Omega)$ of (7.10). Moreover, u can be obtained as the solution of the minimum problem*

$$\min_{v\in H^1(\Omega)} \left\{ \frac{1}{2}\int_\Omega (|\nabla v|^2 + \alpha v^2) - \int_\Omega fv - \langle g, \gamma_0 v\rangle \right\}.$$

Proof. The proof follows by applying the Lax-Milgram Theorem (Corollary 1.7.4) in the Hilbert space $H^1(\Omega)$. □

The constraint $\alpha > 0$ introduced in Theorem 7.1.12 is equivalent to the restriction $\alpha > -\lambda_1$ in Theorem 7.1.8 and it is essential to guarantee the existence and uniqueness of the solution to the Neumann problem (7.10). According to Theorem 3.6.7, we know that, if $\alpha = 0$, lack of existence or uniqueness of solutions may occur. This is due to the fact that 0 is the smallest eigenvalue of the operator $-\Delta$ under homogeneous Neumann boundary conditions, namely it is the smallest real number λ such that the problem

$$\int_\Omega \nabla u \nabla v = \lambda \int_\Omega uv \qquad \forall v \in H^1(\Omega) \tag{7.13}$$

admits a non trivial solution $u \in H^1(\Omega)$. We will denote by $\sigma_N(-\Delta)$ the spectrum, namely the set of eigenvalues. Problem (7.13) is the variational version of the problem

$$\begin{cases} -\Delta u = \lambda u & \text{in } \Omega \\ \dfrac{\partial u}{\partial \nu} = 0 & \text{on } \partial\Omega. \end{cases}$$

From the compactness of the embedding $H^1(\Omega) \subset L^2(\Omega)$ and from Theorem 1.8.10 it follows that the eigenvalues of $-\Delta$ can be ordered in an increasing sequence diverging to $+\infty$, and each eigenvalue has finite dimensional eigenspace. Moreover, the smallest eigenvalue is $\lambda = 0$, associated to the eigenfunctions that are constant over the connected components of Ω.

Remark 7.1.13. In view of Theorem 1.8.10, we know that the eigenvalues have finite dimensional eigenspaces. In particular, since the dimension of the eigenspace associated to $\lambda = 0$ is given by the number of connected components of Ω, this means that such number must be finite. This property is guaranteed by the boundedness and the C^1 regularity of Ω. In absence of one of these hypotheses (as it happens if Ω has infinitely many connected components) the embedding $H^1(\Omega) \subset L^2(\Omega)$ is not compact and Theorem 1.8.10 cannot be applied. □

An immediate consequence of Theorem 1.8.10 is the following

Theorem 7.1.14. *The following assertions hold true:*

(i) if $-\alpha \notin \sigma_N(-\Delta)$, then for every $f \in L^2(\Omega)$ and for every $g \in H^{-1/2}(\partial\Omega)$, problem (7.2) *admits a unique solution $u \in H^1(\Omega)$;*

(ii) if $-\alpha = \lambda \in \sigma_N(-\Delta)$ then (7.2) *admits a solution if and only if $f \in L^2(\Omega)$ and $g \in H^{-1/2}(\partial\Omega)$ satisfy the condition*

$$\int_\Omega fw + \langle g, \gamma_0 w \rangle = 0$$

for every eigenfunction w of λ; such solution is not unique: if $u \in H^1(\Omega)$ is a solution, then the set of solutions can be written as

$$\{u + w \in H^1(\Omega)\ ;\ w \text{ eigenfunction of } \lambda\}.$$

The case in which $\alpha = 0$ and Ω is connected is particularly significant. Since $0 \in \sigma_N(-\Delta)$ and the associated eigenfunctions are precisely the constant functions, by Theorem 7.1.14, we deduce that the problem

$$\begin{cases} -\Delta u = f & \text{in } \Omega \\ \dfrac{\partial u}{\partial \nu} = g & \text{on } \partial\Omega \end{cases} \tag{7.14}$$

admits a weak solution if and only if the functions $f \in L^2(\Omega)$ and $g \in H^{-1/2}(\partial\Omega)$ satisfy the compatibility condition

$$\int_\Omega f + \langle g, 1\rangle = 0. \tag{7.15}$$

Thus, Theorem 7.1.14 is a generalization of Theorem 3.6.7.

7.1.4 Stokes equation

The Helmholtz-Weyl decomposition

In this section Ω denotes an open bounded (connected) set of $\mathbb{R}^n$ with boundary $\partial\Omega \in C^2$; in some cases, less regularity is sufficient, but here we are not interested in such details. Capital bold letters ($\mathbf{L}^2$, $\mathbf{H}^1$,...) denote spaces of vector valued functions (with values in $\mathbb{R}^n$), while italic small letters (L^2, H^1,...) denote spaces of scalar functions: for brevity, we set $L^2 := L^2(\Omega)$,.... and we will precise the set only when it is different from Ω.

We shall denote by $\nabla \cdot u$ the divergence of the field u. The spaces $\mathbf{E}$ (introduced in (6.15)) and $\mathbf{H}_0^1$ are Hilbert spaces if equipped with the scalar products

$$(u,v)_\mathbf{E} = (u,v)_{\mathbf{L}^2} + (\nabla\cdot u, \nabla\cdot v)_{L^2} \qquad \text{and} \qquad (u,v)_{\mathbf{H}_0^1} := (\nabla u, \nabla v)_{\mathbf{L}^2}\ ,$$

where the last parentheses mean

$$(\nabla u, \nabla v)_{\mathbf{L}^2} = \int_\Omega \nabla u : \nabla v$$

and $\nabla u : \nabla v$ stands for the "Euclidean scalar product" of the Jacobian matrices of u and v. Here we consider also some spaces of divergence free functions. First of all, let us introduce

$$\mathbf{V} := \{u \in \mathbf{H}_0^1;\ \nabla\cdot u = 0 \text{ in } \Omega\}$$

which is a closed subspace of $\mathbf{H}_0^1$ and hence, endowed with the scalar product

$$(u,v)_\mathbf{V} := (\nabla u, \nabla v)_{\mathbf{L}^2}\ ,$$

it is a Hilbert space. Recalling that γ_ν is defined according to Theorem 6.7.4, we consider the spaces

$$\mathbf{G}_1 := \{f \in \mathbf{L}^2;\ \nabla\cdot f = 0,\ \gamma_\nu f = 0\}\ ,$$
$$\mathbf{G}_2 := \{f \in \mathbf{L}^2;\ \nabla\cdot f = 0,\ \exists g \in H^1,\ f = \nabla g\}\ ,$$
$$\mathbf{G}_3 := \{f \in \mathbf{L}^2;\ \exists g \in H_0^1,\ f = \nabla g\}\ .$$

Clearly, $\mathbf{V} \subset \mathbf{G}_1 \subset \mathbf{E}_0$ and $\mathbf{G}_1 \oplus \mathbf{G}_2 \subset \mathbf{E}$ where with $\mathbf{E}_0$ we denoted the subspace of $\mathbf{E}$ introduced in Definition 6.7.5. Furthermore, $\mathbf{V}$ is dense in $\mathbf{G}_1$.

Exercise 7.1.15. Show that in dimension $n = 1$ there hold $G_1 = \{0\}$, $G_2 \equiv \mathbb{R}$ and $G_3 = \{f \in L^2;\ \int_\Omega f = 0\}$. □

A famous result by Helmholtz and Weyl [40, 74] states that the spaces $\mathbf{G}_i$ are mutually orthogonal:

Theorem 7.1.16. (Helmholtz-Weyl Theorem). *The spaces $\mathbf{G}_i$ $(i = 1, 2, 3)$ are mutually orthogonal in $\mathbf{L}^2$ and $\mathbf{L}^2 = \mathbf{G}_1 \oplus \mathbf{G}_2 \oplus \mathbf{G}_3$ (this means that every $f \in \mathbf{L}^2$ can be uniquely written as the sum of three functions $f_i \in \mathbf{G}_i$ for $i = 1, 2, 3$).*

Proof. Here we follow [68, Theorem I.1.5].
Step 1. Let us show that $\mathbf{G}_i \perp \mathbf{G}_j$ when $i \neq j$.

Given $f_1 \in \mathbf{G}_1$ and $f_2 \in \mathbf{G}_2$, it holds

$$\nabla \cdot f_1 = \nabla \cdot f_2 = 0 \text{ in } \Omega\,, \quad \gamma_\nu f_1 = 0 \text{ on } \partial\Omega\,, \quad \exists g \in H^1 \text{ s.t. } f_2 = \nabla g\ .$$

But then, exploiting (6.16),

$$\int_\Omega f_1 \cdot f_2 = \int_\Omega f_1 \cdot \nabla g = -\int_\Omega g(\nabla \cdot f_1) + \langle \gamma_\nu f_1, \gamma_0 g\rangle = 0\ .$$

Let now $f_2 \in \mathbf{G}_2$ and $f_3 \in \mathbf{G}_3$, that is

$$\nabla \cdot f_2 = 0 \text{ in } \Omega\,, \quad \exists g \in H^1 \text{ s.t. } f_2 = \nabla g\,, \quad \exists h \in H^1_0 \text{ s.t. } f_3 = \nabla h\ .$$

Again by (6.16), we get

$$\int_\Omega f_2 \cdot f_3 = \int_\Omega f_2 \cdot \nabla h = -\int_\Omega h(\nabla \cdot f_2) + \langle \gamma_\nu f_2, \gamma_0 h\rangle = 0\ .$$

Let $f_1 \in \mathbf{G}_1$ and $f_3 \in \mathbf{G}_3$; then

$$\nabla \cdot f_1 = 0 \text{ in } \Omega\,, \quad \gamma_\nu f_1 = 0 \text{ on } \partial\Omega\,, \quad \exists g \in H^1_0,\ f_3 = \nabla g\ .$$

Finally, again by (6.16), we conclude that

$$\int_\Omega f_1 \cdot f_3 = \int_\Omega f_1 \cdot \nabla g = -\int_\Omega g(\nabla \cdot f_1) + \langle \gamma_\nu f_1, \gamma_0 g\rangle = 0\ .$$

Step 2. Let us prove that every $f \in \mathbf{L}^2$ can be written as the sum of three functions $f_i \in \mathbf{G}_i$ for $i = 1, 2, 3$.

Let $f \in \mathbf{L}^2$ and let $p \in H^1_0$ be the unique weak solution of the Dirichlet problem

$$\begin{cases} \Delta p = \nabla \cdot f & \text{in } \Omega \\ p = 0 & \text{on } \partial\Omega\ . \end{cases}$$

The last problem makes sense thanks to Proposition 6.2.19 and to Remark 7.1.4. Once p is determined, we set $f_3 = \nabla p$.

Note that $\nabla \cdot (f - \nabla p) = 0$ a.e. in Ω (actually, the equality holds true in the sense of distributions in H^{-1}), so that $f - \nabla p \in \mathbf{E}$ and $\gamma_\nu(f - \nabla p) \in H^{-1/2}(\partial\Omega)$. Therefore, once we have checked the validity of the compatibility condition for the Neumann problem (see (7.15)), we can choose $q \in H^1$ among the weak solutions of the problem

$$\begin{cases} \Delta q = 0 & \text{in } \Omega \\ \dfrac{\partial q}{\partial \nu} = \gamma_\nu(f - \nabla p) & \text{on } \partial\Omega \, . \end{cases}$$

The compatibility condition of this problem reads $\langle \gamma_\nu(f - \nabla p), 1\rangle = 0$; choosing $w = 1$ and $u = f - \nabla p$ in (6.16), we obtain

$$\langle \gamma_\nu(f - \nabla p), 1\rangle = \int_\Omega \nabla \cdot (f - \nabla p) = 0 \, ,$$

being $\nabla \cdot (f - \nabla p) = 0$ a.e. in Ω. Hence, the solution q to the above Neumann problem exists and is unique up to an additive constant. We set $f_2 = \nabla q$.

Finally, setting $f_1 = f - f_2 - f_3$, we have to prove that $f_1 \in \mathbf{G}_1$. To this aim, we compute $\nabla \cdot f_1 = \nabla \cdot (f - f_2 - f_3) = \nabla \cdot f - \Delta q - \Delta p = 0$. Moreover, $\gamma_\nu f_1 = \gamma_\nu(f - f_2 - f_3) = \gamma_\nu f - \frac{\partial q}{\partial \nu} - \frac{\partial p}{\partial \nu} = 0$. This completes the proof. □

From now on we denote by P_i $(i = 1, 2, 3)$ the orthogonal projections of $\mathbf{L}^2$ on $\mathbf{G}_i$.

Remark 7.1.17. (i) Theorem 7.1.16 admits an $\mathbf{L}^s$ version for every $s \in (1, \infty)$, see [25, 43].

(ii) If $n = 3$ and $\Omega \subset \mathbb{R}^3$ is connected but not simply connected, then the space $\mathbf{G}_1$ can be decomposed as the direct sum of $\mathbf{G}_1' = \ker(\text{curl}) \cap \mathbf{G}_1$ and of its orthogonal complement $\mathbf{G}_1''$ in $\mathbf{G}_1$. If N is the order of connection of Ω, it is proved that $\dim(\mathbf{G}_1') = N$ (see [23]). Therefore, if Ω is simply connected, $\mathbf{G}_1 = \mathbf{G}_1''$, $\text{curl}(\mathbf{H}^1) = \text{curl}(\mathbf{H}^1 \cap \mathbf{G}_1) = \mathbf{G}_1 \oplus \mathbf{G}_2$ and $[\text{curl}(\mathbf{H}^1)]^\perp = \mathbf{G}_3$. □

Exercise 7.1.18. (i) Without using Theorem 7.1.16, prove that $\mathbf{G}_i \cap \mathbf{G}_j = \{0\}$ if $i \neq j$.

(ii) Prove that $\mathbf{G}_2 \cap \mathbf{E}_0 = \{0\}$, $\mathbf{G}_3 \cap \mathbf{E} = \{f \in \mathbf{L}^2;\ \exists g \in H_0^1 \cap H^2,\ f = \nabla g\}$, $\mathbf{G}_3 \cap \mathbf{E}_0 = \{f \in \mathbf{L}^2;\ \exists g \in H_0^2,\ f = \nabla g\}$. □

Variational formulation of the Stokes problem

The Stokes equation represents the linearized stationary form of the Navier-Stokes equation that appears in the mechanics of incompressible fluids, see [43, 68]. Let $\Omega \subset \mathbb{R}^n$ be an open bounded domain with $\partial\Omega \in C^2$, and let $f \in \mathbf{L}^2$ be a given vector function in Ω (representing the density of the external force applied to the incompressible fluid): the unknowns are the vector function $u = (u_1, ..., u_n)$ and the scalar function p, which represent the fluid velocity and pressure respectively. The functions u and p satisfy (in a suitable sense) the system

$$\begin{cases} -\eta\Delta u + \nabla p = f & \text{in} \quad \Omega \\ \nabla \cdot u = 0 & \text{in} \quad \Omega\,, \end{cases} \tag{7.16}$$

where $\eta > 0$ is the fluid viscosity. In view of (6.16) (with $w \equiv 1$) we have a necessary condition for the solvability of (7.16): $\langle \gamma_\nu u, 1 \rangle = 0$. The simplest example in which such condition holds true is the so called **non-slip boundary condition**

$$u = 0 \qquad \text{on} \quad \partial\Omega\,. \tag{7.17}$$

From a physical view point, (7.17) represents the case in which the fluid is adherent to the boundary of Ω. Note that (7.16) is a system of $n+1$ scalar equations in the $n+1$ unknowns $u_1, ..., u_n, p$, while (7.17) gives just n boundary conditions. The reason is that the pressure p appears just with its first order derivatives, therefore it is determined up to an additive constant. For this reason, we introduce the space $H^1/\mathbb{R}$ of the H^1 functions defined up to an additive constant.

In order to arrive to the variational formulation of (7.16), we proceed as in the previous sections. Multiplying both sides of the first vector equation by a test function $v \in \mathbf{V}$ and applying (6.16), we obtain

$$\eta \int_\Omega \nabla u : \nabla v + \int_\Omega \nabla p \cdot v = \int_\Omega fv.$$

But $\nabla p \in \mathbf{G}_2 \oplus \mathbf{G}_3$ and $\mathbf{V} \subset \mathbf{G}_1$: hence, by Theorem 7.1.16, there holds

$$\int_\Omega \nabla p \cdot v = 0 \qquad \forall v \in \mathbf{V}\,,$$

from which we derive

$$\eta \int_\Omega \nabla u : \nabla v = \int_\Omega fv \qquad \forall v \in \mathbf{V}.$$

Next we observe that for every $f \in \mathbf{L}^2$ the function $v \mapsto \int_\Omega fv$ is linear and continuous over $\mathbf{V}$. Therefore, in view of Lax-Milgram Theorem (Corollary 1.7.4), we infer that

$$\forall f \in \mathbf{L}^2 \quad \exists! u \in \mathbf{V} \quad \text{s.t.} \quad \eta \int_\Omega \nabla u : \nabla v = \int_\Omega fv \quad \forall v \in \mathbf{V}\,. \tag{7.18}$$

Applying the elliptic regularity theory (see [68, Proposition I.2.2] and Section 7.1.5 within few pages), we also deduce that $u \in \mathbf{H}^2$. Thus we may rewrite (7.18) as

$$\forall f \in \mathbf{L}^2 \quad \exists! u \in \mathbf{H}^2 \cap \mathbf{V} \quad \text{s.t.} \quad -\eta \int_\Omega \Delta u \cdot v = \int_\Omega fv \quad \forall v \in \mathbf{V}\,.$$

Thanks to the density of $\mathbf{V}$ in $\mathbf{G}_1$, we immediately obtain

$$\forall f \in \mathbf{L}^2 \quad \exists! u \in \mathbf{H}^2 \cap \mathbf{V} \quad \text{s.t.} \quad \int_\Omega (\eta\Delta u + f)v = 0 \quad \forall v \in \mathbf{G}_1 \ .$$

By Theorem 7.1.16, this means that $(\eta\Delta u + f) \in (\mathbf{G}_2 \oplus \mathbf{G}_3)$ and hence that there exists a unique $p \in H^1/\mathbb{R}$ such that $\eta\Delta u + f = -\nabla p$. We have then proved the following

Theorem 7.1.19. *For every $f \in \mathbf{L}^2$ there exists a unique $(u, p) \in (\mathbf{H}^2 \cap \mathbf{V}) \times (H^1/\mathbb{R})$ which satisfies (7.16) a.e. in Ω and (7.17) in the sense of traces.*

From Theorem 7.1.19, we have the following obvious consequence:

Corollary 7.1.20. *For every $f \in \mathbf{L}^2$ there exists a unique $v \in \mathbf{H}^2 \cap \mathbf{V}$ such that $P_1(\Delta v) = P_1 f$. In particular, if $P_1 f = 0$ then $v = 0$.*

The spectrum of the Stokes operator

The linear operator Δ turns $(\mathbf{H}^2 \cap \mathbf{V}) \subset \mathbf{G}_1$ into $\mathbf{G}_1 \oplus \mathbf{G}_2$; indeed $\nabla \cdot \Delta u = 0$ for every $u \in \mathbf{H}^2 \cap \mathbf{V}$. Here we deal with some properties of its eigenfunctions and eigenvalues. This problem appears quite different from the elliptic eigenvalue problems considered in the previous sections. Indeed, we want to determine the real values of λ for which the following problem admits non trivial solutions $u \in \mathbf{H}^2 \cap \mathbf{V}$ for some $p \in H^1$:

$$\begin{cases} -\Delta u + \nabla p = \lambda u & \text{in } \Omega \\ \nabla \cdot u = 0 & \text{in } \Omega \\ u = 0 & \text{on } \partial\Omega \ . \end{cases} \tag{7.19}$$

For the sake of simplicity, here we have assumed that $\eta = 1$. As in (7.18), the weak formulation of (7.19) reads

$$\int_\Omega \nabla u : \nabla v = \lambda \int_\Omega uv \qquad \forall v \in \mathbf{V} \ . \tag{7.20}$$

By Theorem 7.1.16, we have that $\nabla p \in \mathbf{G}_2 \oplus \mathbf{G}_3$, hence it is clear that in (7.19) it holds $-P_1(\Delta u) = \lambda u$. On the other hand, being $\Delta u \in \mathbf{G}_1 \oplus \mathbf{G}_2$ and $u \in \mathbf{G}_1$, we infer $P_3(\nabla p) = P_3(\Delta u + \lambda u) = 0$ and hence $\nabla p \in \mathbf{G}_2$. The last condition implies that necessarily $P_2(\Delta u) = \nabla p$; this means that ∇p appears in (7.19) in order that the equation projected on $\mathbf{G}_2$ is satisfied. Finally, we observe that such function p is weakly harmonic ($\Delta p = 0$).

Choosing $v = u$ in (7.20), we obtain

$$\int_\Omega |\nabla u|^2 = \lambda \int_\Omega u^2 \ .$$

This shows that all the eigenvalues λ of (7.19) are strictly positive. We are interested in determining the asymptotic behavior of

$$N(\lambda) = \sum_{\lambda_j \le \lambda} 1 \ ,$$

namely the number of eigenvalues that are less or equal to λ. We summarize some results in the following statement:

Theorem 7.1.21. (Métivier Theorem). *Problem* (7.19) *admits a sequence of positive eigenvalues* $\{\lambda_m\}$ *which diverges to infinity. Each eigenvalue has finite multiplicity and there exists a constat* $C = C(\Omega) > 0$ *such that*

$$N(\lambda) \sim C\,\lambda^{n/2} \ .$$

Moreover, the eigenfunctions can be chosen so that they form a complete orthonormal system in $\mathbf{V}$ *(and in* $\mathbf{G}_1$*).*

Proof. The first part of the proof follows from Theorem 1.8.10, but the remaining part is quite long and delicate; for this reason we refer to [49], where the reader may also find the explicit value of $C(\Omega)$. □

Theorem 7.1.21 states that we can associate with each eigenvalue λ_m an eigenfunction e_m such that

$$\|e_m\|_{\mathbf{L}^2} = 1\,, \quad -P_1(\Delta e_m) = \lambda_m e_m\,, \quad \int_\Omega \nabla e_m : \nabla e_k = \int_\Omega e_m e_k = 0 \text{ if } m \neq k\ . \tag{7.21}$$

In the following, we denote by

$$\mathcal{P}_m \text{ the orthogonal projection on } \mathrm{span}\{e_1, ..., e_m\} \qquad \text{and} \qquad \mathcal{Q}_m = I - \mathcal{P}_m\ . \tag{7.22}$$

Consequently, there holds the following generalized Poincaré inequality (compare with Theorem 6.2.18):

$$\lambda_{m+1}^{1/2} \|\mathcal{Q}_m v\|_{\mathbf{L}^2} \le \|\mathcal{Q}_m v\|_{\mathbf{V}} \qquad \forall v \in \mathbf{V}\ . \tag{7.23}$$

Theorem 7.1.21 has several important consequences. One of these will be enlightened in Theorem 8.2.10.

7.1.5 Regularity of weak solutions

In Section 6.1 we have motivated the introduction of Sobolev spaces as a tool to derive solutions of partial differential equations. In the previous sections we have obtained weak solutions by recasting the equation as a variational problem over

Sobolev spaces. The solutions so found are indeed weak solutions. It remains to pass from weak solutions to classical solutions, namely (see the scheme proposed in Section 6.1) we have to show that a weak solution is regular. A complete reference about the regularity theory for elliptic problems can be found in [2], but we warn the bravest reader that it is a very delicate and complicated matter, involving an incredible amount of estimates and technicalities. In this section we limit ourselves to a brief summary (without proofs!) of the results in [2].

The regularity of a weak solution of an elliptic problem depends basically on two parameters: the regularity of the data and the regularity of the boundary of the domain. Basically, we can say that *if the boundary is regular enough, then the solution gains two derivatives with respect to the source and half a derivative with respect to the boundary data.* This can be seen (in a slightly different way) also in the Sobolev spaces $W^{m,p}(\Omega)$ of functions whose generalized derivatives up to the order m belong to $L^p(\Omega)$ (not necessarily $p = 2$): if $p \neq 2$, these are not Hilbert spaces! But let us analyze separately all the problems studied in the previous sections.

Proposition 7.1.22. *Let $k \in \mathbb{N}$ and $p \in (1, \infty)$. Assume that $\alpha \geq 0$, $\partial\Omega \in C^{k+2}$, $f \in W^{k,p}(\Omega)$. Then the weak solution u of* (7.1) *satisfies $u \in W^{k+2,p}(\Omega)$.*

Remark 7.1.23. In the case $p = 2$, this proposition states that if $\alpha \geq 0$, $\partial\Omega \in C^{k+2}$, $f \in H^k(\Omega)$, then $u \in H^{k+2}(\Omega)$; therefore, by Theorem 6.6.3, if $k > \frac{n}{2} + 2$ we infer that $u \in C^2(\overline{\Omega})$, namely u is a classical solution of (7.1). □

For the non homogeneous Dirichlet problem, we limit ourselves to consider Hilbertian Sobolev spaces:

Proposition 7.1.24. *Let $k \in \mathbb{N}$. Assume that $\alpha \geq 0$, $\partial\Omega \in C^{k+2}$, $f \in H^k(\Omega)$, $g \in H^{k+\frac{3}{2}}(\partial\Omega)$. Then the weak solution u of* (7.7) *satisfies $u \in H^{k+2}(\Omega)$.*

Proposition 7.1.24 reveals that also the trace of g influences the regularity of the solution: the solution gains "one derivative and a half" with respect to g. The same holds true for the Neumann problem, but clearly, the gain is just of half a derivative, since the trace affects the derivative of the solution:

Proposition 7.1.25. *Let $k \in \mathbb{N}$. Assume that $\alpha > 0$, $\partial\Omega \in C^{k+2}$, $f \in H^k(\Omega)$, $g \in H^{k+\frac{1}{2}}(\partial\Omega)$. Then the weak solution u of* (7.10) *satisfies $u \in H^{k+2}(\Omega)$.*

Clearly, Remark 7.1.23 applies also to Propositions 7.1.24 and 7.1.25. The reader may wonder whether it is possible to achieve classical regularity starting from continuous data, for instance taking $f \in C^0(\overline{\Omega})$ in the easiest problem (7.1). The answer is negative! To ensure that the solution of (7.1) belongs to $C^2(\overline{\Omega})$, one needs more regularity for f, e.g. f Lipschitz. This is a further reason to introduce Sobolev spaces.

We conclude with the regularity issue for the Stokes problem:

Proposition 7.1.26. *Let $k \in \mathbb{N}$ and $p \in (1, \infty)$. Assume that $\partial\Omega \in C^{k+2}$ and $f \in \mathbf{W}^{k,p}$. Then the weak solution (u, p) of (7.16) – (7.17) satisfies $u \in \mathbf{W}^{k+2,p}$ and $p \in W^{k+1,p}$.*

This result can be also found in [68, Proposition I.2.2].

7.2 Parabolic equations

In this section we deal with existence and uniqueness of weak solutions of parabolic equations. Parabolic equations are the equations that behave as the heat equation. First, we present an abstract reasoning and then we apply it to concrete cases, as the heat equation itself and the Stokes equation.

7.2.1 The Bochner integral

In this section we introduce some new functional spaces. Let X be a Banach space and assume to have a function $u = u(t)$ defined in an interval $I \subset \mathbb{R}$, such that for a.e. $t \in I$ it holds $u(t) \in X$. The simplest example that comes into mind is a function of two real variables:

Example 7.2.1. Let $f \in C^0(\mathbb{R}^2)$. Then $u(t) := f(s,t)$ belongs to $C^0(\mathbb{R})$ for every $s \in \mathbb{R}$. Basically, the variable s is "frozen" and considered as a parameter. Another example of this type has already been introduced in Theorem 2.3.6. □

If we choose the interval I as $I = (0, T)$ with $T > 0$, for a generic Banach space X we can define

$$L^p(0,T;X) = \left\{ u : (0,T) \to X;\ \int_0^T \|u(t)\|_X^p \, dt < \infty \right\} \tag{7.24}$$

for $1 \le p < \infty$ and

$$L^\infty(0,T;X) = \left\{ u : (0,T) \to X;\ \operatorname*{ess\,sup}_{t\in(0,T)} \|u(t)\|_X < \infty \right\}.$$

Provided that we identify functions equal a.e. , we can endow these spaces with the norms

$$\|u\|_{L^p(0,T;X)} := \left(\int_0^T \|u(t)\|_X^p dt \right)^{1/p} \qquad (1 \le p < \infty)$$

$$\|u\|_{L^\infty(0,T;X)} := \operatorname*{ess\,sup}_{t\in(0,T)} \|u(t)\|_X$$

that generate a structure of Banach space. If $p = 2$ and X is a Hilbert space, then also $L^2(0,T;X)$ is a Hilbert space with the scalar product

$$(u,v)_{L^2(0,T;X)} = \int_0^T (u(t), v(t))_X \, dt \ .$$

We give the following

Definition 7.2.2. *A function $s : (0,T) \to X$ is said to be simple if there exists a finite number of measurable sets $A_1, ..., A_n \subset (0,T)$ and of vectors $x_1, ..., x_n \in X$ such that*

$$s(t) = \sum_{k=1}^{n} \chi_{A_k}(t) x_k ,$$

where χ_{A_k} are real functions which equal 1 in A_k and 0 in $(0,T) \setminus A_k$.

If s is a simple function, it is natural to define

$$\int_0^T s(t)dt := \sum_{k=1}^{n} |A_k| x_k \in X ,$$

where we recall that by $|A_k|$ we denote the Lebesgue measure of the set A_k.

A further characterization of the elements of $L^1(0,T;X)$ is stated in the

Proposition 7.2.3. *Let $u \in L^1(0,T;X)$. Then there exists a sequence of simple functions $\{s_n\}$ such that $s_n(t) \to u(t)$ in X for a.e. $t \in (0,T)$ and*

$$\int_0^T \|s_n(t) - u(t)\|_X \, dt \to 0$$

as $n \to \infty$.

Proof. The characterization of $L^1(0,T;X)$ introduced in this proposition is often given as the definition of $L^1(0,T;X)$ itself and (7.24) is interpreted as a consequence. However, in order to make the understanding easier also to the unexperienced reader, we prefer to consider (7.24) as the definition. For more details we refer to [77, Section 5, Chapter V]. □

In view of Proposition 7.2.3, it is reasonable the

Definition 7.2.4. *Let $u \in L^1(0,T;X)$ and let $\{s_n\}$ be a sequence of simple functions as in Proposition 7.2.3. Then we define*

$$\int_0^T u(t)dt := \lim_{n\to\infty} \int_0^T s_n(t)dt \in X \tag{7.25}$$

where the limit is meant with respect to the convergence in X.

It is easy to prove that given $u \in L^1(0,T;X)$ and $\{s_n\}$ as in Proposition 7.2.3, the limit appearing in (7.25) exists and does not depend on the *approximating* sequence $\{s_n\}$, but only on u.

Besides the spaces $L^p(0,T;X)$, we can also define the space

$$C^0([0,T];X) = \left\{ u : [0,T] \to X;\ \lim_{t\to t_0} \|u(t) - u(t_0)\|_X = 0 \quad \forall t_0 \in [0,T] \right\},$$

which is a Banach space as well, equipped with the norm $\|\cdot\|_{L^\infty(0,T;X)}$.

We will widely use the

Theorem 7.2.5. (Bochner Theorem). *Let X be a Banach space and let $L \in X'$. If $u \in L^1(0,T;X)$, then the scalar function $t \mapsto \langle L, u(t)\rangle$ belongs to $L^1(0,T)$. Moreover*

$$\int_0^T \langle L, u(t)\rangle dt = \langle L, \int_0^T u(t)dt\rangle.$$

Proof. See [77, Corollary 2, Section 5, Chapter V]. □

We conclude this section with a few words about distributions with values in a Banach space.

Definition 7.2.6. *Let X be a Banach space. We define $\mathcal{D}'(0,T;X)$ as the space $\mathcal{L}(\mathcal{D}(0,T),X)$ of linear and continuous operators $\Lambda : \mathcal{D}(0,T) \to X$, namely satisfying*

$$\Lambda(\varphi_k) \to \Lambda(\varphi) \quad \text{in } X$$

for every sequence such that $\varphi_k \to \varphi$ in $\mathcal{D}(0,T)$ (for the definition of $\mathcal{D}(0,T)$ and the corresponding convergence we refer to Section 2.6). Given $\Lambda \in \mathcal{D}'(0,T;X)$ and $\varphi \in \mathcal{D}(0;T)$, we set

$$\langle \Lambda, \varphi\rangle := \Lambda(\varphi).$$

Let us define the derivative of a distribution in $\mathcal{D}'(0,T;X)$.

Definition 7.2.7. *Let $u \in \mathcal{D}'(0,T;X)$. We define the distribution u' as*

$$\langle u', \varphi\rangle := -\langle u, \varphi'\rangle \qquad \forall \varphi \in \mathcal{D}(0,T). \tag{7.26}$$

We leave to the reader to check that (7.26) actually defines an element of $\mathcal{D}'(0,T;X)$. For more details we refer to Definition 2.6.11.

Example 7.2.8. *It is possible to associate with each function $u \in L^1(0,T;X)$ the distribution $\Lambda_u \in \mathcal{D}'(0,T;X)$ defined by*

$$\langle \Lambda, \varphi\rangle := \int_0^T \varphi(t)u(t)\,dt \qquad \forall \varphi \in \mathcal{D}(0,T).$$

Arguing as in the scalar case, it is possible to prove that, actually, Λ_u is an element of $\mathcal{D}'(0,T;X)$.

7.2.2 Abstract equations

In this section V and H represent two separable Hilbert spaces such that V is dense in H and the embedding $V \subset H$ is continuous. Identifying H with its dual and denoting by V' the dual of V, we obtain a Hilbert triplet $V \subset H \subset V'$ with continuous dense inclusions, see Chapter 1. We denote by $\langle \cdot, \cdot \rangle$ the duality between V' and V; since we are dealing with a Hilbert triplet, when the element of V' belongs also to H, the duality is meant as a scalar product:

$$\langle w, v \rangle = (w, v)_H \qquad \forall w \in H\ , \quad \forall v \in V. \tag{7.27}$$

The following result holds:

Theorem 7.2.9. *(i) If $u \in L^2(0,T;V)$ and $u' \in L^2(0,T;V)$, then $u \in C^0([0,T];V)$, the function $t \mapsto (u(t), v)_H$ admits a weak derivative for every $v \in V$ and there holds*

$$\frac{d}{dt}(u(t), v)_H = (u'(t), v)_H \qquad \forall v \in V \qquad \text{in } \mathcal{D}'(0,T)\ . \tag{7.28}$$

(ii) If $u \in L^2(0,T;V)$ and $u' \in L^2(0,T;V')$, then $u \in C^0([0,T];H)$, the functions $t \mapsto (u(t), v)_H$ $(v \in V)$ and $t \mapsto \|u(t)\|_H^2$ admit weak derivatives and there hold

$$\frac{d}{dt}(u(t), v)_H = \langle u'(t), v \rangle \qquad \forall v \in V \qquad \text{in } \mathcal{D}'(0,T)\ , \tag{7.29}$$

$$\frac{d}{dt}\|u(t)\|_H^2 = 2\langle u'(t), u(t) \rangle \qquad \text{in } \mathcal{D}'(0,T)\ . \tag{7.30}$$

Proof. See [68, Lemmas 1.1 and 1.2 of Capitolo III]. □

We point out that in the statement of Theorem 7.2.9 the derivative u' is meant in the sense of Definition 7.2.7: by $u' \in L^2(0,T;V)$ we mean that there exists $w \in L^2(0,T;V)$ such that

$$-\int_0^T \varphi'(t)u(t)\,dt = \int_0^T \varphi(t)w(t)\,dt$$

for every $\varphi \in \mathcal{D}(0,T)$. In other words, the distributional derivative $u' \in \mathcal{D}'(0,T;V)$ can be represented in terms of w and in this case we will set $u' := w$.

On the other hand, according to (7.27), by $u' \in L^2(0,T;V')$ we mean that there exists $w \in L^2(0,T;V')$ such that

$$(-\int_0^T \varphi'(t)u(t)\,dt, v)_H = \langle \int_0^T \varphi(t)w(t)\,dt, v \rangle \qquad \forall v \in V.$$

Also in this case we will set $u' := w$.

Let us now comment other aspects of the statement of Theorem 7.2.9. Clearly, $u \in C^0$ is to be read as u *equals a.e. a continuous function*, while (7.28), (7.29), (7.30) are to be meant respectively as

$$\int_0^T (u(t),v)_H\,\varphi'(t)\,dt = -\int_0^T (u'(t),v)_H\,\varphi(t)\,dt \qquad \forall\varphi\in\mathcal{D}(0,T),\ \forall v\in V$$

$$\int_0^T (u(t),v)_H\,\varphi'(t)\,dt = -\int_0^T \langle u'(t),v\rangle\,\varphi(t)\,dt \qquad \forall\varphi\in\mathcal{D}(0,T),\ \forall v\in V$$

$$\int_0^T \|u(t)\|_H^2\varphi'(t)\,dt = -2\int_0^T \langle u'(t),u(t)\rangle\,\varphi(t)\,dt \qquad \forall\varphi\in\mathcal{D}(0,T)\ .$$

Theorem 7.2.9 is not surprising. Indeed, with some imagination, the hypotheses of assertion (i) can be rephrased as $u \in H^1(0,T;V)$ and, being the *domain* of $u = u(t)$ the interval $(0,T) \subset \mathbb{R}$, Theorem 6.2.5 states that u is continuous. Furthermore, (7.28) states that the derivative of the product of a function and a constant equals the product of the derivative of the function and the constant itself. Imagination helps to "turn a blind eye" to the fact that the *codomain* of $u(t)$ is an abstract space, almost mysterious. With regard to the second assertion of Theorem 7.2.9, we need even more imagination, since the codomains of u and u' are distinct. The continuity is then guaranteed in a third "intermediate" space. Finally, (7.30) can be seen as a generalization of the differentiation formula for the square of a function.

Let us assume that over $V \times V$ it is defined a coercive continuous bilinear form $a(\cdot,\cdot)$. We intend to solve the following abstract problem:
given $u_0 \in H$ and $f \in L^2(0,T;V')$, find $u \in L^2(0,T;V)\cap C^0([0,T];H)$ such that

$$\langle u'(t),v\rangle + a(u(t),v) = \langle f(t),v\rangle \qquad \forall v\in V \quad \text{in } \mathcal{D}'[0,T) \tag{7.31}$$

$$u(0) = u_0 \tag{7.32}$$

where the derivative $u'(t)$ is to be meant in the sense of Definition 7.2.7, while, by definition, problem (7.31)-(7.32) can be rephrased as

$$-\int_0^T (u(t),v)_H\varphi'(t)\,dt - \varphi(0)(u_0,v)_H + \int_0^T a(u(t),v)\varphi(t)\,dt = \int_0^T \langle f(t),v\rangle\varphi(t)\,dt$$

$$\forall v\in V\ ,\quad \forall\varphi\in\mathcal{D}[0,T)\ . \tag{7.33}$$

We observe that if $\varphi \in \mathcal{D}[0,T)$ then φ and all its derivatives vanish at T, while *not necessarily* $\varphi(0) = 0$. In this way, the initial condition $u(0) = u_0$ appearing in (7.32) is "encoded" in formula (7.33).

Exercise 7.2.10. Using the energy functional (3.114), prove that if there exists a solution to (7.31), then it is unique. □

To show the existence of a solution to (7.31), let us use the Galerkin method illustrated in Section 6.8.

Step 1. Let $\{V_m\}$ be a Galerkin sequence for V: then it is a Galerkin sequence for H as well, being $\overline{V} = H$. Thus there exists a sequence $\{u_0^m\} \subset V$ such that $u_0^m \to u_0$ in H. For sake of convenience, let $m = \dim V_m$. Let then $\{w_k^m\}_{k=1}^m$ be a basis of V_m, orthonormal in H. We look for $u_m = u_m(t)$ such that

$$u_m(t) = \sum_{k=1}^{m} g_k^m(t)\, w_k^m$$

and that, for $1 \leq k \leq m$,

$$\begin{cases} (u_m'(t), w_k^m)_H + a(u_m(t), w_k^m) = \langle f(t), w_k^m \rangle & \forall t \in (0,T) \\ u_m(0) = u_0^m. \end{cases} \tag{7.34}$$

This problem can also be recast in terms of the unknowns g_k^m:

$$\begin{cases} g_m'(t) + B_m\, g_m(t) = \gamma_m(t) \\ g_m(0) = g_0^m \end{cases} \tag{7.35}$$

where $g_m = (g_1^m, \cdots, g_m^m)^T$, B_m is the matrix whose elements are $\alpha_{kh}^m = a(w_h^m, w_k^m)$, $\gamma_m(t) = (\langle f(t), w_1^m \rangle, \cdots, \langle f(t), w_m^m \rangle)$ and g_0^m is the m-tuple of the components of u_0^m with respect to the basis $\{w_1^m, ..., w_m^m\}$.

Thus we are led to solve a linear $(m \times m)$ system of ordinary differential equations with constant coefficients: such system admits a unique solution $g_m = g_m(t)$ which uniquely determines the solution u_m of (7.34).

Step 2. Now, our goal is to estimate some suitable norms of the sequence $\{u_m\}$. Multiplying by $g_k^m(t)$ the equation in (7.34) and summing for $k = 1, \ldots, m$, by (7.30) we obtain:

$$\frac{1}{2}\frac{d}{dt}\|u_m(t)\|_H^2 + a(u_m(t), u_m(t)) = \langle f(t), u_m(t) \rangle \qquad \text{a.e. in } (0,T). \tag{7.36}$$

We observe that, since we are in a finite dimensional space, the functions u_m enjoy much more regularity than the one we expect for u. This allows us to use formula (7.30) without any problem.

Choosing arbitrarily $t \in (0,T)$, we integrate (7.36) over $(0,t)$, obtaining

$$\frac{1}{2}\|u_m(t)\|_H^2 + \int_0^t a(u_m(\tau), u_m(\tau))\, d\tau = \frac{1}{2}\|u_0^m\|_H^2 + \int_0^t \langle f(\tau), u_m(\tau) \rangle\, d\tau.$$

Thanks to (1.5) and the coercivity of $a(\cdot,\cdot)$, there exists $\alpha > 0$ such that

$$\frac{1}{2}\|u_m(t)\|_H^2 + \alpha \int_0^t \|u_m(\tau)\|_V^2\, d\tau \leq \frac{1}{2}\|u_0^m\|_H^2 + \int_0^t \|f(\tau)\|_{V'}\|u_m(\tau)\|_V\, d\tau.$$

Now we observe that if u_0^m tends to u_0 in H, then there exists $C > 0$ such that

$$\frac{1}{2}\|u_0^m\|_H \leq C \qquad \forall m.$$

Thus, exploiting the Young inequality (2.18), we obtain that for every $t \in [0,T]$)

$$\frac{1}{2}\|u_m(t)\|_H^2 + \frac{\alpha}{2}\int_0^t \|u_m(\tau)\|_V^2 \, d\tau \leq C + \frac{1}{2\alpha}\int_0^T \|f(\tau)\|_{V'}^2 \, d\tau = C'$$

which allows to conclude that

$$\|u_m\|_{L^\infty(0,T;H)} \leq C_1 \qquad \|u_m\|_{L^2(0,T;V)} \leq C_2. \tag{7.37}$$

Step 3. According to (7.37) and Theorem 2.4.17, there exists $u \in L^2(0,T;V) \cap L^\infty(0,T;H)$ such that

$$\begin{aligned} &\int_0^T (u_m(t), v(t))_H dt \to \int_0^T (u(t), v(t))_H dt \qquad \forall v \in L^1(0,T;H) \\ &\int_0^T (u_m(t), v(t))_V dt \to \int_0^T (u(t), v(t))_V dt \qquad \forall v \in L^2(0,T;V)\ , \end{aligned} \tag{7.38}$$

up to subsequences.

For every $m \in \mathbb{N}$ and for every $t \in [0,T]$ we define the functional $A_m(t) \in V'$ as follows:

$$\langle A_m(t), v\rangle = a(u_m(t), v) \qquad \forall v \in V.$$

By (7.37) and by the continuity of a, we infer that there exists $M > 0$ such that

$$|\langle A_m(t), v\rangle| \leq M\|u_m(t)\|_V\|v\|_V \qquad \forall v \in V, \quad \forall m \in \mathbb{N}\ ,$$

therefore, recalling the definition in (1.4), we get

$$\|A_m(t)\|_{V'} \leq M\|u_m(t)\|_V.$$

Thus, integrating, we obtain

$$\int_0^T \|A_m(t)\|_{V'}^2 dt \leq M^2 \int_0^T \|u_m(t)\|_V^2 dt$$

and by (7.37) we deduce that $\{A_m\}$ is bounded in $L^2(0,T;V')$. In view of Theorem 1.4.13 we can affirm that, up to subsequences, there exists $A \in L^2(0,T;V')$ such that

$$A_m \rightharpoonup A \qquad \text{in } L^2(0,T;V') \tag{7.39}$$

where $A = A(t)$ is implicitly defined by

$$\langle A(t), v\rangle = a(u(t), v) \qquad \forall v \in V \quad \text{a.e. in } (0, T).$$

Step 4. Let us consider $\varphi \in \mathcal{D}[0, T)$ and $v \in V$; since V_m is a Galerkin sequence, there exists $\{v_m\}$ such that $v_m \in V_m$ and $v_m \to v$ in V. Setting $\psi_m(t) = \varphi(t)v_m$ and $\psi(t) = \varphi(t)v$, we have

$$\begin{cases} \psi_m \to \psi & \text{in } L^2(0, T; V) \\ \psi'_m \to \psi' & \text{in } L^2(0, T; H); \end{cases} \tag{7.40}$$

actually, the convergences in (7.40) both occur in $L^\infty(0, T; V)$, but we do not need such a strong convergence. Using (7.34), we derive

$$-\int_0^T (u_m(t), \psi'_m(t))_H \, dt - \varphi(0)(u_0^m, v_m)_H + \int_0^T a(u_m(t), \psi_m(t))\, dt$$
$$= \int_0^T \langle f(t), \psi_m(t)\rangle\, dt.$$

By (7.40) we get

$$\int_0^T \langle f(t), \psi_m(t)\rangle \, dt \to \int_0^T \langle f(t), \psi(t)\rangle\, dt.$$

In view of Theorem 1.4.14, by (7.38) and (7.40) we deduce that

$$\int_0^T (u_m(t), \psi'_m(t))_H \, dt \to \int_0^T (u(t), \psi'(t))_H\, dt.$$

Similarly, by (7.38) and (7.39) we obtain

$$\int_0^T a(u_m(t), \psi_m(t)) \, dt \to \int_0^T a(u(t), \psi(t))\, dt.$$

Finally, using again Theorem 1.4.14, we derive

$$(u_0^m, v_m) \to (u_0, v).$$

Thus we have proved the existence of $u \in L^2(0, T; V) \cap L^\infty(0, T; H)$ such that, for every $v \in V$ and for every $\varphi \in \mathcal{D}[0, T)$, equality (7.33) holds true. Hence, from Theorem 7.2.5 it follows that

$$\Big(\int_0^T \varphi'(t)u(t)dt, v\Big)_H = \Big\langle \int_0^T \varphi(t)A(t)dt, v\Big\rangle - \Big\langle \int_0^T \varphi(t)f(t)dt, v\Big\rangle \qquad \forall \varphi \in \mathcal{D}(0, T)$$

(note that now $\varphi(0) = 0!$) and, by the arbitrariness of $v \in V$, we conclude that

$$\int_0^T \varphi'(t)u(t)dt = \int_0^T \varphi(t)[A(t) - f(t)]\,dt \qquad \text{in } V'.$$

This means that u admits a weak derivative and that

$$u'(t) = -A(t) + f(t) \qquad \text{in } V'$$

which implies $u' \in L^2(0,T;V')$. Then, by Theorem 7.2.9, we infer that $u \in C^0([0,T];H)$. Thus we have shown the following

Theorem 7.2.11. *Let $u_0 \in H$ and $f \in L^2(0,T;V')$; then (7.31) admits a unique solution $u \in L^2(0,T;V) \cap C^0([0,T];H)$ such that $u' \in L^2(0,T;V')$.*

7.2.3 The heat equation

Under homogeneous Dirichlet conditions

Let $\Omega \subset \mathbb{R}^n$ be a bounded open set with Lipschitz boundary and consider the problem

$$\begin{cases} u_t - \Delta u = f & \text{in } \Omega \times (0,T) \\ u = 0 & \text{on } \partial\Omega \times (0,T) \\ u(0) = u_0 & \text{in } \Omega. \end{cases} \tag{7.41}$$

We set $V = H_0^1(\Omega)$, $H = L^2(\Omega)$, $V' = H^{-1}(\Omega)$ and

$$a(u,v) = \int_\Omega \nabla u \nabla v = (u,v)_V \qquad \forall u,v \in V; \tag{7.42}$$

it is immediate to check that the bilinear form $a(\cdot,\cdot)$ is continuous and coercive, since

$$|a(u,v)| \le \|u\|_V\,\|v\|_V \quad \forall u,v \in V\,, \qquad a(u,u) = \|u\|_V^2 \quad \forall u \in V\,.$$

Moreover, according to Proposition 2.4.18 and Theorem 6.2.21, we know that the embedding $H_0^1(\Omega) \subset L^2(\Omega)$ is continuous and dense.

Assume that $f \in L^2(0,T;H^{-1}(\Omega))$ and $u_0 \in L^2(\Omega)$. Then the weak solutions of (7.33) are the functions $u \in L^2(0,T;H_0^1(\Omega)) \cap C^0([0,T];L^2(\Omega))$ such that

$$-\int_0^T \int_\Omega u(t)v\,dx\,\varphi'(t)\,dt - \varphi(0)\int_\Omega u_0 v\,dx + \int_0^T \int_\Omega \nabla u(t)\nabla v\,dx\,\varphi(t)\,dt$$
$$= \int_0^T \langle f(t),v\rangle \varphi(t)\,dt \qquad \forall v \in H_0^1(\Omega)\,, \quad \forall \varphi \in \mathcal{D}[0,T)\,, \tag{7.43}$$

which represents the weak formulation of (7.41). By Theorem 7.2.11 we may deduce the

Theorem 7.2.12. *For every $f \in L^2(0,T;H^{-1}(\Omega))$ and for every $u_0 \in L^2(\Omega)$, problem* (7.41) *admits a unique weak solution $u \in L^2(0,T;H_0^1(\Omega)) \cap C^0([0,T];L^2(\Omega))$ such that $u' \in L^2(0,T;H^{-1}(\Omega))$; such solution satisfies* (7.43).

Under homogeneous Neumann conditions

Let $\Omega \subset \mathbb{R}^n$ be a bounded open set with Lipschitz boundary and consider the problem

$$\begin{cases} u_t - \Delta u = f & \text{in } \Omega \times (0,T) \\ \dfrac{\partial u}{\partial \nu} = 0 & \text{on } \partial\Omega \times (0,T) \\ u(0) = u_0 & \text{in } \Omega. \end{cases} \tag{7.44}$$

We set $V = H^1(\Omega)$, $H = L^2(\Omega)$, $V' = (H^1(\Omega))'$; here the problem is that the natural bilinear form associated to (7.44) defined in (7.42) is not coercive over $H^1(\Omega)$ (indeed, if u is a non trivial constant function then $a(u,u) = 0$ but $\|u\|_V > 0$). However, setting $u(x,t) = e^t w(x,t)$ it turns out that w solves

$$\begin{cases} w_t - \Delta w + w = e^{-t} f & \text{in } \Omega \times (0,T) \\ \dfrac{\partial w}{\partial \nu} = 0 & \text{on } \partial\Omega \times (0,T) \\ w(0) = u_0 & \text{in } \Omega \end{cases} \tag{7.45}$$

if and only if u solves (7.44). The bilinear form associated to problem (7.45) is

$$a(u,v) = \int_\Omega (\nabla u \nabla v + uv) \qquad \forall u, v \in H^1(\Omega)$$

which turns out to be continuous and coercive (it is the scalar product!).

Assume that $f \in L^2(0,T;(H^1(\Omega))')$ and $u_0 \in L^2(\Omega)$, then, according to (7.33), the weak solutions w of (7.45) satisfy $w \in L^2(0,T;H^1(\Omega)) \cap C^0([0,T];L^2(\Omega))$ and

$$-\int_0^T \int_\Omega w(t) v \, dx \, \varphi'(t) \, dt - \varphi(0) \int_\Omega u_0 v \, dx + \int_0^T \int_\Omega (\nabla w(t) \nabla v + w(t) v) \, dx \, \varphi(t) \, dt$$

$$= \int_0^T \langle f(t), v \rangle \varphi(t) \, dt \qquad \forall v \in H^1(\Omega) \,, \quad \forall \varphi \in \mathcal{D}[0,T) \,;$$

similarly, the weak solutions u of (7.44) satisfy $u \in L^2(0,T;H^1(\Omega)) \cap C^0([0,T];L^2(\Omega))$ and

$$-\int_0^T \int_\Omega u(t) v \, dx \, \varphi'(t) \, dt - \varphi(0) \int_\Omega u_0 v \, dx + \int_0^T \int_\Omega \nabla u(t) \nabla v \, dx \, \varphi(t) \, dt$$

$$= \int_0^T \langle f(t), v \rangle \varphi(t)\, dt \qquad \forall v \in H^1(\Omega)\ , \quad \forall \varphi \in \mathcal{D}[0,T)\ .$$

By Theorem 7.2.11 (applied to (7.45)!) we deduce the

Theorem 7.2.13. *For every $f \in L^2(0,T;(H^1(\Omega))')$ and for every $u_0 \in L^2(\Omega)$, problem (7.44) admits a unique weak solution $u \in L^2(0,T;H^1(\Omega)) \cap C^0([0,T];L^2(\Omega))$ such that $u' \in L^2(0,T;(H^1(\Omega))')$.*

Exercise 7.2.14. Using the trick of the exponential illustrated above, prove existence and uniqueness of the weak solution of the equation $w_t - \Delta w + \lambda w = f$, for every $\lambda \in \mathbb{R}$, under Dirichlet or Neumann conditions. □

7.2.4 The evolution Stokes equation

Stokes problem (7.16)-(7.17) is nothing else but the stationary version of the problem

$$\begin{cases} u_t - \eta \Delta u + \nabla p = f & \text{in } \Omega \times (0,T) \\ \nabla \cdot u = 0 & \text{in } \Omega \times (0,T) \\ u = 0 & \text{on } \partial\Omega \times (0,T) \\ u(0) = u_0 & \text{in } \Omega. \end{cases} \tag{7.46}$$

Considering the spaces studied in Section 7.1.4, we set

$$V = \mathbf{V} = \{u \in \mathbf{H}^1_0;\ \nabla \cdot u = 0\} \qquad H = \mathbf{G}_1 = \{u \in \mathbf{L}^2; \nabla \cdot u = 0,\ \gamma_\nu u = 0\},,$$

being V' the dual of V. In weak form, problem (7.46) reads

$$-\int_0^T \int_\Omega u(t) \cdot v\, dx\, \varphi'(t)\, dt - \varphi(0) \int_\Omega u_0 \cdot v\, dx + \eta \int_0^T \int_\Omega \nabla u(t) : \nabla v\, dx\, \varphi(t)\, dt$$

$$= \int_0^T \langle f(t), v \rangle \varphi(t)\, dt \qquad \forall v \in V\ , \quad \forall \varphi \in \mathcal{D}[0,T)\ .$$

Since the bilinear form $a(u,v) = \eta(u,v)_V$ is continuous and coercive, by Theorem 7.2.11 we deduce the

Theorem 7.2.15. *For every $f \in L^2(0,T;\mathbf{V}')$ and for every $u_0 \in \mathbf{G}_1$, problem (7.46) admits a unique weak solution $u \in L^2(0,T;\mathbf{V}) \cap C^0([0,T];\mathbf{G}_1)$ such that $u' \in L^2(0,T;\mathbf{V}')$.*

7.3 Wave equation

The Galerkin method can be applied, with some slight modifications, to the wave equation. Let us consider a Hilbert triplet $V \subset H \subset V'$ of separable spaces with dense and compact embeddings. For simplicity, we will not consider an arbitrary continuous symmetric bilinear form defined over $V \times V$, but we shall directly take the scalar product $(\cdot,\cdot)_V$.

Before describing the problem which we intend to study, we give the following

Definition 7.3.1. *Let H be a Hilbert space; we say that u is* **weakly continuous** *over $[0,T]$ with values in H if $u \in C^0_w([0,T];H)$, where $C^0_w([0,T];H)$ is the space*

$$\left\{ u \in L^\infty(0,T;H);\, \lim_{t\to t_0} (u(t)-u(t_0),v)_H = 0, \forall t_0 \in [0,T], \forall v \in H \right\}.$$

By the definition, it is clear that a weakly continuous function is defined *for every* $t \in [0,T]$ and not just a.e. Here we purpose to solve the following problem:

given $u_0 \in V$, $u_1 \in H$ and $f \in L^2(0,T;H)$, find $u \in C^0_w([0,T];V)$ such that $u' \in C^0_w([0,T];H)$, $u'' \in L^2(0,T;V')$ and

$$\langle u''(t),v\rangle + (u,v)_V = (f(t),v)_H \qquad \forall v \in V \quad \text{in } \mathcal{D}'[0,T)\,, \tag{7.47}$$

$$u(0) = u_0\,, \qquad u'(0) = u_1\,. \tag{7.48}$$

The derivatives $u'(t)$ and $u''(t)$ are to be meant in the sense of Definition 7.2.7. By definition, problem (7.47)-(7.48) can be rephrased as

$$\int_0^T (u(t),v)_H\, \varphi''(t)\,dt - \varphi(0)(u_1,v)_H + \varphi'(0)(u_0,v)_H + \int_0^T (u(t),v)_V\,\varphi(t)\,dt \tag{7.49}$$
$$= \int_0^T (f(t),v)_H\,\varphi(t)\,dt \qquad \forall v \in V, \quad \forall \varphi \in \mathcal{D}[0,T)\,.$$

Thanks to the weak continuity of u and u', the initial conditions (7.48) make sense:

$$\lim_{t\to 0^+} (u(t)-u_0,v)_V = 0 \quad \forall v \in V\,, \qquad \lim_{t\to 0^+} (u'(t)-u_1,h)_H = 0 \quad \forall h \in H\,.$$

Existence of solutions

Step 1. Let $\{V_m\}$ be a Galerkin sequence for V and for H. Hence there exist two sequences $\{u_0^m\} \subset V$ and $\{u_1^m\} \subset V$ such that $u_0^m \to u_0$ in V and $u_1^m \to u_1$ in H. Assume that $m = \dim V_m$ and consider a basis $\{w_k^m\}_{k=1}^m$ of V_m, orthonormal in H. We look for $u_m = u_m(t)$ such that

$$u_m(t) = \sum_{k=1}^m g_k^m(t)\; w_k^m$$

and that, for $1 \leq k \leq m$,

$$\begin{cases} (u_m''(t), w_k^m)_H + (u_m(t), w_k^m)_V = (f(t), w_k^m)_H & \forall t \in (0,T) \\ u_m(0) = u_0^m, \qquad u_m'(0) = u_1^m. \end{cases} \tag{7.50}$$

This can be rewritten as

$$\begin{cases} g_m''(t) + B_m\, g_m(t) = \gamma_m(t) \\ g_m(0) = g_0^m, \qquad g_m'(0) = g_1^m \end{cases} \tag{7.51}$$

with g_m and B_m as in (7.35) and $\gamma_m(t) = ((f(t), w_1^m)_H, \cdots, (f(t), w_m^m)_H)$; g_0^m and g_1^m are m-tuples of the components of u_0^m and u_1^m respectively, with respect to the basis $\{w_1^m, ..., w_m^m\}$. The linear problem (7.51) admits a unique solution $g_m = g_m(t)$, then problem (7.50) admits a unique solution $u_m = u_m(t)$ as well.

Step 2. As already noticed, the functions u_m enjoy much more regularity than the one we expect for u, thus we are allowed to use (7.30).

Multiplying by $(g_k^m)'(t)$ both sides of (7.50) and summing for $k = 1, \ldots, m$, thanks to (7.30) applied to both u' and u, we obtain

$$\frac{1}{2}\frac{d}{dt}\|u_m'(t)\|_H^2 + \frac{1}{2}\frac{d}{dt}\|u_m(t)\|_V^2 = (f(t), u_m'(t))_H \qquad \text{a.e. in } (0,T).$$

Choosing arbitrarily $t \in (0,T)$ and integrating over $(0,t)$, by the Schwarz inequality (1.6), we get

$$\frac{1}{2}\left[\|u_m'(t)\|_H^2 + \|u_m(t)\|_V^2\right] = \frac{1}{2}\left[\|u_1^m\|_H^2 + \|u_0^m\|_V^2\right] + \int_0^t (f(\tau), u_m'(\tau))_H \, d\tau$$

$$\leq C + \int_0^t \|f(\tau)\|_H \|u_m'(\tau)\|_H \, d\tau\,, \tag{7.52}$$

for a suitable $C > 0$ which encodes the boundedness of the sequences $\{u_0^m\}$ and $\{u_1^m\}$. Thus, by the Young and Hölder inequalities (2.18) and (2.17), we infer that for every $t \in [0,T)$

$$\|u_m'(t)\|_H^2 + \|u_m(t)\|_V^2 \leq 2C + \int_0^T \|f(\tau)\|_H^2 \, d\tau + \int_0^t \|u_m'(\tau)\|_H^2 \, d\tau$$

$$\leq C' + \int_0^t \|u_m'(\tau)\|_H^2 \, d\tau. \tag{7.53}$$

Setting

$$\psi(t) := \int_0^t \|u_m'(\tau)\|_H^2 \, d\tau \geq 0\,,$$

(7.53) gives

$$\psi'(t) \le C' + \psi(t) \quad \Longrightarrow \quad \frac{d}{dt}[e^{-t}\psi(t)] \le C'e^{-t}$$

which, integrated over $[0,T]$ yields $\psi(T) \le C'(e^T - 1)$ (recall that $\psi(0) = 0$), which implies

$$\int_0^T \|u'_m(\tau)\|_H^2 \, d\tau \le C'' \qquad \forall m \, .$$

Inserting it into (7.53), we deduce that there exist two constants C_1, C_2 such that

$$\|u_m\|_{L^\infty(0,T;V)} \le C_1 \qquad \text{and} \qquad \|u'_m\|_{L^\infty(0,T;H)} \le C_2. \tag{7.54}$$

Step 3. By the estimate (7.54) and by Theorem 2.4.17, we may conclude that there exist $u \in L^\infty(0,T;V)$ and $w \in L^\infty(0,T;H)$ such that

$$\int_0^T (u_m(t) - u(t), v(t))_V \, dt \to 0 \quad \forall v \in L^1(0,T;V) \, , \tag{7.55}$$

$$\int_0^T (u'_m(t) - w(t), h(t))_H \, dt \to 0 \quad \forall h \in L^1(0,T;H) \, ,$$

up to subsequences.

Thanks to these convergences, Theorem 7.2.5 and the definition of distributional derivative imply $w = u'$. Thus we deduce that there exists $u \in L^\infty(0,T;V)$ such that $u' \in L^\infty(0,T;H)$, (7.55) holds true and

$$\int_0^T (u'_m(t) - u'(t), h(t))_H \, dt \to 0 \quad \forall h \in L^1(0,T;H). \tag{7.56}$$

Step 4. Now we have to prove that u is a solution of (7.47)-(7.48) or equivalently of (7.49). Let $\varphi \in \mathcal{D}[0,T)$, $v \in V$ and let $\{v_m\}$ be a sequence such that $v_m \in V_m$ and $v_m \to v$ in V. Setting $\psi_m(t) = \varphi(t)\ v_m$ and $\psi(t) = \varphi(t)\ v$, (7.40) still holds and it also holds

$$\psi''_m \to \psi'' \qquad \text{in } L^2(0,T;V'); \tag{7.57}$$

as already noticed for (7.40), the convergence (7.57) occurs also in $L^\infty(0,T;V)$.

By (7.50) we obtain

$$\int_0^T (u_m(t), \psi''_m(t))_H \, dt - \varphi(0)(u_1^m, v_m)_H + \varphi'(0)(u_0^m, v_m)_H + \int_0^T (u_m(t), \psi_m(t))_V \, dt$$

$$= \int_0^T (f(t), \psi_m(t))_H \, dt.$$

Conditions (7.40) imply

$$\int_0^T (f(t), \psi_m(t))_H \, dt \to \int_0^T (f(t), \psi(t))_H \, dt.$$

According to (7.55), we know that $u_m \rightharpoonup u$ in $L^2(0,T;V)$, while in view of (7.57) we have that $\psi_m \to \psi$ in $L^2(0,T;V)$. Therefore, by Theorem 1.4.14, we derive

$$\int_0^T (u_m(t), \psi_m''(t))_H \, dt \to \int_0^T (u(t), \psi''(t))_H \, dt.$$

Similarly, again by Theorem 1.4.14, by combining (7.40) and (7.55), we deduce

$$\int_0^T (u_m(t), \psi_m(t))_V \, dt \to \int_0^T (u(t), \psi(t))_V \, dt.$$

Finally, there hold also the convergences

$$(u_1^m, v_m)_H \to (u_1, v)_H \qquad (u_0^m, v_m)_H \to (u_0, v)_H.$$

These five passages to the limit show that u solves (7.49) and hence (7.47)-(7.48), provided that u, u' and u'' satisfy the required properties.

Step 5. We have just shown that for every $\varphi \in \mathcal{D}(0,T)$ and for every $v \in V$

$$\int_0^T (u(t), \varphi''(t)v)_H \, dt + \int_0^T (u(t), v)_V \varphi(t) \, dt = \int_0^T (f(t), v)_H \varphi(t) \, dt \qquad (7.58)$$

or, equivalently,

$$-\int_0^T (\varphi'(t)u'(t), v)_H \, dt = -\int_0^T \langle \varphi(t)A(t), v\rangle \, dt + \int_0^T (\varphi(t)f(t), v)_H \, dt,$$

where for every $t \in [0,T]$ the functional $A(t) \in V'$ is defined by

$$\langle A(t), v\rangle = (u(t), v)_V \qquad \forall v \in V.$$

From Theorem 7.2.5, it follows that

$$-\Big(\int_0^T \varphi'(t)u'(t)dt, v\Big)_H = -\Big\langle \int_0^T \varphi(t)A(t)dt, v\Big\rangle + \Big(\int_0^T \varphi(t)f(t)dt, v\Big)_H$$

and, by the arbitrariness of $v \in V$, we have that

$$-\int_0^T \varphi'(t)u'(t)dt = -\int_0^T \varphi(t)A(t)dt + \int_0^T \varphi(t)f(t)dt.$$

This means that u' admits a weak derivative and that $u''(t) = -A(t) + f(t)$ in V', whence

$$u'' \in L^2(0,T;V'). \qquad (7.59)$$

Let us now show that u enjoys the weak continuities expected. First, we observe that by Theorem 7.2.9 we infer $u \in C^0([0,T];H)$, that is

$$\lim_{t\to t_0} \|u(t) - u(t_0)\|_H = 0 \qquad \forall t_0 \in [0,T]\,. \qquad (7.60)$$

Furthermore, since $u \in L^\infty(0,T;V)$, as $t \to t_0 \in [0,T]$ the "sequence" $\{u(t)\}$ is bounded in V. According to Theorem 1.4.13, there exist $w \in V$ and a subsequence $\{u(t_n)\}$ such that $u(t_n) \rightharpoonup w$ as $t_n \to t_0$. In view of the compactness of the embedding $V \subset H$ and (7.60), it necessarily holds $w = u(t_0)$. Repeating the procedure for all the sequences $t_n \to t_0$ we obtain that $u \in C^0_w([0,T];V)$. Similarly, by (7.59) and by applying the first part of Theorem 7.2.9 to u' replacing V by V', we conclude that $u' \in C^0([0,T];V')$. In addition, since $u' \in L^\infty(0,T;H)$, as $t \to t_0 \in [0,T]$ the sequence $\{u'(t)\}$ is bounded in H and, according to Theorem 1.4.13, there exist $w \in H$ and a subsequence $\{u'(t_n)\}$ such that $u'(t_n) \rightharpoonup w$ as $t_n \to t_0$. In view of the compactness of the embedding $H \subset V'$, we deduce that $w = u'(t_0)$ and, subsequently, that $u' \in C^0_w([0,T];H)$. Finally, the initial conditions in (7.48) (obtained as weak limits) can be derived integrating by parts the equality (7.49).

Uniqueness of the solution

Let u and w be two solutions to the problem (7.47)-(7.48) and define $z = u - w$. Then z is a solution to the homogeneous problem

$$z(0) = z'(0) = 0 \ , \qquad \langle z''(t), v\rangle + (z,v)_V = 0 \qquad \forall v \in V \quad \text{in } \mathcal{D}'[0,T) \ .$$

Applying to the function z the estimate (7.52), we readily obtain that $z \equiv 0$ and hence $u \equiv w$. Therefore, the solution is unique.

Hence it is proved the following

Theorem 7.3.2. *Given $u_0 \in V$, $u_1 \in H$ and $f \in L^2(0,T;H)$, problem (7.47) – (7.48) admits a unique weak solution $u \in C^0_w([0,T];V) \cap C^0([0,T];H)$ such that $u' \in C^0_w([0,T];H)$ and $u'' \in L^2(0,T;V')$.*

Let us apply Theorem 7.3.2 to the wave equation under homogeneous Dirichlet conditions:

$$\begin{cases} u_{tt} - \Delta u = f & \text{in } \Omega \times (0,T) \\ u = 0 & \text{on } \partial\Omega \times (0,T) \\ u(0) = u_0 & \text{in } \Omega \\ u_t(0) = u_1 & \text{in } \Omega, \end{cases} \tag{7.61}$$

whose weak solutions satisfy the following variational problem:
given $u_0 \in H^1_0(\Omega)$, $u_1 \in L^2(\Omega)$ and $f \in L^2(0,T;L^2(\Omega))$ find $u \in C^0_w([0,T];H^1_0(\Omega)) \cap C^0([0,T];L^2(\Omega))$ such that $u' \in C^0_w([0,T];L^2(\Omega))$, $u'' \in L^2(0,T;H^{-1}(\Omega))$ and

$$\int_0^T \int_\Omega u(t)v\,dx\,\varphi''(t)\,dt - \varphi(0)\int_\Omega u_1 v\,dx + \varphi'(0)\int_\Omega u_0 v\,dx + \int_0^T \int_\Omega \nabla u(t)\nabla v\,dx\,\varphi(t)\,dt$$

$$= \int_0^T \int_\Omega f(t) v \, \varphi(t) \, dx \, dt \qquad \forall v \in H_0^1(\Omega) \quad \forall \varphi \in \mathcal{D}[0,T) \, .$$

By Theorem 7.3.2, we derive the

Theorem 7.3.3. *Let* $u_0 \in H_0^1(\Omega)$, $u_1 \in L^2(\Omega)$ *and* $f \in L^2(0,T;L^2(\Omega))$; *then* (7.61) *admits a unique weak solution.*

Exercise 7.3.4. Write the weak formulation of the wave equation under homogeneous Neumann conditions. Furthermore, prove the existence and the uniqueness of the solutions. □

Chapter 8

Some nonlinear partial differential equations

Several natural phenomena cannot be described by linear differential equations. Therefore, the theory developed in Chapters 3 and 7 cannot be considered comprehensive. Nonlinear differential equations can be extremely different from each other and their analysis requires techniques almost ad hoc for each type. For this reason, there does not exist a general theory that allows to treat them irrespective of their nature. In this chapter, we limit ourselves to explain how to analyze three specific kinds of nonlinear equations: first order quasilinear equations, quasilinear equations coming from fluid dynamics (called Navier-Stokes equations) and semilinear equations of the type "stationary heat with temperature-depending source".

8.1 First order quasilinear equations

This section is devoted to the study of first order quasilinear equations of the type

$$u_t + A(u)u_x = 0\,, \tag{8.1}$$

with $x \in \mathbb{R}$, $t \geq 0$, $u = u(x,t)$ a real valued function and $A : \mathbb{R} \to \mathbb{R}$ sufficiently smooth. In order to understand the importance of such equation, we present a couple of applications.

We start describing a physical model of gas dynamics, being the gas ideal. Consider a tube with infinite length, filled with a gas, and denote by v the gas velocity, by p the gas pressure and by ρ the gas density. Assume that the physical quantities that we have just introduced are constant on every cross section of the tube, so that they can be considered functions depending only on the spatial variable $x \in \mathbb{R}$ and on the time $t \geq 0$. Furthermore, assume that the gas velocity is parallel to the axis of the tube, so that v can be considered as a scalar quantity.

Under these conditions, the equation (3.46) becomes

$$\rho_t + v\rho_x + \rho v_x = 0. \tag{8.2}$$

On the other hand, by (3.49) and by (3.50), we obtain

$$\rho v_t + \rho v v_x + p_x = 0. \tag{8.3}$$

Under the assumptions of constant entropy and constant C_V (specific heat at constant volume), the pressure turns out to be a function of the sole density, namely $p = f(\rho) = K_0\rho^\gamma$ with $\gamma = C_p/C_V$ (C_p specific heat at constant pressure) and K_0 a suitable constant. Thus, by (8.2) and (8.3), we obtain the following system of two equations in the unknowns ρ and v:

$$\begin{cases} \rho_t + v\rho_x + \rho v_x = 0 \\ v_t + v v_x + \frac{f'(\rho)}{\rho}\rho_x = 0. \end{cases} \tag{8.4}$$

Once we set

$$U = \begin{pmatrix} \rho \\ v \end{pmatrix}, \qquad A(U) = A\begin{pmatrix} \rho \\ v \end{pmatrix} = \begin{pmatrix} v & \rho \\ \frac{f'(\rho)}{\rho} & v \end{pmatrix},$$

the system (8.4) can be rewritten in the form

$$U_t + A(U)U_x = 0 \tag{8.5}$$

which is the vector version of the equation (8.1).

In the sequel we will deal only with the scalar equation (8.1); then, for more details about gas dynamics and quasilinear systems like (8.5), we refer the reader to [66].

The second application of equation (8.1) concerns a model representing the road traffic. If we consider a straight road with infinite length, then the traffic flow may be seen as a fluid with density $\rho = \rho(x,t)$. In this model, we consider a road without sources or sinks, namely without exits and entrances. Moreover, the velocity v of the cars is assumed to depend only on the density ρ.

Following the same arguments of Section 3.2, we arrive to equation (8.2). Calling ρ_m the maximum allowed density and v_m the maximum allowed velocity, it is reasonable to assume that

$$v(\rho) = \left(1 - \frac{\rho}{\rho_m}\right) v_m. \tag{8.6}$$

This constitutive relation shows that when the density is very low, the velocity is close to the maximum allowed velocity, while when the density is close to the maximum allowed density, the velocity is close to zero.

Inserting (8.6) into (8.2) we obtain the equation

$$\rho_t + \left(1 - \frac{2\rho}{\rho_m}\right) v_m \, \rho_x = 0$$

which is of type (8.1) with $A(\rho) = \left(1 - \frac{2\rho}{\rho_m}\right) v_m$.

8.1.1 Classical solutions of the Cauchy problem

We analyze the following Cauchy problem associated to (8.1):

$$\begin{cases} u_t + A(u)u_x = 0 & (x,t) \in \mathbb{R} \times (0,T) \\ u(x,0) = g(x) & x \in \mathbb{R}, \end{cases} \tag{8.7}$$

with $0 < T \leq +\infty$.

As already done in Section 3.2.2, we adopt the method of characteristics, namely we try to determine some curves lying in the half-plane $\mathbb{R} \times (0,\infty)$ along which a solution to (8.7) is constant. If the characteristic curves admit a representation of the form $x = x(t)$, $t \in \mathbb{R}$, then by (8.7) we deduce that, in order that u is constant along a characteristic, it must be

$$x'(t) = A(u(x(t),t)).$$

On the other hand, being u constant along a characteristic, for every $t > 0$ there holds

$$A(u(x(t),t)) = A(u(x(0),0)) = A(g(x(0))).$$

We draw the conclusion that if $x = x(t)$ is the equation of a characteristic passing through the point $(x_0, 0)$, then

$$x(t) = x_0 + A(g(x_0))t\,. \tag{8.8}$$

Thus we can affirm that the characteristic curves are straight lines, as we have already seen for the linear problem (3.4), but unlike that case, they are not necessarily parallel, since each slope coincides with $1/A(g(x_0))$, where $(x_0, 0)$ is the point of intersection between the considered characteristic and the x axis.

For this reason, for (x_0, t_0) fixed with $x_0 \in \mathbb{R}$ and $t_0 > 0$, it is natural to ask whether a characteristic curve passing through (x_0, t_0) is unique. Actually, in some cases lack of uniqueness may occur. It suffices to observe that if there exist $x_1, x_2 \in \mathbb{R}$ and $t_0 > 0$ such that

$$x_1 + A(g(x_1))t_0 = x_2 + A(g(x_2))t_0\;,$$

then the characteristics passing through $(x_1, 0)$ and $(x_2, 0)$ intersect each other in the half-plane $\mathbb{R} \times (0,\infty)$. In this case, we say that the solution u to (8.7) displays the **shock waves** phenomenon. As we will see later, in such situation the solution u may display some discontinuities.

However, under suitable assumptions, this phenomenon cannot occur for times close to $t = 0$. Assume for instance that $A \in C^1(\mathbb{R})$, $g \in C^1(\mathbb{R})$ and that the function $\xi \mapsto A'(g(\xi))g'(\xi)$ is bounded from below in $\mathbb{R}$:

$$\inf_{\xi \in \mathbb{R}} A'(g(\xi))g'(\xi) > -\infty\,. \tag{8.9}$$

In order to prove existence and uniqueness of the classical solution for times sufficiently small, consider the family of straight lines in the plane (x, t) with equation $x = \xi + A(g(\xi))t$, with ξ varying in $\mathbb{R}$. Then define the function

$$F(x, t, \xi) = \xi + A(g(\xi))t - x \qquad \forall (x, t, \xi) \in \mathbb{R}^3\,. \tag{8.10}$$

Let $T > 0$ be defined as

$$T := \begin{cases} -\dfrac{1}{\inf\limits_{\xi \in \mathbb{R}} A'(g(\xi))g'(\xi)} & \text{if } \inf\limits_{\xi \in \mathbb{R}} A'(g(\xi))g'(\xi) < 0 \\ +\infty & \text{if } \inf\limits_{\xi \in \mathbb{R}} A'(g(\xi))g'(\xi) \geq 0. \end{cases} \tag{8.11}$$

For $(x, t) \in \mathbb{R} \times [0, T)$ fixed, we can affirm that there exists a unique $\xi = \xi(x, t) \in \mathbb{R}$ such that $F(x, t, \xi) = 0$. In fact, note that for every $(x, t) \in \mathbb{R} \times [0, T)$

$$\frac{\partial F}{\partial \xi}(x, t, \xi) = 1 + A'(g(\xi))g'(\xi)t \geq 1 - \frac{t}{T} > 0 \qquad \forall \xi \in \mathbb{R}\,, \tag{8.12}$$

where $\frac{t}{T} = 0$ in case $T = +\infty$. Hence, once we fix $(x, t) \in \mathbb{R} \times [0, T)$, we infer that the map $\xi \mapsto F(x, t, \xi)$ is strictly increasing and furthermore, by (8.12), we obtain

$$\lim_{\xi \to -\infty} F(x, t, \xi) = -\infty, \qquad \lim_{\xi \to +\infty} F(x, t, \xi) = +\infty,$$

from which the existence of $\xi = \xi(x, t)$ follows. Thanks to the implicit function Theorem (applied in an open set containing $\mathbb{R} \times [0, T)$) we also deduce that the map $(x, t) \mapsto \xi(x, t)$ is of class C^1 in $\mathbb{R} \times [0, T)$.

In conclusion, for every point $(x_0, t_0) \in \mathbb{R} \times [0, T)$, there exists a unique characteristic passing through (x_0, t_0) and its equation is given by

$$x = \xi(x_0, t_0) + A(g(\xi(x_0, t_0)))t.$$

If u is a solution to (8.7), then u must be constant along the characteristic with equation $x = \xi(x_0, t_0) + A(g(\xi(x_0, t_0)))t$ and, consequently,

$$u(x, t) = u(\xi(x, t), 0) = g(\xi(x, t)) \qquad \forall (x, t) \in \mathbb{R} \times [0, T). \tag{8.13}$$

We have just proved the following

Theorem 8.1.1. *Let $A, g \in C^1(\mathbb{R})$ satisfy* (8.9) *and let $T \in (0, \infty]$ be defined in* (8.11). *Then problem* (8.7) *admits a unique solution $u \in C^1(\mathbb{R} \times [0, T))$ given by formula* (8.13).

Imposing suitable monotonicity conditions on A and g, thanks to Theorem 8.1.1, it is possible to guarantee the existence of solutions for problem (8.7) in the whole $\mathbb{R} \times (0, \infty)$. In fact, there holds the following

Corollary 8.1.2. *Let $A, g \in C^1(\mathbb{R})$ be two non decreasing (resp. non increasing) functions. Then problem* (8.7) *admits a unique solution $u \in C^1(\mathbb{R} \times [0, \infty))$ given by formula* (8.13).

In absence of the monotonicity hypotheses introduced in Corollary 8.1.2, it is not possible to ensure the existence of solutions for every $t > 0$. In order to better explain how this may happen, let us assume that $A, g \in C^1(\mathbb{R})$ and that $A' > 0$ in $\mathbb{R}$. We set $\xi = \xi(x, t)$ the function implicitly defined by the equation $\xi + A(g(\xi))t - x = 0$. For what we have shown formerly, as long as the solution u to (8.7) exists, it is given by $u(x, t) = g(\xi(x, t))$; moreover, its partial derivatives are implicitly given by

$$u_t = -\frac{A(u)g'(\xi)}{1 + A'(u)g'(\xi)t} \quad \text{and} \quad u_x = \frac{g'(\xi)}{1 + A'(u)g'(\xi)t}. \tag{8.14}$$

Let us assume that there exists $\xi_0 \in \mathbb{R}$ such that $g'(\xi_0) < 0$ and let us consider the characteristic passing through $(\xi_0, 0)$:

$$\begin{cases} x(s) = \xi_0 + A(g(\xi_0))s \\ t(s) = s \end{cases} \qquad (s \in \mathbb{R}).$$

Since $\xi(x(s), s) = \xi_0$ and $u(x(s), s) = g(\xi_0)$ for every s, it holds

$$u_t(x(s), s) = -\frac{A(g(\xi_0))g'(\xi_0)}{1 + A'(g(\xi_0))g'(\xi_0)s} \quad \text{and} \quad u_x(x(s), s) = \frac{g'(\xi_0)}{1 + A'(g(\xi_0))g'(\xi_0)s}.$$

We immediately notice that the partial derivatives of u tend to infinity as $s \to S := -1/(A'(g(\xi_0))g'(\xi_0)) > 0$. Therefore, the solution u cannot be extended as a C^1 function up to $t = S$.

8.1.2 Weak solutions of the Cauchy problem

In the previous section we have noticed that it is not possible to guarantee the existence of classical solutions of (8.7) for all the times $t > 0$. However, from a mathematical but also physical point of view, it is worth to take into account also non classical solutions, whence the necessity of a notion of weak solutions. We proceed formally, assuming that u is a classical solution of (8.7), namely $u \in C^1(\mathbb{R} \times [0, \infty))$

verifying (8.7). Let $\varphi \in C^1(\mathbb{R} \times [0,\infty))$ have compact support in $\mathbb{R} \times [0,\infty)$: be careful, it may happen that $\varphi(x,0) \neq 0$! Let $a < b$ and $T > 0$ be such that $\text{supp}\varphi \subset [a,b] \times [0,T]$. Multiplying by φ both sides of the equation and integrating, we obtain

$$0 = \int_{-\infty}^{+\infty} \int_0^{+\infty} (u_t + A(u)u_x)\varphi \, dx \, dt = \int_{[a,b]\times[0,T]} (u_t + A(u)u_x)\varphi \, dx \, dt. \tag{8.15}$$

Integrating by parts we get

$$\int_{[a,b]\times[0,T]} u_t\varphi \, dx \, dt = \int_a^b \left(\int_0^T u_t\varphi \, dt \right) dx$$

$$= \int_a^b [u(x,T)\varphi(x,T) - u(x,0)\varphi(x,0)]dx - \int_a^b \left(\int_0^T u\varphi_t \, dt \right) dx$$

$$= -\int_{\mathbb{R}} g(x)\varphi(x,0) \, dx - \int_{-\infty}^{+\infty} \int_0^{+\infty} u\varphi_t \, dx \, dt. \tag{8.16}$$

Setting $B(s) = \int_0^s A(t) \, dt$, the equation in (8.7) can be rewritten as

$$u_t + (B(u))_x = 0$$

and hence, integrating by parts, we have

$$\int_{[a,b]\times[0,T]} A(u)u_x\varphi \, dx \, dt = \int_{[a,b]\times[0,T]} (B(u))_x\varphi \, dx \, dt = \int_0^T \left(\int_a^b (B(u))_x\varphi \, dx \right) dt$$

$$= \int_0^T [B(u(b,t))\varphi(b,t) - B(u(a,t))\varphi(a,t)]dt - \int_0^T \left(\int_a^b B(u)\varphi_x \, dx \right) dt$$

$$= -\int_{-\infty}^{+\infty} \int_0^{+\infty} B(u)\varphi_x \, dx \, dt. \tag{8.17}$$

Inserting (8.16) and (8.17) into (8.15), we find

$$\int_{-\infty}^{+\infty} \int_0^{+\infty} (u\varphi_t + B(u)\varphi_x) dx \, dt + \int_{\mathbb{R}} g(x)\varphi(x,0) \, dx = 0. \tag{8.18}$$

We can give the following

Definition 8.1.3. *Given $g \in L^\infty(\mathbb{R})$ and $A \in C^0(\mathbb{R})$, we say that a function $u \in L^\infty(\mathbb{R} \times (0,\infty))$ is a* **weak solution** *of the Cauchy problem (8.7) if (8.18) holds true for every function $\varphi \in C^1(\mathbb{R} \times [0,\infty))$ compactly supported in $\mathbb{R} \times [0,\infty)$.*

In view of what we have shown above, it is clear that a classical solution of (8.7) is also a weak solution. Conversely, let us assume that $u \in C^1(\mathbb{R} \times [0, \infty))$ is a weak solution of (8.7). The identity (8.18) holds true, in particular, for the functions $\varphi \in C^1(\mathbb{R} \times [0, \infty))$ with $\text{supp}\varphi \subset \mathbb{R} \times (0, \infty)$. Proceeding as in (8.16) and in (8.17), we obtain

$$\int_{-\infty}^{+\infty} \int_0^{+\infty} [u_t + (B(u))_x]\varphi \, dx \, dt = 0.$$

Therefore, by Proposition 2.4.22, we infer that

$$u_t + (B(u))_x = 0 \qquad \text{a.e. in } \mathbb{R} \times (0, \infty) . \tag{8.19}$$

In addition, being u of class C^1, the previous identity is valid for every $(x, t) \in \mathbb{R} \times (0, \infty)$. Let now $\varphi \in C^1(\mathbb{R} \times [0, \infty))$ be compactly supported in $\mathbb{R} \times [0, \infty)$ (so it is not necessarily zero on the x axis). Integrating by parts as in (8.16) and in (8.17), by (8.18) we get

$$\int_{-\infty}^{+\infty} \int_0^{+\infty} [u_t + (B(u))_x]\varphi \, dx \, dt + \int_{\mathbb{R}} [u(x, 0) - g(x)]\varphi(x, 0) \, dx = 0$$

and, keeping in mind (8.19), we conclude that

$$\int_{\mathbb{R}} [u(x, 0) - g(x)]\varphi(x, 0) \, dx = 0$$

for every $\varphi \in C^1(\mathbb{R} \times [0, \infty))$ compactly supported in $\mathbb{R} \times [0, \infty)$. In view of Proposition 2.4.22, we have $u(x, 0) = g(x)$ for a.e. $x \in \mathbb{R}$. Actually, since u is continuous, we infer $u(x, 0) = g(x)$ for every $x \in \mathbb{R}$. Thus we have proved that u is a classical solution of (8.7). For the existence and uniqueness of the weak solution of (8.7), see Theorem 8.1.9 at the end of this section.

Let us now consider a weak solution u of (8.7) and let us assume that such solution is discontinuous along an arc Γ of a regular curve with equation $x = \eta(t)$. Analogous results hold for regular curves in parametric form: $x = x(s)$, $t = t(s)$. We will show that not all the types of discontinuities are admissible. Let us assume that u undergoes a jump discontinuity along Γ, namely that u is regular outside Γ and admits limits from "both sides" of Γ. To be more rigorous, for every $(x_0, t_0) \in \Gamma$ let $D \subset \mathbb{R} \times (0, \infty)$ be an open disk centered at (x_0, t_0) with a suitable radius and let

$$D_1 = \{(x, t) \in D : x < \eta(t)\} , \qquad D_2 = \{(x, t) \in D : x > \eta(t)\} . \tag{8.20}$$

Assume that for every $(x, t) \in \Gamma \cap D$ the following limits exist:

$$u_-(x, t) := \lim_{(y,\tau) \to (x,t)} u_{|D_1}(y, \tau) , \qquad u_+(x, t) := \lim_{(y,\tau) \to (x,t)} u_{|D_2}(y, \tau).$$

In addition, assume that $u \in C^1(\overline{D}_1)$ and $u \in C^1(\overline{D}_2)$, where in the first case we mean $u = u_-$ on $\Gamma \cap D$ while in the second case $u = u_+$ on $\Gamma \cap D$.

Let now $\varphi \in C^1(\mathbb{R} \times [0, \infty))$ be compactly supported in D. Then by (8.18) we infer

$$0 = \int_D (u\varphi_t + B(u)\varphi_x) dx\, dt = \int_{D_1} (u\varphi_t + B(u)\varphi_x) dx\, dt + \int_{D_2} (u\varphi_t + B(u)\varphi_x) dx\, dt. \tag{8.21}$$

Since $u \in C^1(\overline{D}_i)$ and it solves the equation $u_t + (B(u))_x = 0$ in D_i for $i = 1, 2$, thanks to the Green formula in the plane (see (6.16)), we have

$$\int_{D_i} (u\varphi_t + B(u)\varphi_x) dx\, dt = \int_{D_i} [(u\varphi)_t + (B(u)\varphi)_x] dx\, dt = \int_{\partial^+ D_i} (-u\varphi\, dx + B(u)\varphi\, dt) \tag{8.22}$$

where by $\partial^+ D_i$ we mean the boundary of D_i, run in counterclockwise sense. But since $\varphi \equiv 0$ on ∂D, the last line integral reduces to the integral along the sole Γ, whence

$$\int_{\partial^+ D_1} (-u\varphi\, dx + B(u)\varphi\, dt) = \int_{t_1}^{t_2} [-u_-(\eta(t), t)\eta'(t) + B(u_-(\eta(t), t))]\varphi(\eta(t), t)\, dt \tag{8.23}$$

and

$$\int_{\partial^+ D_2} (-u\varphi\, dx + B(u)\varphi\, dt) = \int_{t_1}^{t_2} [u_+(\eta(t), t)\eta'(t) - B(u_+(\eta(t), t))]\varphi(\eta(t), t)\, dt. \tag{8.24}$$

By (8.21)-(8.24) we deduce

$$\int_{t_1}^{t_2} \{[u(\eta(t), t)]\eta'(t) - [B(u(\eta(t), t))]\}\varphi(\eta(t), t)\, dt = 0\,,$$

where we have set $[u] = u_+ - u_-$ and $[B(u)] = B(u_+) - B(u_-)$ on Γ. By the arbitrariness of φ and of $(x_0, t_0) \in \Gamma$, we conclude that

$$\eta' \cdot [u] = [B(u)] \qquad \text{on } \Gamma. \tag{8.25}$$

The function η' is called **shock speed** and the equation (8.25) is called **jump condition**. We can then state the following

Proposition 8.1.4. *Let $A, g \in C^1(\mathbb{R})$ and let u be a weak solution of* (8.7). *Let Γ, η and D be as above. Assume that $u \in C^1(D \setminus \Gamma)$ and that $u \notin C^0(D)$. Introducing the sets D_1, D_2 according to* (8.20), *suppose that the functions*

$$u_-(x, t) := \lim_{(y,\tau)\to(x,t)} u_{|D_1}(y, \tau) \qquad \forall (x, t) \in \overline{D}_1,$$

$$u_+(x, t) := \lim_{(y,\tau)\to(x,t)} u_{|D_2}(y, \tau) \qquad \forall (x, t) \in \overline{D}_2$$

are well defined and that $u_- \in C^1(\overline{D}_1)$, $u_+ \in C^1(\overline{D}_2)$. *Setting*

$$[u] = u_+ - u_-, \qquad [B(u)] = B(u_+) - B(u_-)$$

on Γ, *then the function* η *satisfies* (8.25).

In order to better understand how there can exist discontinuous solutions of (8.7), we propose the following model problem (with $A(u) = u$)

$$\begin{cases} u_t + uu_x = 0 & (x,t) \in \mathbb{R} \times (0,\infty) \\ u(x,0) = g(x) & x \in \mathbb{R}. \end{cases} \tag{8.26}$$

The type of nonlinearity appearing in (8.26) comes out in a natural way from fluid dynamics equations: it is enough to observe the second equation in system (8.4). Let us see some examples (essentially taken from [66]) of initial conditions for problem (8.26), which produce respectively the appearance of shock waves and the opposite phenomenon, known as **rarefaction waves**. In the latter case, the characteristics go farther and farther apart as t increases and admit intersection points only for $t < 0$.

We start with an example displaying shock waves.

Example 8.1.5. Consider problem (8.26) with

$$g(x) = \begin{cases} 1 & \text{if } x < 0 \\ 1 - x & \text{if } 0 \le x \le 1 \\ 0 & \text{if } x > 1. \end{cases}$$

By (8.8), the Cartesian equations of the characteristics passing through $(x_0, 0)$ are

$$\begin{cases} x = x_0 + t & \text{if } x_0 < 0 \\ x = x_0 + (1 - x_0)t & \text{if } 0 \le x_0 \le 1 \\ x = x_0 & \text{if } x_0 > 1. \end{cases}$$

We readily observe that for $0 < t < 1$ the characteristics do not intersect, as shown in the next picture.

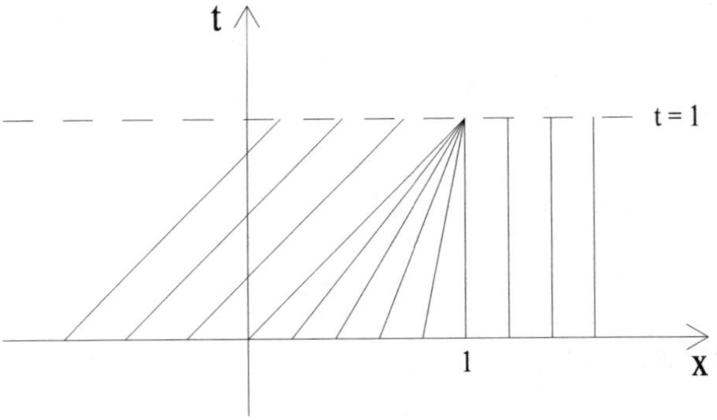

Characteristic curves in Example 8.1.5

Therefore, for $(x,t) \in \mathbb{R} \times [0,1)$, we have

$$u(x,t) = \begin{cases} 1 & \text{if } x < t \\ \dfrac{1-x}{1-t} & \text{if } t \le x \le 1 \\ 0 & \text{if } x > 1. \end{cases} \tag{8.27}$$

The function u so defined is continuous in $\mathbb{R} \times [0,1)$. It is possible to define u also for $t \ge 1$, on condition that we allow u to undergo jump discontinuities. If $x = \eta(t)$ is the Cartesian equation of the curve along which u displays a jump discontinuity, then it must be

$$u(x,t) = \begin{cases} u_-(x,t) = 1 & \text{if } x < \eta(t) \\ u_+(x,t) = 0 & \text{if } x > \eta(t). \end{cases}$$

Consequently, we have that $[u] = u_+ - u_- = -1$ and $[\frac{1}{2}u^2] = \frac{1}{2}(u_+^2 - u_-^2) = -\frac{1}{2}$ and by (8.25) we infer $\eta'(t) = \frac{1}{2}$. The last condition implies $\eta(t) = \frac{1}{2}t + C$, for some $C \in \mathbb{R}$ to be determined. On the other hand, with an extension by continuity, thanks to (8.27), we can define

$$u(x,1) = \begin{cases} 1 & \text{if } x < 1 \\ 0 & \text{if } x > 1 \end{cases}$$

and hence the curve of equation $x = \eta(t)$ must pass through the point $(1,1)$. Therefore it must be

$$\eta(t) = \frac{1}{2}(t+1) \qquad \forall t \ge 1.$$

In conclusion, for $(x,t) \in \mathbb{R} \times [1,\infty)$ it holds

$$u(x,t) = \begin{cases} 1 & \text{if } x < \frac{1}{2}(t+1) \\ 0 & \text{if } x > \frac{1}{2}(t+1), \end{cases}$$

which is a discontinuous function. □

We now pass to an example that displays rarefaction waves.

Example 8.1.6. Consider problem (8.26) with

$$g(x) = \begin{cases} 0 & \text{if } x < 0 \\ x & \text{if } 0 \le x \le 1 \\ 1 & \text{if } x > 1. \end{cases}$$

In view of (8.8), the characteristics through $(x_0, 0)$, as x_0 varies in $\mathbb{R}$, are

$$\begin{cases} x = x_0 & \text{if } x_0 < 0 \\ x = x_0 + x_0 t & \text{if } 0 \le x_0 \le 1 \\ x = x_0 + t & \text{if } x_0 > 1. \end{cases}$$

As we can see from the picture below, the characteristic curves "go farther and farther apart" as t increases.

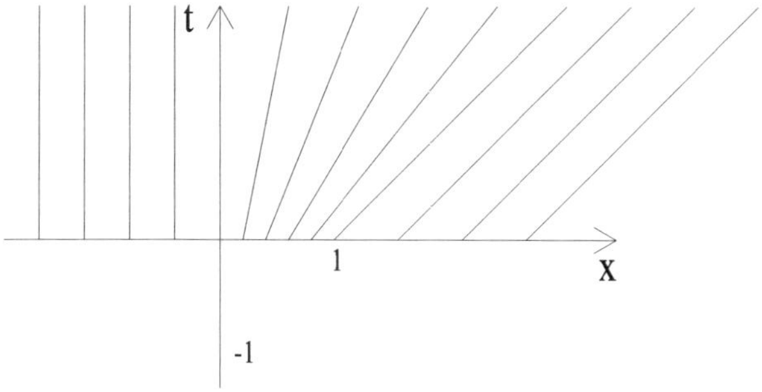

Characteristic curves in Example 8.1.6

The solution u of the Cauchy problem is continuous in the whole half-plane $\mathbb{R} \times [0, \infty)$ and is given by

$$u(x,t) = \begin{cases} 0 & \text{if } x < 0 \\ \dfrac{x}{1+t} & \text{if } 0 \le x \le t+1 \\ 1 & \text{if } x > t+1. \end{cases}$$

The possible discontinuity would be at $t = -1 < 0$. □

The next example shows how a Cauchy problem with a discontinuous initial datum may admit a continuous solution for $t > 0$, furthermore there may exist several solutions of the same problem.

Example 8.1.7. Consider problem (8.26) with

$$g(x) = \begin{cases} 0 & \text{if } x < 0 \\ 1 & \text{if } x > 0. \end{cases}$$

We look for a solution u_1 which undergoes a discontinuity along a curve of equation $x = \eta(t)$ and such that

$$u_1(x,t) = \begin{cases} 0 & \text{if } x < \eta(t) \\ 1 & \text{if } x > \eta(t). \end{cases}$$

By (8.25), we deduce that necessarily $\eta'(t) = \frac{1}{2}$ and hence $\eta(t) = \frac{1}{2}t$.

But it is easy to check that also the function

$$u_2(x,t) = \begin{cases} 0 & \text{if } x < 0 \\ \dfrac{x}{t} & \text{if } 0 \le x \le t \\ 1 & \text{if } x > t \end{cases}$$

is a solution of the Cauchy problem and, in addition, it is continuous in $\mathbb{R} \times (0, \infty)$.
□

The previous example reveals that the uniqueness of the weak solution of (8.7) is not guaranteed. It is natural to ask which weak solutions are physically relevant. In order to answer this question, let us assume that $A, g \in C^1(\mathbb{R})$, $\inf_{\mathbb{R}} A' > 0$ and that $g' > 0$ in $\mathbb{R}$. The same procedure applies also in the case $A' < 0$ and $g' < 0$.

Thanks to Corollary 8.1.2, it is possible to affirm that (8.7) admits a unique classical solution $u \in C^1(\mathbb{R} \times [0, \infty))$ and, in view of (8.14), we have

$$u_x = \frac{g'(\xi)}{1 + A'(u)g'(\xi)t} \le \frac{1}{tA'(u)} \le \frac{E}{t},$$

where we have set $E = \frac{1}{\inf_{\mathbb{R}} A'} \in (0, \infty)$. By the Lagrange Theorem, we obtain

$$u(x+a,t) - u(x,t) \le \frac{E}{t}a \qquad \forall x \in \mathbb{R}, \forall a > 0, \forall t > 0. \tag{8.28}$$

We can give the following

Definition 8.1.8. *Let $A \in C^1(\mathbb{R})$, $g \in L^\infty(\mathbb{R})$ and assume that $A' > 0$ in $\mathbb{R}$. We say that a weak solution u of (8.7) satisfies the* **entropy condition** *if for every $x \in \mathbb{R}$, for every $a > 0$ and for every $t > 0$, the inequality (8.28) holds true.*

Going back to Example 8.1.7, we readily observe that between the two found solutions only u_2 satisfies the entropy condition. Indeed, it is enough to choose $x = \frac{t}{2} - 1$ and $a = 2$ in (8.28) to see that

$$u_2\left(\frac{t}{2} + 1, t\right) - u_2\left(\frac{t}{2} - 1, t\right) = 1 \qquad \forall t > 0.$$

Example 8.1.7 suggests that problem (8.7), even if it has several solutions, may admit only one solution satisfying the entropy condition. In fact, the following result of existence and uniqueness holds true:

Theorem 8.1.9. *Let $A \in C^1(\mathbb{R})$, $g \in L^\infty(\mathbb{R})$ and assume that $A' > 0$ in $\mathbb{R}$. Then (8.7) admits a unique weak solution satisfying the entropy condition.*

For the proof of this result, see [66, Chapter 16].

8.1.3 Cauchy problem for more general equations

Problem (8.7) is a particular case of Cauchy problem associated to a first order quasilinear equation. As already explained in Section 3.3, we can consider more general problems of the form

$$\begin{cases} F(\nabla u(x), u(x), x) = 0 & x \in \Omega \subset \mathbb{R}^n \\ u(x) = g(x) & x \in \Gamma\,, \end{cases} \tag{8.29}$$

where $F : \mathbb{R}^{2n+1} \to \mathbb{R}$ is a linear function with respect to the first n variables, Ω is an open domain and Γ is an $(n-1)$-dimensional surface contained into Ω. As we have seen in Section 3.3 for the linear case, also for problem (8.29), which is a generalization of (3.8)-(3.9), neither the existence nor the uniqueness is guaranteed.

In this section we will deal with Cauchy problems in dimension $n = 2$ of the type

$$\begin{cases} a(x,y,u)u_x + b(x,y,u)u_y = c(x,y,u) & \text{in } \Omega \\ u = g \quad \text{on } \Gamma\,, \end{cases} \tag{8.30}$$

where Γ is the curve with parametric equations

$$\begin{cases} x = \varphi(s) \\ y = \psi(s) \end{cases} \qquad (s \in I)$$

and I is an interval of $\mathbb{R}$. The condition $u = g$ on Γ is to be read as

$$u(\varphi(s), \psi(s)) = h(s) := g(\varphi(s), \psi(s)) \qquad \forall s \in I. \tag{8.31}$$

We shall assume that the functions a, b, c and φ, ψ, h are of class C^1 in the respective sets of definition.

In analogy with the linear case, we will call **characteristic curves** the curves in $\mathbb{R}^3$ that solve the following system of ordinary differential equations:

$$\begin{cases} x' = a(x, y, z) \\ y' = b(x, y, z) \\ z' = c(x, y, z). \end{cases} \tag{8.32}$$

As it happened for the characteristics introduced in (3.14) for problem (3.12), also the characteristics defined in (8.32) are tangent at every point to the graph of the solution of (8.30).

Let us assume that $0 \in I$ and let us set $(x_0, y_0, z_0) = (\varphi(0), \psi(0), h(0))$. We may conclude by stating the following

Theorem 8.1.10. *Let a, b, c be functions of class C^1 in a neighborhood of (x_0, y_0, z_0) and let φ, ψ, h be of class C^1 in a neighborhood of $s = 0$. If*

$$J := b(x_0, y_0, z_0)\varphi'(0) - a(x_0, y_0, z_0)\psi'(0) \neq 0\,, \tag{8.33}$$

then there exists a neighborhood U of (x_0, y_0) such that problem (8.30) *admits a unique solution of class C^1 defined in U.*

For the proof, see [41, Paragrafo 1.5].

8.2 Navier-Stokes equations

8.2.1 Description of the physical model

Navier-Stokes equations describe the motion of a viscous incompressible fluid in the space $\mathbb{R}^n$. Let us go back to the equation (3.80) and let us assume that in place of the gravity $\mathbf{g}$ it acts a generic force $\mathbf{f}$ per unit volume; then, calling p the fluid pressure and $\mathbf{v}$ the fluid velocity, equation (3.80) becomes

$$\rho \frac{d\mathbf{v}}{dt} = -\nabla p + \mathbf{f}$$

and, using (3.50), we obtain

$$\frac{\partial \mathbf{v}}{\partial t} + (\mathbf{v} \cdot \nabla)\mathbf{v} = -\frac{1}{\rho}\nabla p + \frac{\mathbf{f}}{\rho}. \tag{8.34}$$

This is the equation which governs the motion of a non viscous fluid. Now we aim to derive the equation for a viscous fluid. Exploiting the continuity equation (3.46) and (8.34), we infer

$$\frac{\partial}{\partial t}(\rho \mathbf{v}) = \rho \frac{\partial \mathbf{v}}{\partial t} + \frac{\partial \rho}{\partial t}\mathbf{v} = -\nabla p - \rho(\mathbf{v}\cdot\nabla)\mathbf{v} - \mathrm{div}(\rho\mathbf{v})\mathbf{v} + \mathbf{f}$$

which, written component by component, reads

$$\frac{\partial}{\partial t}(\rho v_i) = -\frac{\partial p}{\partial x_i} - \sum_{j=1}^{n} \frac{\partial}{\partial x_j}(\rho v_i v_j) + f_i = -\sum_{j=1}^{n} \frac{\partial \Pi_{ij}}{\partial x_j} + f_i \qquad (i = 1, ..., n) \quad (8.35)$$

where Π is the tensor with components $\Pi_{ij} = p\delta_{ij} + \rho v_i v_j$ (with $\delta_{ij} = 1$ if $i = j$ and $\delta_{ij} = 0$ if $i \neq j$). The physical meaning of the tensor Π will be clear in a moment.

Consider the portion of fluid which flows in a fixed domain $\Omega \subset \mathbb{R}^n$; by (8.35) we derive

$$\frac{d}{dt}\int_\Omega \rho\mathbf{v}\, dx = -\int_\Omega \mathrm{div}\Pi\, dx + \int_\Omega \mathbf{f}\, dx$$

and, by the divergence Theorem, we deduce

$$\frac{d}{dt}\int_\Omega \rho\mathbf{v}\, dx = -\int_{\partial\Omega} \Pi\nu\, dS + \int_\Omega \mathbf{f}\, dx.$$

We have obtained a balance of the linear momentum in which the tensor Π measures the flux of the linear momentum passing through the surface $\partial\Omega$ in unit time. In the case of a viscous fluid, it is necessary to add to the tensor Π another term which, in accordance to [44], we shall denote by $-\sigma'_{ij}$. In this way, the tensor Π becomes

$$\Pi_{ij} = p\delta_{ij} + \rho v_i v_j - \sigma'_{ij} = -\sigma_{ij} + \rho v_i v_j\,,$$

where we have set $\sigma_{ij} = -p\delta_{ij} + \sigma'_{ij}$. The tensor σ_{ij} is called **stress tensor** and σ'_{ij} **viscous stress tensor**. The internal friction phenomena due to the fluid viscosity occur when portions of fluid move with different velocities, producing a sort of sliding of the stream threads over each other. From a mathematical point of view, this translates to a dependence of σ'_{ij} on the sole spatial derivatives of the velocity $\mathbf{v}$. If the gradient of the velocity is small, we may assume that the viscous stress tensor only depends on the first derivatives of $\mathbf{v}$. Note that, in the case in which the fluid rotates uniformly as in Example 3.4.5, it holds

$$\frac{\partial v_i}{\partial x_j} + \frac{\partial v_j}{\partial x_i} = 0.$$

Imposing that the tensor σ'_{ij} vanishes, in such situation we have to assume that it contains these symmetric combinations of the derivatives of $\mathbf{v}$. We will assume that σ'_{ij} is of the form

$$\sigma'_{ij} = a\left(\frac{\partial v_i}{\partial x_j} + \frac{\partial v_j}{\partial x_i}\right) + b(\mathrm{div}\mathbf{v})\delta_{ij}\,,$$

with $a, b \in \mathbb{R}$ independent of $\mathbf{v}$, or of the more particular kind

$$\sigma'_{ij} = \eta \left(\frac{\partial v_i}{\partial x_j} + \frac{\partial v_j}{\partial x_i} - \frac{2}{3}(\mathrm{div}\mathbf{v})\delta_{ij} \right) + \zeta(\mathrm{div}\mathbf{v})\delta_{ij}. \tag{8.36}$$

The coefficients η and ζ are called respectively **dynamic viscosity** and **second viscosity** and are both positive; they are independent of the fluid velocity, but only depend on the pressure and the temperature. In many cases, however, η and ζ do not vary their value significantly, therefore they may be considered constant, see [44]. To obtain the equation of motion, it remains to insert (8.36) into the equation (8.35) that can be rewritten as

$$\frac{\partial(\rho\mathbf{v})}{\partial t} = -\mathrm{div}\Pi + \mathbf{f}.$$

After some calculations, we obtain

$$\frac{\partial(\rho\mathbf{v})}{\partial t} - \eta\Delta\mathbf{v} + \nabla p + \rho(\mathbf{v}\cdot\nabla)\mathbf{v} + \mathrm{div}(\rho\mathbf{v})\mathbf{v} - \left(\frac{1}{3}\eta + \zeta\right)\nabla(\mathrm{div}\mathbf{v}) = \mathbf{f}. \tag{8.37}$$

In the case in which the fluid density ρ is constant (incompressible fluid), by the continuity equation (3.46) we obtain $\mathrm{div}\mathbf{v} = 0$, which inserted into (8.37) gives the Navier-Stokes equation

$$\frac{\partial\mathbf{v}}{\partial t} - \frac{\eta}{\rho}\Delta\mathbf{v} + \frac{1}{\rho}\nabla p + (\mathbf{v}\cdot\nabla)\mathbf{v} = \frac{\mathbf{f}}{\rho}. \tag{8.38}$$

In the sequel we will take $\rho = 1$ in a suitable unit of measure, since this does not affect the nature of the equation. We shall then denote by u the generic solution of the Navier-Stokes equation (8.38) and we will denote the divergence of a vector function w by $\nabla \cdot w$.

8.2.2 Stationary equation

Variational formulation

In this section we analyze the stationary problem

$$\begin{cases} -\eta\Delta u + \nabla p + (u\cdot\nabla)u = f & \text{in } \Omega \\ \nabla\cdot u = 0 & \text{in } \Omega \\ u = 0 & \text{on } \partial\Omega \end{cases} \tag{8.39}$$

where the unknowns are the vector function u (velocity field) and the scalar function p (pressure), and Ω is an open bounded domain of $\mathbb{R}^n$ $(2 \le n \le 4)$ with C^2 boundary $\partial\Omega$. Compared to (7.16), the equation in (8.39) possesses in addition the term $(u\cdot\nabla)u$

and hence it is a *quasilinear* elliptic equation. This nonlinearity deserves a particular attention. Let us consider the spaces $\mathbf{G}_i$ and $\mathbf{V}$ introduced in Section 7.1.4. First, we prove a result which explains the reason why we need the restriction $2 \le n \le 4$ on the space dimension:

Lemma 8.2.1. *Assume that* $u \in \mathbf{V}$, *then* $(u \cdot \nabla)u \in \mathbf{H}^{-1}$.

Proof. By Theorem 6.2.21 we deduce that

– if $n = 2$ then $\mathbf{H}^1 \subset \mathbf{L}^q$ for every $q < \infty$, so that $(u \cdot \nabla)u \in \mathbf{L}^r$ for every $r < 2$ and hence $(u \cdot \nabla)u \in \mathbf{H}^{-1}$;

– if $n = 3$ then $\mathbf{H}^1 \subset \mathbf{L}^6$, so that $(u \cdot \nabla)u \in \mathbf{L}^{3/2} \subset \mathbf{L}^{6/5} \subset \mathbf{H}^{-1}$;

– if $n = 4$ then $\mathbf{H}^1 \subset \mathbf{L}^4$, so that $(u \cdot \nabla)u \in \mathbf{L}^{4/3} \subset \mathbf{H}^{-1}$.

For the aforementioned inclusions we have used the trivial implication $\mathbf{H}_0^1 \subset \mathbf{L}^q \Rightarrow \mathbf{L}^{q'} \subset \mathbf{H}^{-1}$. □

Then, we prove the

Lemma 8.2.2. *The trilinear function* $b : \mathbf{V} \times \mathbf{V} \times \mathbf{V} \to \mathbb{R}$ *defined by*

$$b(u,v,w) \;:=\; \int_\Omega (u \cdot \nabla)v \cdot w \;=\; \sum_{i,j=1}^n \int_\Omega u_i \frac{\partial v_j}{\partial x_i} w_j$$

has the following properties:

(i) it is continuous in $\mathbf{V} \times \mathbf{V} \times \mathbf{V}$;

(ii) for every $u, v, w \in \mathbf{V}$ *there hold* $b(u,v,v) = 0$ *and* $b(u,v,w) = -b(u,w,v)$;

(iii) if $\{u^m\} \subset \mathbf{V}$ *satisfies* $u^m \rightharpoonup u$ *in* $\mathbf{V}$, *then* $b(u^m, u^m, v) \to b(u,u,v)$ *for every* $v \in \mathbf{V}$.

Proof. (i) Since $2 \le n \le 4$, by Theorem 6.2.21, we have the continuous embedding $\mathbf{V} \subset \mathbf{L}^4$. Therefore, by the Hölder inequality (2.17) we deduce

$$|b(u,v,w) - b(u',v,w)| \;\le\; \int_\Omega |u - u'| \cdot |\nabla v| \cdot |w|$$

$$\le \|u - u'\|_{\mathbf{L}^4} \cdot \|\nabla v\|_{\mathbf{L}^2} \cdot \|w\|_{\mathbf{L}^4} \le c_n \|u - u'\|_{\mathbf{V}} \cdot \|v\|_{\mathbf{V}} \cdot \|w\|_{\mathbf{V}} \,,$$

where the constant c_n depends only on the dimension and on the domain Ω; it is the constant for the embedding $\mathbf{V} \subset \mathbf{L}^4$ (actually, c_4 is independent of $\Omega \subset \mathbb{R}^4$). The previous estimate shows that $b(u_k, v, w) \to b(u,v,w)$ if $u_k \to u$ in $\mathbf{V}$. To complete the proof of (i) we can proceed in a similar way for v and w.

(ii) By assertion (i) and by density, to prove the first property we may assume that $u, v \in \mathbf{V} \cap \mathcal{D}$. For such u and v an integration by parts yields

$$\int_\Omega u_i \frac{\partial v_j}{\partial x_i} v_j = \frac{1}{2} \int_\Omega u_i \frac{\partial v_j^2}{\partial x_i} = -\frac{1}{2} \int_\Omega v_j^2 \frac{\partial u_i}{\partial x_i} \qquad \forall i, j = 1, ..., n$$

so that, summing over i,

$$b(u,v,v) = -\frac{1}{2}\sum_{j=1}^{n}\int_\Omega (\nabla\cdot u)\, v_j^2 = 0 ,$$

being $\nabla\cdot u = 0$. The second property follows from the first one by replacing v with $v+w$.

(iii) First we take $v \in \mathbf{V}\cap\mathcal{D}$. According to assertion (ii), we know that

$$b(u^m,u^m,v) = -b(u^m,v,u^m) = -\sum_{i,j=1}^{n}\int_\Omega u_i^m \frac{\partial v_j}{\partial x_i} u_j^m .$$

Since $\frac{\partial v_j}{\partial x_i} \in L^\infty$ and $u_i^m \to u_i$ in L^2 for every i, the result is achieved for every $v\in\mathbf{V}\cap\mathcal{D}$. Using a standard density argument, the same conclusion holds true also for a generic $v\in\mathbf{V}$. □

Remark 8.2.3. Lemma 8.2.2 (i) states that if $u^m \to u$ in $\mathbf{V}$, $v^m\to v$ in $\mathbf{V}$ and $w^m \to w$ in $\mathbf{V}$, then $b(u^m,v^m,w^m)\to b(u,v,w)$. Lemma 8.2.2 (iii) *does not* affirm that if $u^m \rightharpoonup u$ in $\mathbf{V}$ and $v^m \rightharpoonup v$ in $\mathbf{V}$, then $b(u^m,v^m,w)\to b(u,v,w)$. □

Once these properties are proved, we can give the *variational formulation of* (8.39): given $f\in\mathbf{L}^2$, find $u\in\mathbf{V}$ such that

$$\eta\int_\Omega \nabla u : \nabla v + b(u,u,v) = \int_\Omega fv \qquad \forall v\in\mathbf{V} . \tag{8.40}$$

By Theorem 7.1.16, it is clear that if we multiply (8.39) by $v\in\mathbf{V}\subset\mathbf{G}_1$ and we integrate, then the term containing ∇p vanishes and hence a solution $u\in\mathbf{V}$ satisfies (8.40). Viceversa, if $u\in\mathbf{V}$ satisfies (8.40), then

$$\langle -\eta\Delta u + (u\cdot\nabla)u - f, v\rangle = 0 \qquad \forall v\in\mathbf{V}, \tag{8.41}$$

with $\Delta u\in\mathbf{H}^{-1}$, see Proposition 6.2.19. Furthermore, by Lemma 8.2.1, $(u\cdot\nabla)u\in\mathbf{H}^{-1}$ as well. Therefore, all the terms tested with v in (8.41) belong to $\mathbf{H}^{-1}$.

Now we show that the solvability of problem (8.40) implies the solvability of problem (8.39). To this aim, we use the following result:

Proposition 8.2.4. *Let $q=(q_1,...,q_n)\in\mathbf{H}^{-1}$ be a vector field. A necessary and sufficient condition for the existence of $p\in L^2$ such that $\nabla p = q$ is that $\langle q,v\rangle = 0$ for every $v\in\mathbf{V}$.*

Proof. See [68, Propositions I.1.1 and I.1.2]. □

By combining (8.41) with Proposition 8.2.4, we see that there exists $p\in L^2$ such that $-\eta\Delta u + (u\cdot\nabla)u - f = -\nabla p$ in $\mathbf{H}^{-1}$, that is $\langle -\eta\Delta u+\nabla p+(u\cdot\nabla)u-f,v\rangle=0$ for every $v\in\mathbf{H}_0^1$. We have then shown that the solution of (8.39) may be reduced to the solution of (8.40), which we discuss below.

Remark 8.2.5. In view of the above mentioned results, it is clear that the weak formulation is still valid if $f \in \mathbf{H}^{-1}$. Here we limit ourselves to the case $f \in \mathbf{L}^2$ because we want to emphasize the role of the Helmholtz-Weyl decomposition (Theorem 7.1.16). □

Existence of solutions

We prove the following

Theorem 8.2.6. *For every $f \in \mathbf{L}^2$ there exists $u \in \mathbf{V}$ such that* (8.40) *holds true. Moreover, there exists $p \in L^2$ such that* (8.39) *holds in the $\mathbf{H}^{-1}$ sense.*

Proof. In view of what we have seen formerly, we have just to prove the existence of u. To this purpose, we follow [68] and use the Galerkin method, cf. Section 6.8. We consider the sequence $\{e_m\} \subset \mathbf{V} \cap \mathbf{C}^\infty$ of eigenvectors of the Stokes operator, see Theorem 7.1.21. For every $m \in \mathbb{N}$ we define an approximate solution $u^m \in \mathbf{V}_m := \mathrm{span}\{e_1, ..., e_m\}$ of (8.40) given by

$$u^m = \sum_{i=1}^m \xi_{i,m} e_i\,, \qquad \xi_{i,m} \in \mathbb{R}\,,$$

$$\eta \int_\Omega \nabla u^m : \nabla e_i + b(u^m, u^m, e_i) = \int_\Omega f e_i \qquad \forall i = 1, ..., m\,. \tag{8.42}$$

The existence of a solution u^m to (8.42) follows from the next statement:

Lemma 8.2.7. *Let X be a finite dimensional Hilbert space, with scalar product $[\cdot,\cdot]$ and norm $\|\cdot\|$. Let $P : X \to X$ be a continuous function such that $[P(x), x] > 0$ for every $\|x\| = k > 0$. Then there exists $x \in X$ such that $\|x\| \le k$ and $P(x) = 0$.*

Proof. By contradiction, suppose that P does not have any zero in $B := \{x \in X; \|x\| \le k\}$. Then we can define the continuous function $S : B \to B$ as

$$S(x) = -k\,\frac{P(x)}{\|P(x)\|}\,.$$

According to the Brouwer fixed point Theorem [78, Proposition 2.6], we know that S has a fixed point at some $x_0 \in B$, that is

$$x_0 = -k\,\frac{P(x_0)}{\|P(x_0)\|}\,.$$

In particular, $\|x_0\| = k$; thus, multiplying (scalar product) by x_0 both sides of the previous equation, we obtain

$$k^2 = -k\,\frac{[P(x_0), x_0]}{\|P(x_0)\|}\,,$$

in contrast with the starting assumption $[P(x),x]>0$ if $\|x\|=k$. □

We apply Lemma 8.2.7 as follows. Let $X=\mathbf{V}_m$ with $[u,v]=\int_\Omega \nabla u:\nabla v$ and let $P=P_m$ be implicitly defined by

$$[P(u),v]=\eta[u,v]+b(u,u,v)-\int_\Omega fv \qquad \forall u,v\in X\ .$$

The continuity of P follows from Lemma 8.2.2 (i). Moreover, by Lemma 8.2.2 (ii) we infer that

$$[P(u),u]=\eta\|u\|_{\mathbf{V}}^2-\int_\Omega fu\geq \eta\|u\|_{\mathbf{V}}^2-\|f\|_{\mathbf{L}^2}\|u\|_{\mathbf{L}^2}\geq \|u\|_{\mathbf{V}}(\eta\|u\|_{\mathbf{V}}-c\|f\|_{\mathbf{L}^2})\,,$$

where c is the constant coming from the Poincaré inequality ($\|v\|_{\mathbf{L}^2}\leq c\|v\|_{\mathbf{V}}$ for every $v\in X\subset\mathbf{V}$, see Theorem 6.2.18). Therefore, if $\|u\|_{\mathbf{V}}=k$ with k large enough, then $[P(u),u]>0$. Then the hypotheses of Lemma 8.2.7 are satisfied, from which we deduce that *there exists a solution* u^m *of* (8.42). Since (8.42) holds for every e_i, it holds for every linear combination of the e_i as well. We choose as test function the solution u^m. Again by Hölder and Poincaré inequalities, we find

$$\eta\|u^m\|_{\mathbf{V}}^2=\int_\Omega fu^m\leq\|u^m\|_{\mathbf{L}^2}\cdot\|f\|_{\mathbf{L}^2}\leq c\|u^m\|_{\mathbf{V}}\cdot\|f\|_{\mathbf{L}^2}$$

so that the sequence $\{u^m\}$ is bounded in $\mathbf{V}$ and, up to subsequences, $u^m\rightharpoonup u$ in $\mathbf{V}$, due to Theorem 1.4.13. For every $\ell\in\mathbb{N}$, let us choose $v\in\mathbf{V}_\ell\subset\mathbf{V}$ and $m\geq\ell$; applying Lemma 8.2.2 (iii), we obtain

$$0=\eta\int_\Omega\nabla u^m:\nabla v+b(u^m,u^m,v)-\int_\Omega fv\to\eta\int_\Omega\nabla u:\nabla v+b(u,u,v)-\int_\Omega fv\ ,$$

namely $u\in\mathbf{V}$ is a solution to (8.40). This completes the proof of Theorem 8.2.6. □

Regularity of solutions

In Theorem 8.2.6 we have assumed that $f\in\mathbf{L}^2$. In view of the elliptic regularity, we should then expect that the solution $u\in\mathbf{V}$ of (8.40) belongs to $\mathbf{H}^2$; let us discuss here this problem. To this purpose, we need to recall the following Sobolev embedding [1], which is a generalization of Theorem 6.6.1:

$$\forall m\in\mathbb{N}\,,\quad\forall r\geq 1\qquad \mathbf{W}^{m,r}\subset\begin{cases}\mathbf{L}^q \text{ for every } 1\leq q\leq\frac{nr}{n-mr} & \text{if } mr<n\\ \mathbf{L}^q \text{ for every } 1\leq q<\infty & \text{if } mr=n\\ \mathbf{L}^\infty & \text{if } mr>n\ ,\end{cases} \tag{8.43}$$

where $\mathbf{W}^{m,r}$ denotes the Sobolev space of functions whose generalized derivatives up to the order m belong to L^r (not necessarily $r = 2$).

Let us now exploit the regularity results for (7.16), see Proposition 7.1.26. First of all, we rewrite (8.39) as

$$\begin{cases} -\eta\Delta u + \nabla p = f - (u \cdot \nabla)u & \text{in } \Omega \\ \nabla \cdot u = 0 & \text{in } \Omega \\ u = 0 & \text{on } \partial\Omega \end{cases} \tag{8.44}$$

so that the regularity of the solution of (8.39) depends on the new function $g := f - (u \cdot \nabla)u$. In view of the proof of Lemma 8.2.1, we know that

$$(u \cdot \nabla)u \in \begin{cases} \mathbf{L}^r \text{ for every } r < 2 & \text{if } n = 2 \\ \mathbf{L}^{3/2} & \text{if } n = 3 \\ \mathbf{L}^{4/3} & \text{if } n = 4\,. \end{cases}$$

Therefore, if we assume that $f \in \mathbf{L}^2$, we have

$$g \in \begin{cases} \mathbf{L}^r \text{ for every } r < 2 & \text{if } n = 2 \\ \mathbf{L}^{3/2} & \text{if } n = 3 \\ \mathbf{L}^{4/3} & \text{if } n = 4\,. \end{cases}$$

Since u solves (8.44), by applying Proposition 7.1.26, we deduce that

$$u \in \begin{cases} \mathbf{W}^{2,r} \text{ for every } r < 2 & \text{if } n = 2 \\ \mathbf{W}^{2,3/2} & \text{if } n = 3 \\ \mathbf{W}^{2,4/3} & \text{if } n = 4\,. \end{cases}$$

If $n = 2$, since $u \in \mathbf{W}^{2,3/2}$ by (8.43), we deduce that $u \in \mathbf{L}^\infty$ so that $(u \cdot \nabla)u \in \mathbf{L}^2$. Thus $g \in \mathbf{L}^2$ and $u \in \mathbf{H}^2$ by Proposition 7.1.26.

If $n = 3$, by $u \in \mathbf{W}^{2,3/2}$, first we obtain that $u \in \mathbf{L}^r$ for every $r \in [1, \infty)$, and then that $\nabla u \in \mathbf{W}^{1,3/2} \subset \mathbf{L}^3$ (in the latter, $\mathbf{W}^{1,3/2}$ denotes a Sobolev space of matrix valued functions, being ∇u the *Jacobian* matrix of u); consequently, $(u \cdot \nabla)u \in \mathbf{L}^q$ for every $q \in [1, 3)$. Hence $g \in \mathbf{L}^2$ and we have again $u \in \mathbf{H}^2$.

If $n = 4$, by $u \in \mathbf{W}^{2,4/3}$ we have that $u \in \mathbf{L}^4$ and that $\nabla u \in \mathbf{W}^{1,4/3} \subset \mathbf{L}^2$ and there is *no improvement* with respect to the already known condition $u \in \mathbf{V}$; thus, in this case we do not achieve the $\mathbf{H}^2$ regularity by means of this simple trickery.

Nevertheless, in dimension $n = 4$ there are still regularity results, but they are obtained with more advanced tools, see [34].

In a similar way, further regularity of the solution (assuming f more regular) may be obtained.

Remark 8.2.8. The theory developed so far for (8.39) applies also to more general problems.

– One may consider *non homogeneous* Dirichlet problems for (8.39), replacing the boundary condition with $u = \varphi$ on $\partial\Omega$, see [68, Section II.1.4].

– One may consider the same problem (8.39), but set in *unbounded* domains; here some embedding problems arise and the variational formulation changes drastically, see [68, Section II.1.3]. □

Structure of the set of solutions

Theorem 8.2.6 says nothing about the uniqueness of the weak solution of (8.40); let us now discuss such a problem. For every $f \in \mathbf{L}^2$ we set

$$S_f \;=\; \text{the set of weak solutions of (8.40).}$$

Let P_1 be the projection on $\mathbf{G}_1$ introduced in Theorem 7.1.16. We prove uniqueness when f is sufficiently small, in a suitable sense:

Theorem 8.2.9. *There exists $\gamma > 0$ such that if $\|P_1 f\|_{\mathbf{L}^2} < \gamma$, then the solution of (8.40) is unique.*

Proof. Let $f \in \mathbf{L}^2$ and take $u \in S_f$. Choosing $u = v$ in (8.40) and exploiting the Poincaré inequality (Theorem 6.2.18), we obtain

$$\eta\|u\|_{\mathbf{V}}^2 = \int_\Omega fu = \int_\Omega P_1 f\ u \le \|P_1 f\|_{\mathbf{L}^2} \cdot \|u\|_{\mathbf{L}^2} \le c\|P_1 f\|_{\mathbf{L}^2} \cdot \|u\|_{\mathbf{V}} \ ,$$

where we have used also Lemma 8.2.2, Theorem 7.1.16 and Hölder inequality (2.17). Therefore, there exists $c_1 > 0$ such that every $u \in S_f$ satisfies the a priori estimate

$$\|u\|_{\mathbf{V}} \le c_1\|P_1 f\|_{\mathbf{L}^2} \qquad \forall u \in S_f \ . \tag{8.45}$$

Let us assume that (8.40) admits two distinct solutions u^1 and u^2. Set $w = u^1 - u^2 \neq 0$; subtracting the equations (8.40) associated to u^1 and u^2, we obtain

$$\eta \int_\Omega \nabla w : \nabla v + b(u^1, w, v) + b(w, u^2, v) = 0 \qquad \forall v \in \mathbf{V} \ .$$

Choosing $v = w$ in the last equation and applying Lemma 8.2.2 (ii), we get

$$\eta\|w\|_{\mathbf{V}}^2 = -b(w, u^2, w) \le c(n)\|w\|_{\mathbf{V}}^2 \cdot \|u^2\|_{\mathbf{V}} \ .$$

Since $w \neq 0$, this shows that $\|u^2\|_{\mathbf{V}} \geq \eta/c(n)$ and, according to (8.45), this implies that $\|P_1 f\|_{\mathbf{L}^2} \geq \eta/(c_1 c(n)) =: \gamma$. Thus we have proved that if (8.40) admits two different solutions, then necessarily $\|P_1 f\|_{\mathbf{L}^2} \geq \gamma$. This concludes the proof. □

In particular, Theorem 8.2.9 states that if $P_1 f = 0$, then $u = 0$. This last claim is similar to the statement in the linear case, see Corollary 7.1.20.

If the hypothesis that $P_1 f$ is "small" fails, nothing can be said in general. However, a profound result by Foias and Temam [22] suggests that, when the space dimension is $n = 2$ or $n = 3$ (namely the dimensions with a physical meaning), there are not "too many" solutions. This restriction on the dimension ($n = 4$ is not admissible) follows not only from the regularity results discussed above, but also from the next inequality, due to Gagliardo [26] and Nirenberg [54]:

$$\|v\|_{\mathbf{L}^4} \leq c\|v\|_{\mathbf{L}^2}^{1-n/4} \cdot \|\nabla v\|_{\mathbf{L}^2}^{n/4} \qquad \forall v \in \mathbf{H}_0^1 \; ; \tag{8.46}$$

clearly, if $n = 4$ the inequality (8.46) reduces to the Sobolev embedding, cf. Theorem 6.2.20. There holds the

Theorem 8.2.10. (Foias-Temam Theorem). *Assume that $n = 2, 3$ and that $\partial\Omega$ has finitely many connected components. Let $f \in \mathbf{L}^2$, then S_f is homeomorphic to a compact subset of $\mathbb{R}^m$, for some m. Moreover, there exists a dense open set $\mathcal{O} \subset \mathbf{L}^2$ such that for every $f \in \mathcal{O}$ the set S_f is finite and the number of solutions is constant on every connected component of $\mathcal{O}$.*

Proof. Here we adopt the notations introduced in (7.21) and (7.22). Let $f \in \mathbf{L}^2$; first, we claim that if m is large enough then there exists $c_m > 0$ such that

$$\|u^1 - u^2\|_{\mathbf{V}} \leq c_m \|\mathcal{P}_m(u^1 - u^2)\|_{\mathbf{L}^2} \qquad \forall u^1, u^2 \in S_f \; . \tag{8.47}$$

In the following $c > 0$ and $c_m > 0$ denote some constants which may vary from line to line, being the second one depending on m. To prove (8.47), consider $u^1, u^2 \in S_f$ and let $w = u^1 - u^2$: subtracting the two relations (8.40) associated to u^1 and u^2, we obtain

$$\eta \int_\Omega \nabla w : \nabla v = -b(u^1, w, v) - b(w, u^2, v) \qquad \forall v \in \mathbf{V} \; ;$$

choosing $v = \mathcal{Q}_m w$ and exploiting Lemma 8.2.2, we get

$$\begin{aligned}
\eta\|\mathcal{Q}_m w\|_{\mathbf{V}}^2 &= -b(u^1, \mathcal{P}_m w, \mathcal{Q}_m w) - b(\mathcal{P}_m w, u^2, \mathcal{Q}_m w) - b(\mathcal{Q}_m w, u^2, \mathcal{Q}_m w) \\
&\leq c\|u^1\|_{\mathbf{V}} \cdot \|\mathcal{P}_m w\|_{\mathbf{V}} \cdot \|\mathcal{Q}_m w\|_{\mathbf{V}} + c\|\mathcal{P}_m w\|_{\mathbf{V}} \cdot \|u^2\|_{\mathbf{V}} \cdot \|\mathcal{Q}_m w\|_{\mathbf{V}} \\
&\quad + c\|u^2\|_{\mathbf{V}} \cdot \|\mathcal{Q}_m w\|_{\mathbf{L}^4}^2 \; .
\end{aligned}$$

By (8.45) and by the equivalence of the norms in finite dimensional spaces, we get

$$c\|u^1\|_{\mathbf{V}} \cdot \|\mathcal{P}_m w\|_{\mathbf{V}} \cdot \|\mathcal{Q}_m w\|_{\mathbf{V}} + c\|\mathcal{P}_m w\|_{\mathbf{V}} \cdot \|u^2\|_{\mathbf{V}} \cdot \|\mathcal{Q}_m w\|_{\mathbf{V}} \leq c_m \|\mathcal{P}_m w\|_{\mathbf{L}^2} \cdot \|\mathcal{Q}_m w\|_{\mathbf{V}} .$$

On the other hand, keeping in mind (7.23), (8.45) and (8.46), we derive

$$c\|u^2\|_{\mathbf{V}} \cdot \|\mathcal{Q}_m w\|^2_{\mathbf{L}^4} \leq c\lambda_{m+1}^{-1+n/4}\|\mathcal{Q}_m w\|^2_{\mathbf{V}} \,.$$

Summarizing, we have

$$\eta\|\mathcal{Q}_m w\|^2_{\mathbf{V}} \leq c_m\|\mathcal{P}_m w\|_{\mathbf{L}^2} \cdot \|\mathcal{Q}_m w\|_{\mathbf{V}} + c\lambda_{m+1}^{-1+n/4}\|\mathcal{Q}_m w\|^2_{\mathbf{V}} \,.$$

Now, instead of discussing whether $\mathcal{Q}_m w = 0$ or $\mathcal{Q}_m w \neq 0$, we prefer to apply the Young inequality (2.18), obtaining

$$\eta\|\mathcal{Q}_m w\|^2_{\mathbf{V}} \leq \left(\frac{\eta}{4} + c\lambda_{m+1}^{-1+n/4}\right)\|\mathcal{Q}_m w\|^2_{\mathbf{V}} + c_m\|\mathcal{P}_m w\|^2_{\mathbf{L}^2} \,.$$

Therefore, if m is large enough (we recall that, according to Theorem 7.1.21, $\lambda_m \to \infty$ as $m \to \infty$), it holds $\|\mathcal{Q}_m w\|^2_{\mathbf{V}} \leq c_m\|\mathcal{P}_m w\|^2_{\mathbf{L}^2}$ and consequently

$$\|w\|^2_{\mathbf{V}} = \|\mathcal{P}_m w\|^2_{\mathbf{V}} + \|\mathcal{Q}_m w\|^2_{\mathbf{V}} \leq c_m\|\mathcal{P}_m w\|^2_{\mathbf{L}^2}$$

which proves (8.47).

If $u^1, u^2 \in S_f$ satisfy $\mathcal{P}_m u^1 = \mathcal{P}_m u^2$, by (8.47) we deduce that $u^1 = u^2$: therefore, S_f is contained into a space isomorphic to span$\{e_1, ..., e_m\}$. Moreover, in view of (8.45) and of the compact embedding $\mathbf{V} \subset \mathbf{L}^2$, we infer that S_f is compact in $\mathbf{L}^2$ (cf. Definition 1.8.3), and hence S_f is homeomorphic to a compact subset of $\mathbb{R}^m$.

The second part of Theorem 8.2.10 is more delicate and can be obtained by applying an appropriate version of Sard-Smale Theorem, see for instance [22, Theorem A]. This requires some knowledge about nonlinear Fredholm operators, see [78, Section 8.4]. One should prove that the operator $A : (\mathbf{H}^2 \cap \mathbf{V}) \to \mathbf{L}^2$ defined by $A(u) = -\eta\Delta u + (u \cdot \nabla)u$ is a particular Fredholm operator of index 0. For a detailed proof we refer to [22, Section 2.2]. □

Some generalizations of Theorem 8.2.10 can be found in [23].

8.2.3 Evolution equation

Here we analyze the evolution equation

$$\begin{cases} u_t - \eta\Delta u + \nabla p + (u \cdot \nabla)u = f & \text{in} \quad \Omega \times (0,T) \\ \nabla \cdot u = 0 & \text{in} \quad \Omega \times (0,T) \\ u = 0 & \text{on} \quad \partial\Omega \times (0,T) \\ u(x,0) = u_0(x) & \text{in} \quad \Omega \end{cases} \tag{8.48}$$

where, as usual, the unknowns are the vector field u and the scalar function p, while Ω is a bounded open subset of $\mathbb{R}^n$ $(2 \leq n \leq 4)$ with C^2 boundary. Here we assume that $f \in L^2(0,T;\mathbf{V}')$ and that $u_0 \in \mathbf{G}_1$. Under such hypotheses, we are interested in finding weak solutions of (8.48), namely

$$u \in L^2(0,T;\mathbf{V}) \quad \text{such that} \quad u(0) = u_0 \quad \text{and}$$
$$\frac{d}{dt}\int_\Omega u(t)v + \eta\int_\Omega \nabla u(t) : \nabla v + b(u(t),u(t),v) = \langle f(t), v\rangle \quad \forall v \in \mathbf{V} \text{ in } \mathcal{D}'[0,T) \ . \tag{8.49}$$

Note that under the sole assumption $u \in L^2(0,T;\mathbf{V})$, the initial condition $u(0) = u_0$ does not make sense. In fact, such a u is not defined pointwise for every t (in particular, for $t = 0$), being just an element of $L^2(0,T)$. To justify this fact, we need the following result:

Lemma 8.2.11. *Let Ω be a bounded open subset of $\mathbb{R}^n$ $(2 \le n \le 4)$ with C^2 boundary. Assume that $u \in L^2(0,T;\mathbf{V})$. Then the function Bu defined by*

$$\langle Bu(t), v\rangle = b(u(t),u(t),v) \quad \forall v \in \mathbf{V} \ \textit{a.e. in } [0,T]$$

belongs to $L^1(0,T;\mathbf{V}')$.

Proof. See [68, Lemma III.3.1]. □

Let us now rewrite the equation in (8.49) as

$$\langle u'(t), v\rangle = -\eta\int_\Omega \nabla u(t) : \nabla v - b(u(t),u(t),v) + \langle f(t), v\rangle \quad \forall v \in \mathbf{V} \ ;$$

in view of Lemma 8.2.11 and by the assumptions made on f (and on u) we derive that $u' \in L^1(0,T;\mathbf{V}')$. Then, by a result analogous to Theorem 7.2.9, we infer that $t \mapsto u(t)$ is weakly continuous from $[0,T]$ to $\mathbf{G}_1$ (see Definition 7.3.1), in the sense that $\Phi_v : [0,T] \to \mathbb{R}$ defined by

$$\Phi_v(t) := \int_\Omega u(t)v \qquad (t \in [0,T])$$

is a continuous (scalar!) function for every $v \in \mathbf{G}_1$. Therefore, the initial condition actually makes sense, since $u(t)$ is defined pointwise as a function with values in $\mathbf{G}_1$. There holds the

Theorem 8.2.12. *Let Ω be a bounded open subset of $\mathbb{R}^n$ $(2 \le n \le 4)$ with C^2 boundary. Assume that $f \in L^2(0,T;\mathbf{V}')$ and $u_0 \in \mathbf{G}_1$. Then there exists at least one solution $u = u(t)$ to (8.49). Moreover, every solution u to (8.49) is weakly continuous from $[0,T]$ to $\mathbf{G}_1$ and satisfies $u \in L^\infty(0,T;\mathbf{G}_1)$.*

Proof. The proof is based on the Galerkin method described in Section 6.8, see [68, Theorem III.3.1]. □

Remark 8.2.13. After having obtained the existence of u, we should deal with the proof of the existence of p. This can be done but it is a difficult task, see [68]. □

Concerning the uniqueness, a complete answer can be given only in dimension $n = 2$:

Theorem 8.2.14. *Let $n = 2$. Under the assumptions of Theorem 8.2.12, the solution $u = u(t)$ of (8.49) is unique and coincides a.e. with a continuous function from $[0, T]$ to $\mathbf{G}_1$.*

Proof. See [68, Theorem III.3.2]. □

As for the stationary equation, the dimensional constraint is due to the regularity of the solution which, in turn, depends on the Sobolev embeddings. Since in low dimensions these embeddings are "better" (see Theorem 6.2.20), in these dimensions one achieves stronger results. For some results relating uniqueness to regularity see [68, Theorem III.3.4] and [69, Theorem I.3.2]. Without any further regularity, until now it is not known any uniqueness result for the weak solution of (8.49) in the case $n \geq 3$. We draw the interested (and avid!) reader's attention to the existence of a 1.000.000 dollars award for him who will be able to prove (or disprove) such result.

8.3 Semilinear elliptic equations

8.3.1 Some physical models described by nonlinear equations

This section is devoted to the study of a class of equations which model a wide set of physical phenomena, from the wave propagation to the heat diffusion. We will treat equations of the form

$$-\Delta u = f(u) \,, \tag{8.50}$$

set in the whole $\mathbb{R}^n$ or in bounded domains under a Dirichlet boundary condition. We will illustrate how, starting from different physical phenomena, it is possible to reduce to an equation of type (8.50).

Let us begin with the nonlinear Klein-Gordon equation:

$$\Phi_{tt} - \Delta\Phi + a^2\Phi = g(\Phi) \,, \tag{8.51}$$

where $\Phi = \Phi(x, t)$ is a function with values in $\mathbb{C}$ defined for $x \in \mathbb{R}^n$ and $t \in \mathbb{R}$, $a \in \mathbb{R}$ and $g : \mathbb{C} \to \mathbb{C}$ is a function satisfying

$$g(\rho e^{i\vartheta}) = f(\rho)e^{i\vartheta}, \qquad \rho, \vartheta \in \mathbb{R} \tag{8.52}$$

for a suitable $f : \mathbb{R} \to \mathbb{R}$ continuous and odd. Let us restrict the search of solutions of (8.51) to the functions with separate variables of the form $\Phi(x, t) = e^{i\omega t}u(x)$, with

$u : \mathbb{R}^n \to \mathbb{R}$ and $\omega \in \mathbb{R}$ (ω represents the angular frequency of oscillation). Then, by (8.52) u must solve the elliptic equation

$$-\Delta u + (a^2 - \omega^2)u = f(u) \qquad \text{in } \mathbb{R}^n. \tag{8.53}$$

Also the search of particular solutions of the nonlinear Schrödinger equation

$$i\Phi_t - \Delta\Phi = g(\Phi) \tag{8.54}$$

can be reduced to the study of equation (8.50). In this model the wave function $\Phi = \Phi(x, t)$ is complex valued and is defined for $x \in \mathbb{R}^n$ and $t \in \mathbb{R}$. As for the Klein-Gordon equation, suppose that the function g is of the form (8.52). The search of solutions of (8.54) with separate variables of the type $\Phi(x, t) = e^{i\omega t}u(x)$ leads to the study of the elliptic equation

$$-\Delta u - \omega u = f(u) \qquad \text{in } \mathbb{R}^n. \tag{8.55}$$

For further details on the models described so far, see [4, 7, 14, 24, 35].

Equation (8.50) can be seen also as the stationary version of the equation

$$u_t - \Delta u = f(u), \tag{8.56}$$

where $u = u(x, t)$ with $x \in \mathbb{R}^n$ and $t \geq 0$. Models of this kind come (for instance) from population dynamics, in which the function u represents the population density, as the position x and the time t vary. For more details, see [5, 20, 21, 73].

On the other hand, the equation (8.56) is also a model of thermal diffusion, in which the heat source $f(\vartheta)$ depends in a nonlinear way on the temperature ϑ. Nonlinear models of this type may appear when considering a combustion process, with a mixture containing a combustible substance of density ρ, cf. [32]. The mixture temperature ϑ and the combustible density ρ affect each other as follows: during the combustion process, a certain amount of heat is released, depending on both ϑ and ρ. On the other hand, a certain amount of combustible is destroyed during the reaction, whose velocity depends on both ϑ and ρ. A phenomenon of this type can be modeled by the system of equations

$$\begin{cases} \vartheta_t - k\Delta\vartheta = f_1(\vartheta, \rho) \\ \rho_t - D\Delta\rho = f_2(\vartheta, \rho), \end{cases} \tag{8.57}$$

where in the heat equation (3.91) we have assumed $C_V = 1$. The function $f_1(\vartheta, \rho)$ is positive since the reaction produces heat, and is increasing with respect to ϑ and ρ since the higher the mixture temperature and the combustible density are, the faster the velocity of the reaction will be. Conversely, the function $f_2(\vartheta, \rho)$ is negative since the reaction destroys the combustible, and is increasing in modulus with respect to ϑ and ρ, still due to the dependence of the velocity of reaction on the temperature and the density. A possible choice of these functions is given by $f_1(\vartheta, \rho) = \varepsilon_1(1+\vartheta)^{p-1}\rho^{q-1}$ and $f_2(\vartheta, \rho) = -\varepsilon_2(1+\vartheta)^{p-1}\rho^{q-1}$ with $\varepsilon_1, \varepsilon_2 > 0$ and $p, q > 1$. The combustion process inside the tank Ω containing the mixture may essentially take place in two different ways. In the first type of reaction, the time is not sufficient to waste the produced heat across the tank walls, thereby increasing the temperature and the velocity of reaction until a lack of combustible occurs. In the second type of reaction, the thermal diffusion coefficient is large enough to allow the produced heat to rapidly leak out through the tank walls. In the latter kind of process, it is reasonable that during the reaction, for times close enough to the initial time, the amount of combustible remains almost constant, so that $\rho = \rho_0$, where the constant ρ_0 is the combustible density at the initial time. The heat equation in (8.57) becomes

$$\vartheta_t - k\Delta\vartheta = f_1(\vartheta, \rho_0). \tag{8.58}$$

Moreover, if the diffusion coefficient k is large enough, for times close to the initial time, the solution of (8.58) does not differ too much from the solution of the stationary equation

$$-k\Delta\vartheta = f_1(\vartheta, \rho_0). \tag{8.59}$$

If we prescribe the temperature on the tank walls, namely $\vartheta = \vartheta_0$ on $\partial\Omega$, then the occurrence or not of the second type of process as k varies is equivalent to the problem of existence or not of a solution of the Dirichlet problem

$$\begin{cases} -k\Delta\vartheta = f_1(\vartheta, \rho_0) & \text{in } \Omega \\ \vartheta = \vartheta_0 & \text{on } \partial\Omega. \end{cases} \tag{8.60}$$

We conclude this part devoted to physical applications by describing the following model, taken from [27]: let Ω be a body which is heated by a constant (in both direction and magnitude) current flow and let I be the current surface density. Assume that the resistivity R of the body depends in a nonlinear way on the temperature, namely let $R = R(\vartheta)$ where $\vartheta = \vartheta(x,t)$ denotes the temperature of the body. It is reasonable to assume that the resistivity increases with temperature, thereby the function $\vartheta \mapsto R(\vartheta)$ is monotone increasing and $R(0) > 0$. A possible choice of the function R is given by $R(\vartheta) = (1+\vartheta)^{p-1}$ for some $p > 1$. Since the amount of heat per unit volume and time which is produced by the current flow equals RI^2, then, setting $k = C_V = 1$, the heat equation (3.91) becomes

$$\vartheta_t - \Delta\vartheta = I^2 R(\vartheta). \tag{8.61}$$

At this point we may ask for which values of I there exist stationary solutions of (8.61) satisfying the Dirichlet condition $\vartheta = \vartheta_0$ on $\partial\Omega$. In conclusion, also in this case we are led to study a Dirichlet problem of the type

$$\begin{cases} -\Delta\vartheta = I^2 R(\vartheta) & \text{in } \Omega \\ \vartheta = \vartheta_0 & \text{on } \partial\Omega. \end{cases} \tag{8.62}$$

8.3.2 Weak solutions and their regularity

We start considering equations of the type (8.50) set on bounded domains. Let then $\Omega \subset \mathbb{R}^n$ be a bounded open set with C^1 boundary and consider the following Dirichlet problem:

$$\begin{cases} -\Delta u = f(u) & \text{in } \Omega \\ u = 0 & \text{on } \partial\Omega\ , \end{cases} \tag{8.63}$$

where $f : \mathbb{R} \to \mathbb{R}$ is a C^1 function. We will say that $u \in H_0^1(\Omega)$ is a **weak solution** of (8.63) if

$$\int_\Omega \nabla u \nabla v = \int_\Omega f(u) v \qquad \forall v \in H_0^1(\Omega)\ .$$

A justification for this kind of definition of solution may be given proceeding as in the case of linear elliptic equations described in Chapter 7. In fact, following the same procedure introduced in Chapter 7, it is possible to verify that classical solutions of (8.63) are weak solutions and, viceversa, that weak solutions of class $C^2(\overline{\Omega})$ are classical solutions.

For this reason it may be important to determine sufficient conditions to ensure that weak solutions are also classical solutions. For this to happen, it is necessary to impose on the function f some growth conditions at infinity; to this purpose we recall that if $n \geq 3$, the number $2^* = 2n/(n-2)$ is the critical Sobolev exponent introduced in Theorem 6.2.21. In fact there holds the following

Theorem 8.3.1. (Brezis-Kato Theorem). *Assume that $n \geq 3$ and that $\Omega \subset \mathbb{R}^n$ is a bounded open set with $\partial\Omega \in C^1$. If there exists a constant $C > 0$ such that*

$$|f(s)| \leq C(1 + |s|^{2^*-1}) \qquad \forall s \in \mathbb{R}\,, \tag{8.64}$$

then every weak solution of (8.63) is a classical solution.

Proof. See [11]. □

The result stated in Theorem 8.3.1 is optimal, since if f violates the growth condition (8.64) then there may exist weak solutions of (8.63) that are non classical solutions, as it is shown in the following

Example 8.3.2. Let $p > \frac{2(n-1)}{n-2}$, let B be the unit ball of $\mathbb{R}^n$ centered at the origin and consider the problem

$$\begin{cases} -\Delta u = \lambda(1+u)^{p-1} & \text{in } B \\ u = 0 & \text{on } \partial B \end{cases} \tag{8.65}$$

with $\lambda = \frac{2}{p-2}(n-2-\frac{2}{p-2}) > 0$. The function $u(x) = |x|^{-2/(p-2)} - 1$ solves (8.65) in $B \setminus \{0\}$. It is easy to check that $u \in H_0^1(B)$ if and only if $p > 2^*$. Therefore u is a weak solution of (8.65) if and only if $p > 2^*$. Nevertheless, it is not a classical solution, since it is not bounded. For $p \le 2^*$, the function u does not belong to $H_0^1(B)$ and hence it is not a weak solution. However it satisfies the equation in (8.65) in the sense of classical derivatives at every point $x \in B \setminus \{0\}$. Moreover, thanks to condition $p > \frac{2(n-1)}{n-2}$, we can affirm that $u \in L^{p-1}(B)$ and that u solves the equation in (8.65) in the sense of distributions, that is

$$-\int_B u\Delta\varphi = \lambda \int_B (1+u)^{p-1}\varphi \qquad \forall \varphi \in \mathcal{D}(B). \qquad \square$$

8.3.3 The role of the critical Sobolev exponent

In order to emphasize the role of the exponent 2^* we need to assume that $n \ge 3$. Let us begin to study the Dirichlet problem

$$\begin{cases} -\Delta u = u^{p-1} & \text{in } \Omega \\ u > 0 & \text{in } \Omega \\ u = 0 & \text{on } \partial\Omega\ . \end{cases} \tag{8.66}$$

The positivity requirement ($u > 0$) is due to physical reasons, see Section 8.3.1; in particular, we are interested in *non trivial* solutions of the homogeneous Dirichlet problem in (8.66).

Before starting the analysis of (8.66), we introduce some notions about functionals (functions with values in $\mathbb{R}$) defined over Banach spaces.

Definition 8.3.3. *Let X be a Banach space and let $J : X \to \mathbb{R}$ be a functional.*

(i) We will say that J is continuous if there holds the implication

$$u_m \to u \text{ in } X \quad \Longrightarrow \quad J(u_m) \to J(u)\ .$$

(ii) We will say in addition that J is Fréchet differentiable in X if for every $u \in X$ there exists a linear continuous functional over X, which we denote by $J'(u)$, such that

$$u_m \to u \text{ in } X \quad \Longrightarrow \quad \frac{|J(u_m) - J(u) - \langle J'(u), u_m - u\rangle|}{\|u_m - u\|_X} \to 0.$$

The functional $J'(u) \in X'$ will be called **derivative** *of J at u.*

(iii) We will say that J is of class C^1 in X if J is Fréchet differentiable in X and if the map $J' : X \to X'$ is continuous, in the sense that there holds the implication

$$u_m \to u \text{ in } X \quad \Longrightarrow \quad J'(u_m) \to J'(u) \text{ in } X'.$$

Exercise 8.3.4. Rewrite Definition 8.3.3 in the case $X = \mathbb{R}^n$ and derive the definition of differentiability of a function with several real variables. □

In the following example we introduce two functionals that will prove useful in the study of problem (8.66).

Example 8.3.5. Let $\Omega \subset \mathbb{R}^n$ be a bounded set and let $X = H_0^1(\Omega)$. Then $J(u) = \|\nabla u\|_{L^2}^2$ and $I_p(u) = \|u\|_{L^p}^p$ with $1 < p \leq 2^*$ are of class C^1. Moreover,

$$\langle J'(u), v\rangle = 2\int_\Omega \nabla u \nabla v \ , \qquad \langle I_p'(u), v\rangle = p\int_\Omega |u|^{p-2}uv. \tag{8.67}$$

We observe that for $p = 1$ the functional I_p is continuous but not differentiable, while for $p > 2^*$, the functional I_p is not well defined in $H_0^1(\Omega)$, since there exist functions of $H_0^1(\Omega)$ which do not belong to $L^p(\Omega)$. □

Definition 8.3.3 and Example 8.3.5 play an important role in the proof of

Theorem 8.3.6. *Let $n \geq 3$ and let $\Omega \subset \mathbb{R}^n$ be a bounded open set with $\partial\Omega \in C^1$. If $2 < p < 2^*$, problem (8.66) admits at least one solution.*

Proof. Instead of proceeding directly with the study of (8.66), we consider the following minimization problem, whose relationship with the solutions of (8.66) will be explained later. We intend to study the best constant S_p in the Sobolev inequality

$$S_p\|u\|_{L^p}^2 \leq \|\nabla u\|_{L^2}^2 \qquad \forall u \in H_0^1(\Omega)\ . \tag{8.68}$$

The inequality (8.68) shows the continuity of the embedding $H_0^1(\Omega) \subset L^p(\Omega)$ (see Theorem 6.2.20). In other words, S_p is the largest constant $C > 0$ such that the following inequality holds true:

$$C\|u\|_{L^p}^2 \leq \|\nabla u\|_{L^2}^2 \qquad \forall u \in H_0^1(\Omega)\ .$$

Note that if $p = 2$ (case not covered by the assumptions), (8.68) reduces to the Poincaré inequality, cf. Theorem 6.2.18.

Setting $B = \{u \in H_0^1(\Omega);\ \|u\|_{L^p} = 1\}$, S_p admits the following characterization:

$$S_p = \inf_{u \in B} \|\nabla u\|_{L^2}^2\ . \tag{8.69}$$

In other words, we are minimizing the functional $J(u) = \|\nabla u\|_{L^2}^2$ under the constraint $I_p(u) = \|u\|_{L^p}^p = 1$. Let us try to establish for which values of p the minimum in (8.69) is achieved. To this aim, let us consider a minimizing sequence for S_p, namely a sequence $\{u_m\} \subset B$ such that $\|\nabla u_m\|_{L^2}^2 \to S_p$ as $m \to \infty$. Such sequence turns out to be bounded in $H_0^1(\Omega)$ and hence, up to subsequences, by Theorem 1.4.13 we can affirm that there exists $\overline{u} \in H_0^1(\Omega)$ such that $u_m \rightharpoonup \overline{u}$. Being $p < 2^*$, thanks to the compactness of the embedding $H_0^1(\Omega) \subset L^p(\Omega)$ (see Theorem 6.6.3) we can affirm that $u_m \to \overline{u}$ strongly in $L^p(\Omega)$: in particular, this entails that $\overline{u} \in B$. On the other hand, again by Theorem 1.4.13, we infer

$$\|\nabla \overline{u}\|_{L^2}^2 \le \liminf_{m\to\infty} \|\nabla u_m\|_{L^2}^2 = S_p \ .$$

Consequently, the function $\overline{u} \in B$ satisfies $\|\nabla \overline{u}\|_{L^2}^2 = S_p$.

It can be verified that if $u \in H_0^1(\Omega)$ then $|u| \in H_0^1(\Omega)$ and $|\nabla u| = |\nabla |u||$ a.e. in Ω. Applying this result to $\overline{u}$ and observing that $\||\overline{u}|\|_{L^p} = \|\overline{u}\|_{L^p}$, we infer that

$$|\overline{u}| \in B \quad \text{and} \quad \|\nabla|\overline{u}|\|_{L^2}^2 = \|\nabla \overline{u}\|_{L^2}^2 = S_p \ .$$

But for $|\overline{u}|$ in place of $\overline{u}$, we can affirm that there exists a non zero function $\overline{u} \ge 0$ a.e. in Ω which satisfies (8.68) with the equality sign. Such $\overline{u}$ is a minimizer for J, constrained to B. Using the infinite dimensional version of Lagrange multipliers Theorem, we deduce that there exists $\lambda > 0$ such that

$$J'(\overline{u}) = \lambda I_p'(\overline{u})$$

with J and I_p as in Example 8.3.5. To check that $\lambda > 0$, it is enough to notice that

$$0 < 2\|\nabla \overline{u}\|_{L^2}^2 = \langle J'(\overline{u}), \overline{u}\rangle = \lambda \langle I_p'(\overline{u}), \overline{u}\rangle = \lambda p \|\overline{u}\|_{L^p}^p = \lambda p.$$

In view of (8.67), we obtain

$$\int_\Omega \nabla \overline{u} \nabla v = \frac{\lambda p}{2} \int_\Omega \overline{u}^{p-1} v \qquad \forall v \in H_0^1(\Omega) \ .$$

Setting $\overline{v} = (\frac{\lambda p}{2})^{1/(p-2)} \overline{u}$, from the previous identity we derive

$$\int_\Omega \nabla \overline{v} \nabla v = \int_\Omega \overline{v}^{p-1} v \qquad \forall v \in H_0^1(\Omega) \ .$$

Applying Theorem 8.3.1 we obtain that $\overline{v} \ge 0$ is a classical solution of the problem

$$\begin{cases} -\Delta \overline{v} = \overline{v}^{p-1} & \text{in } \Omega \\ \overline{v} = 0 & \text{on } \partial\Omega \ . \end{cases}$$

If Ω is connected, since $-\Delta\overline{v} \geq 0$ in Ω, by the maximum principle (see Corollary 3.6.15), we infer that $\overline{v} > 0$ in Ω. If instead Ω is not connected, we need to verify that $\overline{v} > 0$ on each connected component of Ω. We leave this check to the reader: we suggest to argue by contradiction, exploiting the inequality

$$\frac{a+b}{c+d} < \frac{a}{c} + \frac{b}{d} \qquad \forall a, b, c, d > 0$$

and the fact that S_p is a minimum. □

Now we see that the assumption $p < 2^*$ is necessary to guarantee the validity of Theorem 8.3.6. We start remarking that its proof does not work in the case $p \geq 2^*$, since the embedding $H_0^1(\Omega) \subset L^p(\Omega)$ is not compact any more for $p = 2^*$ and is not valid any more for $p > 2^*$. We give the following

Definition 8.3.7. *We say that a set $\Omega \subset \mathbb{R}^n$ is* **star-shaped** *with respect to the point $x_0 \in \Omega$ if for every $x \in \partial\Omega$ it holds $(x - x_0) \cdot \nu_x > 0$, where ν_x is the outer unit normal to $\partial\Omega$ at the point x.*

A star-shaped set is then a set whose boundary "may be illuminated by only one lamp": placing the lamp at the center of star-shapedness x_0, the light rays reach all the points of $\partial\Omega$ with an angle less than 90°. Also, the segment $\overline{x_0 x}$ (the light ray) is entirely contained into Ω.

Then there holds the

Theorem 8.3.8. (Pohozaev Theorem). *Let $n \geq 3$ and let $\Omega \subset \mathbb{R}^n$ be a star-shaped bounded open set with $\partial\Omega \in C^1$. If $p \geq 2^*$, problem* (8.66) *has no solution.*

Proof. We observe that (8.66) is an autonomous problem, that is, both sides of the equation do not directly depend on x, but just indirectly, by means of the function u. Consequently, up to translations of the domain Ω that, for the above mentioned reason, would leave the equation unchanged, we may assume that the center of star-shapedness of Ω is the origin 0, so that

$$x \cdot \nu > 0 \qquad \text{on } \partial\Omega\,. \tag{8.70}$$

Let us now assume that $u > 0$ solves (8.66); thus, being $\partial\Omega$ be a zero level curve for u, the gradient ∇u is parallel to the normal ν to the boundary and hence

$$(x \cdot \nabla u)\frac{\partial u}{\partial \nu} = (x \cdot \nu)\left(\frac{\partial u}{\partial \nu}\right)^2 = (x \cdot \nu)|\nabla u|^2 \qquad \text{on } \partial\Omega\,. \tag{8.71}$$

Then, integrating by parts, we infer

$$\int_\Omega \Delta u(x \cdot \nabla u) = \int_{\partial\Omega} (x \cdot \nabla u)\frac{\partial u}{\partial \nu} - \int_\Omega \nabla u \nabla(x \cdot \nabla u) \tag{8.72}$$

and, observing that

$$\nabla u \nabla(x \cdot \nabla u) = |\nabla u|^2 + \frac{1}{2} \sum_{i,j=1}^{n} x_j \frac{\partial}{\partial x_j} \left[\left(\frac{\partial u}{\partial x_i} \right)^2 \right],$$

by (8.72) we obtain

$$\int_\Omega \Delta u (x \cdot \nabla u) = \int_{\partial\Omega} (x \cdot \nabla u) \frac{\partial u}{\partial \nu} - \int_\Omega |\nabla u|^2 - \frac{1}{2} \sum_{i,j=1}^{n} \int_\Omega x_j \frac{\partial}{\partial x_j} \left[\left(\frac{\partial u}{\partial x_i} \right)^2 \right].$$

A further integration by parts gives

$$= \int_{\partial\Omega} (x \cdot \nabla u) \frac{\partial u}{\partial \nu} - \int_\Omega |\nabla u|^2 - \frac{1}{2} \sum_{i,j=1}^{n} \left[\int_{\partial\Omega} x_j \nu_j \left(\frac{\partial u}{\partial x_i} \right)^2 - \int_\Omega \left(\frac{\partial u}{\partial x_i} \right)^2 \right]$$

$$= \int_{\partial\Omega} (x \cdot \nabla u) \frac{\partial u}{\partial \nu} - \frac{1}{2} \int_{\partial\Omega} (x \cdot \nu) |\nabla u|^2 + \frac{n-2}{2} \int_\Omega |\nabla u|^2$$

and finally, in view of (8.71), we have

$$\int_\Omega \Delta u (x \cdot \nabla u) = \frac{1}{2} \int_{\partial\Omega} (x \cdot \nu) |\nabla u|^2 + \frac{n-2}{2} \int_\Omega |\nabla u|^2. \tag{8.73}$$

Now we observe that

$$u^{p-1} (x \cdot \nabla u) = \frac{1}{p} \operatorname{div}(x u^p) - \frac{n}{p} u^p. \tag{8.74}$$

By (8.74), using the divergence Theorem and exploiting the fact that $u = 0$ on $\partial\Omega$, we obtain

$$\int_\Omega (x \cdot \nabla u) u^{p-1} = \frac{1}{p} \int_{\partial\Omega} (x \cdot \nu) u^p - \frac{n}{p} \int_\Omega u^p = -\frac{n}{p} \int_\Omega u^p. \tag{8.75}$$

Moreover, multiplying by u the equation in (8.66), integrating by parts and using again the fact that $u = 0$ on $\partial\Omega$, we deduce

$$\int_\Omega u^p = -\int_\Omega u \Delta u = \int_\Omega |\nabla u|^2. \tag{8.76}$$

Inserting (8.76) into (8.73), we have

$$\int_\Omega \Delta u (x \cdot \nabla u) = \frac{1}{2} \int_{\partial\Omega} (x \cdot \nu) |\nabla u|^2 + \frac{n-2}{2} \int_\Omega u^p. \tag{8.77}$$

On the other hand, multiplying by $x \cdot \nabla u$ the equation in (8.66) and integrating, by (8.75) we deduce

$$-\int_\Omega \Delta u (x \cdot \nabla u) = \int_\Omega (x \cdot \nabla u) u^{p-1} = -\frac{n}{p} \int_\Omega u^p. \tag{8.78}$$

Finally, by combining (8.77) and (8.78), we obtain

$$\frac{2n-(n-2)p}{2p}\int_{\Omega} u^p = \frac{1}{2}\int_{\partial\Omega} |\nabla u|^2 (x\cdot\nu). \tag{8.79}$$

In view of (8.70), the right hand side of (8.79) is non negative; if it was zero, it would be $|\nabla u| = 0$ on $\partial\Omega$, which is absurd since applying the divergence Theorem it would imply

$$0 = \int_{\partial\Omega} |\nabla u| = -\int_{\partial\Omega} \frac{\partial u}{\partial \nu} = -\int_{\Omega} \Delta u = \int_{\Omega} u^{p-1} > 0\ .$$

In conclusion, the right hand side of (8.79) is strictly positive; but then also the left hand side of (8.79) must be positive, that is $2n-(n-2)p > 0$.

Thus we have proved that if (8.66) admits a solution on a star-shaped set, then necessarily $p < 2^*$. This completes the proof. □

Remark 8.3.9. Actually, Theorem 8.3.8 is still valid under the weaker assumption of weak star-shapedness, namely when the inequality (8.70) becomes $x\cdot\nu \geq 0$, see [58]. In this case, however, one needs to apply the Hopf principle to guarantee that a superharmonic positive function in Ω such that $u = 0$ on $\partial\Omega$ satisfies $|\nabla u| > 0$ on the whole boundary $\partial\Omega$. □

Theorem 8.3.8 is valid in star-shaped sets; without such hypothesis, problem (8.66) may have solutions even if $p \geq 2^*$, see [57]. Nevertheless, even without the star-shapedness assumption, the Theorems 8.3.6 and 8.3.8 show a substantial difference between the subcritical case $p < 2^*$ and the supercritical case $p \geq 2^*$.

The critical exponent 2^* plays a fundamental role also in the case of equations set in the whole $\mathbb{R}^n$. We begin by stating two results that we consider relevant:

Theorem 8.3.10. *Let $n \geq 3$ and for $p > 1$ consider the equation*

$$-\Delta u = u^{p-1}, \quad u \geq 0 \qquad \text{in } \mathbb{R}^n\ . \tag{8.80}$$

(i) If $p < 2^*$ then the unique solution to (8.80) is $u \equiv 0$.
(ii) If $p \geq 2^*$ then (8.80) admits infinitely many positive bounded solutions, radially symmetric with respect to the origin.

Proof. See [36] and [53, Theorem 6.4]. □

Therefore, also for the problem (8.80) the critical exponent 2^* equals the power p discriminating between existence and non existence of non trivial solutions. However, compared to the Dirichlet problem on an open bounded set Ω, the situation is inverted, in the sense that the existence of non trivial solutions to (8.80) occurs in the case $p \geq 2^*$, while for problem (8.66) it occurred in the case $2 < p < 2^*$.

The next statement gives a further confirmation of the role played by 2^* in the analysis of semilinear elliptic equations.

Theorem 8.3.11. *Let $n \geq 3$ and for $1 < q < p$ consider the equation*

$$-\Delta u = -u^{q-1} + u^{p-1}, \quad u \geq 0 \qquad \text{in } \mathbb{R}^n . \tag{8.81}$$

(i) If $p < 2^*$ then (8.81) admits a unique positive bounded solution, radially symmetric with respect to the origin.
(ii) If $p \geq 2^*$ then the unique bounded solution of (8.81) is $u \equiv 0$.

Proof. See [7, 30, 52, 64]. □

This time, unlike what happens for problem (8.80), non trivial solutions exist if and only if $p < 2^*$. This phenomenon, apparently inexplicable, is due to the different behavior of the functions $f_1(s) = s^{p-1}$ and $f_2(s) = -s^{q-1} + s^{p-1}$ as $s \to 0^+$. Indeed, since $1 < q < p$, the function f_2 turns out to be negative for $s > 0$ sufficiently close to $s = 0$, while the function f_1 turns out to be positive for every $s > 0$. A detailed explanation of the different behavior of the problems (8.80) and (8.81) would require the introduction of some advanced mathematical tools and this goes beyond the scopes of this book. Therefore, for more details, we refer to [30, 53].

8.3.4 Other critical exponents

Besides the critical Sobolev exponent analyzed in Section 8.3.3, there exist other critical exponents of Functional Analysis which appear in various problems related to the solutions of elliptic equations of the type (8.50).

In order to keep the same structure of the previous section, we begin with an abstract problem of Functional Analysis. Let $\Omega \subset \mathbb{R}^n$ ($n \geq 3$) be a bounded open set with non empty boundary $\partial\Omega$ of class C^1 and let $\gamma_0 : H^1(\Omega) \mapsto L^q(\partial\Omega)$ be the trace operator introduced in Theorem 6.5.3. We wonder which is the largest $q \geq 2$ such that this operator is continuous. According to [1, Theorem 5.22], we know that this value is

$$q = 2_T = \frac{2(n-1)}{n-2} .$$

In other words, there holds the following result:

$$\exists C = C_\Omega > 0 , \qquad \|\gamma_0 u\|_{L^q(\partial\Omega)} \leq C\|u\|_{H^1(\Omega)} \qquad \forall u \in H^1(\Omega)$$

for every $1 \leq q \leq 2_T$, and 2_T is the largest possible exponent. We will call 2_T **critical trace exponent**.

Now, we look for weak solutions of the problem

$$\begin{cases} -\Delta u \geq u^{p-1} & \text{in } \mathbb{R}^n \\ u \geq 0 , \quad u \not\equiv 0 & \text{in } \mathbb{R}^n . \end{cases} \tag{8.82}$$

By weak solution of (8.82) we mean a function $u \in H^1_{loc}(\mathbb{R}^n)$ such that

$$\int_{\mathbb{R}^n} \nabla u \nabla \varphi \geq \int_{\mathbb{R}^n} u^{p-1}\varphi \qquad \forall \varphi \in \mathcal{D}(\mathbb{R}^n)\ , \quad \varphi \geq 0\ ,$$

where $H^1_{loc}(\mathbb{R}^n)$ denotes the set of functions defined on $\mathbb{R}^n$, whose restriction to any bounded open set $\Omega \subset \mathbb{R}^n$ belongs to $H^1(\Omega)$.

The first result which emphasizes the importance of the critical exponent of trace is the

Theorem 8.3.12. *Assume that $n \geq 3$ and $p > 1$.*
(i) If $p \leq 2_T$ then (8.82) has no weak solution $u \in H^1_{loc}(\mathbb{R}^n)$.
(ii) If $p > 2_T$ then (8.82) has infinitely many weak solutions.

Proof. See [51, Theorem 12.1], [51, p.53], [65, Theorem I]. □

The exponent 2_T comes into play also studying the existence of solutions of the problem

$$\Delta u = u^{q-1} \qquad \text{in } \mathbb{R}^n \setminus \{0\} \tag{8.83}$$

under the conditions

$$\lim_{x\to 0} |x|^{n-2}\, u(x) = [n(n-2)]^{(n-2)/2}, \qquad \lim_{|x|\to\infty} u(x) = 0. \tag{8.84}$$

We observe that (8.84) entails that u displays a singularity as $x \to 0$ of the same order of the fundamental solution of $-\Delta u = \delta_0$. There holds the

Theorem 8.3.13. *Assume that $n \geq 3$ and $q > 1$.*
(i) If $q < 2_T$ then (8.83)-(8.84) has a unique radial non negative solution.
(ii) If $q \geq 2_T$ then (8.83)-(8.84) has no solution.

Proof. See [19], [28] and [71, Theorem 1.1]. □

Remark 8.3.14. We observe that 2_T is hidden also in problem (8.65): indeed $\lambda = \frac{2}{p-2}(n - 2 - \frac{2}{p-2}) > 0$ if and only if $p > 2_T$. □

We conclude this section (and the whole book) by introducing a third critical exponent. One of the consequences of Theorem 8.3.8 is that the best Sobolev constant S_{2^*} defined in (8.69) is not attained if Ω is star-shaped, namely the problem

$$S_{2^*} = \inf_{u\in B} \|\nabla u\|^2_{L^2}$$

with B as in (8.69), does not admit a minimizer. In fact, if the best Sobolev constant was attained, arguing as in the proof of Theorem 8.3.6, we would obtain the existence of a solution to (8.66), in contradiction with Theorem 8.3.8. Actually, it is possible to show that S_{2^*} is not attained in any bounded domain Ω. Therefore we may ask if, for some $q \geq 1$, there exist a positive constant C_q (depending on Ω and on q) such that

$$\|\nabla u\|_{L^2}^2 \geq S_{2^*}\|u\|_{L^{2^*}}^2 + C_q\|u\|_{L^q}^2 \qquad \forall u \in H_0^1(\Omega) .$$

In order to face this question, we introduce the exponent

$$2_R = \frac{n}{n-2} = 2_T - 1 ;$$

then the answer to the previous question is given by the

Theorem 8.3.15. (Brezis-Nirenberg Theorem). *Let $n \geq 3$ and let $\Omega \subset \mathbb{R}^n$ be a bounded open set with C^1 boundary. Then for every $q \in [1, 2_R)$ there exists $C_q > 0$ such that*

$$\|\nabla u\|_{L^2}^2 \geq S_{2^*}\|u\|_{L^{2^*}}^2 + C_q\|u\|_{L^q}^2 \qquad \forall u \in H_0^1(\Omega) .$$

Moreover, $\lim_{q\to 2_R^-} \overline{C_q} = 0$, *where*

$$\overline{C_q} \;=\; \inf_{\|u\|_{L^q}=1} \left(\|\nabla u\|_{L^2}^2 - S_{2^*}\|u\|_{L^{2^*}}^2\right) .$$

Proof. See [12]. □

Theorem 8.3.15 states that the Sobolev inequality can be improved by adding an L^q norm if and only if $q < 2_R$. Hence we may call 2_R **critical exponent of the residual term** or simply residual critical exponent.

To conclude, we present a result that relates the three critical exponents introduced so far. Theorem 8.3.11 suggests to analyze the behavior of positive solutions of (8.81) (which exist if $p < 2^*$) as $p \to (2^*)^-$, since for $p = 2^*$ problem (8.81) does not admit positive solutions. In the next (and last!) result we describe the asymptotical behavior of the L^∞ norm.

Theorem 8.3.16. (Ferrero-Gazzola Theorem). *For every $1 < q < p < 2^*$ let u_p be the unique radial positive solution to* (8.81). *Then there exist $\alpha_{n,q}, \beta_n, \gamma_{n,q} > 0$ such that*

$$\lim_{p\to 2^*} (2^* - p)\|u_p\|_{L^\infty}^{2^*-q} = \alpha_{n,q} \qquad \text{if } q > 2_R$$

$$\lim_{p\to 2^*} \frac{2^* - p}{|\log(2^* - p)|}\|u_p\|_{L^\infty}^{\frac{n}{n-2}} = \beta_n \qquad \text{if } q = 2_R$$

$$\lim_{p\to 2^*} (2^* - p)\|u_p\|_{L^\infty}^{\frac{2^*-q}{2_T-q}} = \gamma_{n,q} \qquad \text{if } q < 2_R .$$

Proof. See [19] and also previous results contained in [28, 29]. □

Theorem 8.3.16 shows that $\|u_p\|_{L^\infty} \to +\infty$ as $p \to (2^*)^-$. The fact that the L^∞ norm diverges as $p \to (2^*)^-$ gives an intuitive explanation of the absence of positive solutions of (8.81) for $p = 2^*$.

Glossary of the prerequisites

Did you forget some definitions or properties that we took for granted? Let us try to help you with this list (without any mathematical rigor!):

Analytic function. A function f is analytic if its power series has a non zero radius of convergence and converges to f.

Circle of convergence. A power series over $\mathbb{C}$ has as set of convergence a disk in the complex plane and its radius is the radius of convergence.

Closed differential form. A differential form $\sum_i f_i \, dx_i$ is said to be closed if

$$\frac{\partial f_i}{\partial x_j} = \frac{\partial f_j}{\partial x_i} \qquad \forall i, j.$$

Closure of a set. Given $E \subseteq \mathbb{R}^n$, the closure of E is defined as $E \cup \partial E$ and is denoted by $\overline{E}$.

Connected set. A subset of $\mathbb{R}^n$ is said to be connected if it cannot be represented as the union of two separated nonempty sets.

Differential operators. Let $F = (F_1, ..., F_n)$ be a differentiable vector field in $\Omega \subset \mathbb{R}^n$, let $G = (G_1, G_2, G_3)$ be a differentiable vector field in $\Omega \subset \mathbb{R}^3$, let $u \in C^2(\Omega)$:

$$\text{div} F = \nabla \cdot F = \sum_{i=1}^{n} \frac{\partial F_i}{\partial x_i}, \qquad \nabla F = \left(\frac{\partial F_i}{\partial x_j} \right)_{i,j=1,...,n},$$

$$\text{curl}\, G = \nabla \wedge G = \begin{vmatrix} \mathbf{i} & \mathbf{j} & \mathbf{k} \\ \frac{\partial}{\partial x_1} & \frac{\partial}{\partial x_2} & \frac{\partial}{\partial x_3} \\ G_1 & G_2 & G_3 \end{vmatrix},$$

$$\nabla u = (u_{x_1}, ..., u_{x_n}), \quad D^2 u = \left(u_{x_i x_j} \right)_{i,j=1,...,n}, \quad \Delta u = \text{div} \nabla u = \sum_{i=1}^{n} \frac{\partial^2 u}{\partial x_i^2}.$$

Differentiation under the integral sign. It is the operation which allows to apply this procedure:

$$\frac{d}{dx}\left(\int f(x,y)\,dy\right) = \int f_x(x,y)\,dy.$$

Divergence Theorem. (Which, in dimension $n = 1$, reduces to the fundamental Theorem of the integral calculus) Let F be a vector field defined in a bounded set Ω, then

$$\int_\Omega \nabla \cdot F = \int_{\partial\Omega} F \cdot \nu.$$

Exact differential form. A differential form $\sum_i f_i\, dx_i$ is said to be exact if it is the differential of a potential function F, namely $\nabla F = (f_1, \ldots, f_n)$. Every exact differential form is also closed, the viceversa is in general false.

Extension of a function from real variables to complex variables. Let $f(x)$ be a function of the *real* variable x. By replacing the real variable x with the complex variable $z = x + iy$, we obtain the function $f(z)$ defined as the extension of the function $f(x)$ from the real line to the complex plane.

Gradient rule. Let f be a function which is differentiable at a point $x \in \mathbb{R}^n$ and let v be a vector. The gradient rule says that

$$\frac{\partial f}{\partial v}(x) = \nabla f(x) \cdot v.$$

Isomorphism. It is a linear bijective function.

Limit inferior and limit superior. Given a sequence $\{a_n\} \subset \mathbb{R}$, then:

$$\liminf_{n\to\infty} a_n := \lim_{n\to\infty} \inf_{m\geq n} a_m, \qquad \limsup_{n\to\infty} a_n := \lim_{n\to\infty} \sup_{m\geq n} a_m$$

Map. Continuous function.

Mean value. Let $E \subset \mathbb{R}^n$ be a measurable set with finite measure and let $f \in L^1(E)$. It is defined the mean value of f on E as

$$⨍_E f := \frac{1}{|E|}\int_E f.$$

Normal convergence. The series $\sum f_n$ normally converges on A if:

$$\sum_{n=1}^{\infty} \sup_{x\in A} |f_n(x)| < \infty.$$

Rectifiable curve. It is any measurable curve with finite length.

Restriction of a function. Let $A \subseteq X$ be a set and let $f : X \to Y$ be a function. The restriction of f to A is denoted by $f_{|A}$ and satisfies:

$$f_{|A} : A \to Y, \; f_{|A}(x) = f(x) \quad \forall x \in A\,.$$

Separated sets. Let $A, B \subset \mathbb{R}^n$. They are said to be separated if $\overline{A} \cap B = \emptyset$ and $A \cap \overline{B} = \emptyset$.

Simply connected set. It is a set in which every closed path contained in the set can be contracted (without leaving the set) to a point of the set.

Bibliography

[1] R.A. Adams, *Sobolev spaces*, Academic Press Inc. 1975

[2] S. Agmon, A. Douglis, L. Nirenberg, *Estimates near the boundary for solutions of elliptic partial differential equations satisfying general boundary conditions. I*, Comm. Pure Appl. Math. 12, 1959, 623-727

[3] L. Amerio, *Analisi matematica con elementi di analisi funzionale, Vol.3, parti I e II*, Ed. UTET, Torino, 1982

[4] D. Anderson, G. Derrick, *Stability of time dependent particle like solutions in nonlinear field theories*, J. Math. Phys. 11, 1970, 1336-1346; e 12, 1971, 945-952

[5] D.G. Aronson, H.F. Weinberger, *Nonlinear diffusion in population genetics*, Lecture Notes in Math. 446, Ed. Springer, Berlin, Heidelberg, New York, 1975

[6] G. Belloni, A. Lo Conte, *Costruzione di macchine*, Ulrico Hoepli editore, Milano, 2002

[7] H. Berestycki, P.L. Lions, *Non-linear scalar field equations. I. Existence of a ground state*, Arch. Rational Mech. Anal. 82, 1983, 313-345

[8] A. Bernasconi, M. Filippini, M. Giglio, A. Lo Conte, G. Petrone, M. Sangirardi, *Fondamenti di costruzioni di macchine*, McGraw Hill, Milano, 2002

[9] M. Bramanti, *Esercitazioni di analisi 3*, Ed. CUSL, Milano, 1993

[10] H. Brezis, *Analisi funzionale*, Ed. Liguori, Napoli, 1986

[11] H. Brezis, T. Kato, *Remarks on the Schrodinger operator with singular complex potentials*, J. Math. Pures Appl. 58, 1979, 137-151

[12] H. Brezis, L. Nirenberg, *Positive solutions of nonlinear elliptic equations involving critical Sobolev exponents*, Comm. Pure Appl. Math. 36, 1983, 437-477

[13] M. Capurso, *Lezioni di scienza delle costruzioni*, Ed. Pitagora, Bologna, 1998

[14] S. Coleman, *The fate of the false vacuum*, Phys. Rev. D. 15, 1977, 2929

[15] J.P. Den Hartog, *Mechanical vibrations*, Dover Publications Inc. New York, 1985

[16] G. Diana, F. Cheli, *Dinamica e vibrazione dei sistemi meccanici, Vol. 1*, UTET Libreria, Torino, 1993

[17] G. Diana, F. Cheli, *Dinamica e vibrazione dei sistemi meccanici, Vol. 2*, UTET Libreria, Torino, 1993

[18] L.C. Evans, *Partial differential equations*, Ed. American Mathematical Society, 1998

[19] A. Ferrero, F. Gazzola, *Asymptotic behavior of ground states of quasilinear elliptic problems with two vanishing parameters, Part III*, J. Diff. Eq. 198, 2004, 53-90

[20] P.C. Fife, *Asymptotics states for equations of reaction and diffusion*, Bull. A.M.S. 84, 1978, 693-726

[21] P.C. Fife, L.A. Peletier, *Nonlinear diffusion in population genetics*, Arch. Rat. Mech. Anal. 64, 1977, 93-109

[22] C. Foias, R. Temam, *Structure of the set of stationary solutions of the Navier-Stokes equations*, Comm. Pure Appl. Math. 30, 1977, 149-164

[23] C. Foias, R. Temam, *Remarques sur les équations de Navier-Stokes stationnaires et les phénomènes successifs de bifurcation*, Ann. Sc. Norm. Sup. Pisa - Serie IV, Vol.5, 1978, 29-63

[24] P.H. Frampton, *Consequences of vacuum instability in quantum field theory*, Phys. Rev. D. 15, 1977, 2922-2928

[25] D. Fujiwara, H. Morimoto, *An L_r-theorem of the Helmholtz decomposition of vector fields*, J. Fac. Sci. Univ. Tokyo, 24, 1977, 685-700

[26] E. Gagliardo, *Proprietà di alcune classi di funzioni in più variabili*, Ricerche Mat. 7, 1958, 102-137, & 8, 1959, 24-51

[27] F. Gazzola, A. Malchiodi, *Some remarks on the equation $-\Delta u = \lambda(1+u)^p$ for varying λ, p and varying domains*, Comm. Part. Diff. Eq. 27, 2002, 809-845

[28] F. Gazzola, L. Peletier, P. Pucci, J. Serrin, *Asymptotic behavior of ground states of quasilinear elliptic problems with two vanishing parameters, Part II*, Ann. Inst. H. Poincaré A.N.L. 20, 2003, 947-974

[29] F. Gazzola, J. Serrin, *Asymptotic behavior of ground states of quasilinear elliptic problems with two vanishing parameters*, Ann. Inst. H. Poincaré A.N.L. 19, 2002, 477-504

[30] F. Gazzola, J. Serrin, M. Tang, *Existence of ground states and free boundary problems for quasilinear elliptic operators*, Adv. Diff. Eq. 5, 2000, 1-30

[31] F. Gazzola, F. Tomarelli, M. Zanotti, *Analytic functions, integral transforms, differential equations* Esculapio, 2015, ISBN 978-88-7488-660-9

[32] I.M. Gelfand, *Some problems in the theory of quasi-linear equations*, Amer. Math. Soc. Transl. 29, 1963, 295-381

[33] G. Genta, *Vibrazioni delle strutture e delle macchine*, Levrotto-Bella, Torino 1996

[34] C. Gerhardt, *Stationary solutions to the Navier-Stokes equations in dimension four*, Math. Z. 165, 1979, 193-197

[35] B. Gidas, *Bifurcation phenomena in mathematical physics and related topics*, Ed. C. Bardos & D. Bessis, Dordrecht, Olanda, 1980

[36] B. Gidas, J. Spruck, *Global and local behavior of positive solutions of nonlinear elliptic equations*, Comm. Pure Appl. Math. 34, 1981, 525-598

[37] Y. Giga, *Domains of fractional powers of the Stokes operator in L_r spaces*, Arch. Rat. Mech. Anal. 89, 1985, 251-265

[38] G. Gilardi, *Analisi tre*, McGraw-Hill, Milano, 1994

[39] P. Grisvard, *Elliptic problems in nonsmooth domains*, Pitman 1985

[40] H. Helmholtz, *Über die theorie der elektrodynamic. Erste abhandlung. Über die bewegungsgleichungen der elektricität für ruhende leitende körper*, J. Reine Angew. Math. 72, 1870, 57-129

[41] F. John, *Partial differential equations*, Springer, New York, 1971

[42] P.K. Kundu, I.M. Cohen, *Fluid Mechanics*, Academic Press, 2nd Edition, 2002

[43] O.A. Ladyzhenskaya, *The mathematical theory of viscous incompressible flow*, Gordon and Breach, 2nd. Edition, 1969

[44] L.D. Landau, E.M. Lifshitz, *Fluid Mechanics*, Pergamon Press, Great Britain, 1959

[45] L.D. Landau, E.M. Lifshitz, *Theory of elasticity*, Pergamon Press, Great Britain, 2nd Edition, 1970

[46] J.L. Lions, E. Magenes, *Problèmes aux limites non homogènes et applications, vol.1*, Dunod, Paris 1968

[47] A. Lorenzi, E. Paparoni, *Esercizi e problemi di analisi matematica, funzioni di una variabile complessa*, Ed. CLUED, Milano, 1984

[48] P. Mazzoldi, M. Nigro, C. Voci, *Fisica*, vol. II, EdiSES, Napoli 2002

[49] G. Métivier, *Valeurs propres d'opérateurs définis par la restriction de systèmes variationnels à des sous-espaces*, J. Math. Pures Appl. 57, 1978, 133-156

[50] V.P. Michajlov, *Equazioni differenziali alle derivate parziali*, Ed. Mir 1984

[51] E. Mitidieri, S. Pohozaev, *Apriori estimates and blow-up of solutions to non-linear partial differential equations and inequalities*, Proc. Steklov Inst. Math. 234, 2001, 1-362

[52] W.M. Ni, J. Serrin, *Nonexistence theorems for quasilinear partial differential equations*, Rend. Circolo Mat. Palermo (Centenary Supplement), Series II, 8, 1985, 171-185

[53] W.M. Ni, J. Serrin, *Existence and nonexistence theorems for ground states for quasilinear partial differential equations. The anomalous case*, Accad. Naz. Lincei, Atti dei Convegni 77, 1986, 231-257

[54] L. Nirenberg, *On elliptic partial differential equations*, Ann. Sc. Norm. Sup. Pisa 13, 1959, 115-162

[55] C.D. Pagani, S. Salsa, *Analisi Matematica, Volume 1*, Ed. Zanichelli, Milano, 1999

[56] C.D. Pagani, S. Salsa, *Analisi Matematica, Volume 2*, Ed. Zanichelli, Milano, 1999

[57] D. Passaseo, *Nontrivial solutions of elliptic equations with supercritical exponent in contractible domains*, Duke Math. J. 92, 1998, 429-457

[58] S.J. Pohozaev, *Eigenfunctions of the equation $\Delta u + \lambda f(u) = 0$*, Soviet Math. Doklady 6, 1965, 1408-1411

[59] D. Roux, *Istituzioni di analisi superiore*, Ed. Masson, Milano, 1984

[60] W. Rudin, *Analisi reale e complessa*, Ed. Boringhieri, Torino, 1982

[61] W. Rudin, *Functional Analysis*, Ed. McGraw-Hill, Singapore, 1991

[62] S. Salsa, *Equazioni a derivate parziali*, Ed. Springer, Milano, 2004

[63] E. Sernesi, *Geometria 1*, Ed. Bollati Boringhieri, Torino, 1994

[64] J. Serrin, M. Tang, *Uniqueness of ground states for quasilinear elliptic equations*, Indiana Univ. Math. J. 49, 2000, 897-923

[65] J. Serrin, H. Zou, *Cauchy-Liouville and universal boundedness theorems for quasilinear elliptic equations and inequalities*, Acta Math. 189, 2002, 79-142

[66] J. Smoller, *Shock waves and reaction-diffusion equations*, Springer-Verlag, 1983

[67] M.R. Spiegel, *Trasformate di Laplace*, Collana Schaum, Etas Libri, Milano, 1976

[68] R. Temam, *Navier-Stokes equations, theory and numerical analysis*, North-Holland, Amsterdam, 1979

[69] R. Temam, *Navier-Stokes equations and nonlinear functional analysis*, SIAM, Philadelphia, 1983

[70] F. Tomarelli, *Esercizi di metodi matematici per l'ingegneria*, Ed. CLUED, Milano, 1990

[71] J.L. Vázquez, L. Veron, *Removable singularities of some strongly nonlinear elliptic equations*, Manuscripta Math. 33, 1980, 129-144

[72] P. Villaggio, *Mathematical models for elastic structures*, Cambridge University Press, 2005

[73] M.F. Weinberger, *Asymptotic behavior of a model in population genetics*, Lecture Notes in Math. 648, Ed. Springer, Berlin, Heidelberg, New York, 1978

[74] H. Weyl, *The method of orthogonal projection in potential theory*, Duke Math. J. 7, 1940, 411-444

[75] R.L. Wheeden, A. Zygmund, *Measure and integral*, Marcel Dekker Inc., 1977

[76] J. Wloka, *Partial differential equations*, Cambridge University Press, 1987

[77] K. Yosida, *Functional Analysis*, Springer-Verlag, 1980

[78] E. Zeidler, *Nonlinear functional analysis and its applications*, Vol. I, Springer-Verlag, 1986

Index